353
Topics in Current Chemistry

Aims and Scope

The series Topics in Current Chemistry presents critical reviews of the present and future trends in modern chemical research. The scope of coverage includes all areas of chemical science including the interfaces with related disciplines such as biology, medicine and materials science.

The goal of each thematic volume is to give the non-specialist reader, whether at the university or in industry, a comprehensive overview of an area where new insights are emerging that are of interest to larger scientific audience.

Thus each review within the volume critically surveys one aspect of that topic and places it within the context of the volume as a whole. The most significant developments of the last 5 to 10 years should be presented. A description of the laboratory procedures involved is often useful to the reader. The coverage should not be exhaustive in data, but should rather be conceptual, concentrating on the methodological thinking that will allow the non-specialist reader to understand the information presented.

Discussion of possible future research directions in the area is welcome.

Review articles for the individual volumes are invited by the volume editors.

Readership: research chemists at universities or in industry, graduate students.

More information about this series at
http://www.springer.com/series/128

Kenneth M. Nicholas
Editor

Selective Catalysis for Renewable Feedstocks and Chemicals

With contributions by

C. Boucher-Jacobs · J.J. Bozell · E.Y-X Chen · M. Dusselier ·
S. Dutta · M. Mascal · Y. Nakagawa · K.M. Nicholas ·
B.F. Sels · M. Tamura · K. Tomishige · Y. Zhang

Springer

Editor
Kenneth M. Nicholas
Department of Chemistry and Biochemistry
University of Oklahoma
Norman
Oklahoma
USA

ISSN 0340-1022 ISSN 1436-5049 (electronic)
ISBN 978-3-319-08653-8 ISBN 978-3-319-08654-5 (eBook)
DOI 10.1007/978-3-319-08654-5
Springer Cham Heidelberg New York Dordrecht London

Library of Congress Control Number: 2014950413

Printed on acid-free paper

Springer is part of Springer Science+Business Media (www.springer.com)

Preface

Organic biomass from plants provided the energy and chemical resources for mankind for centuries, until the widespread discovery and mining of fossil resources, particularly coal and petroleum, fueled worldwide industrialization during the last 150 years. The burgeoning world population and its increasing demand for energy and materials/chemicals, the unequally distributed and non-renewability of the fossil resources, and their environmental and climatological impacts have stimulated renewed interest in the discovery and development of new processes for the conversion of renewable, biomass-derived, and carbon-neutral feedstocks into useful fuels and value-added chemicals. The latter is the focus of the present volume.

The purpose of this volume is to highlight several recent research efforts to discover and to develop *selective, catalytic* reactions and processes for the conversion of ligno-cellulosic and plant oil-derived feedstocks to value-added chemicals. The value of catalytic chemical processes lies in their reduced energy consumption and potentially increased efficiency and product selectivity, which minimize environmental and societal impacts. Selectivity is especially important for the preparation of commodity and specialty chemical and materials where single compounds are usually needed, unlike fuel production where mixtures of high energy compounds (e.g., gasoline and diesel) are useable.

The challenge to chemists seeking to reconstitute/refunctionalize biomass resources is their polyoxygenated nature $(CHO)_n$, which requires the transformation of C–O bonds for oxygen removal or substitution, whereas the functionalization of hydrocarbon-rich fossil resources focuses on C–C and C–H activation. Different chemistries and types of catalysts are thus needed to effect the selective, catalytic conversions desired for refunctionalizing typical biomass feedstocks.

In this volume we begin with an introduction to identify and assess the value of potential chemical conversions of biomass feedstocks. This is then followed by several chapters from leaders in the field covering various catalytic methods for oxygen-removal and refunctionalization of polyoxygenates, including selective dehydration to multipurpose furan derivatives, carbohydrate retroaldol/

dehydrations/isomerization to the platform chemical lactic acid, polyol hydrogenolysis to partially deoxygenated products, and deoxydehydration to unsaturated products. Finally, we conclude with two chapters focusing on biomass reconstruction and deconstruction processes – the polymerization of biomass-derived monomers to macromolecules (polyesters, etc.) and lignin deconstruction reactions.

In is our hope that this volume will help to stimulate continued interest and developments in biomass conversion catalysis for the biorefineries of the future.

Norman, OK, USA
May 2014

Kenneth M. Nicholas

Contents

Top Curr Chem (2014) 353: 1–40
DOI: 10.1007/128_2014_544

Published online: 20 May 2014

Top Chemical Opportunities from Carbohydrate Biomass: A Chemist's View of the Biorefinery

Michiel Dusselier, Mark Mascal, and Bert F. Sels

Abstract Cheap fossil oil resources are becoming depleted and crude oil prices are rising. In this context, alternatives to fossil fuel-derived carbon are examined in an effort to improve the security of carbon resources through the development of novel technologies for the production of chemicals, fuels, and materials from renewable feedstocks such as biomass. The general concept unifying the conversion processes for raw biomass is that of the biorefinery, which integrates biofuels with a selection of pivot points towards value-added chemical end products via so-called "platform chemicals". While the concept of biorefining is not new, now more than ever there is the motivation to investigate its true potential for the production of carbon-based products. A variety of renewable chemicals have been proposed by many research groups, many of them being categorized as drop-ins, while others are novel chemicals with the potential to displace petrochemicals across several markets. To be competitive with petrochemicals, carbohydrate-derived products should have advantageous chemical properties that can be profitably exploited, and/or their production should offer cost-effective benefits. The production of drop-ins will likely proceed in short term since the markets are familiar, while the commercial introduction of novel chemicals takes longer and demands more technological and marketing effort.

Rather than describing elaborate catalytic routes and giving exhaustive lists of reactions, a large part of this review is devoted to creating a guideline for the selection of the most promising (platform) chemicals derived via chemical-catalytic reaction routes from lignocellulosic biomass. The major rationale behind our recommendations is a maximum conservation of functionality, alongside a high atom economy. Nature provides us with complex molecules like cellulose and hemicellulose, and it should be possible to transform them into chemical products while

M. Dusselier and B.F. Sels (✉)
Center for Surface Chemistry and Catalysis, KU Leuven, Leuven, Belgium
e-mail: michiel.dusselier@biw.kuleuven.be; bert.sels@biw.kuleuven.be

M. Mascal (✉)
Department of Chemistry, University of California Davis, Davis, CA 95616, USA
e-mail: mjmascal@ucdavis.edu

maintaining aspects of their original structure, rather than taking them completely apart only to put them back together again in a different order, or turning them into metabolites and CO_2. Thus, rather than merely pursuing energy content as in the case of biofuels, the chemist sees atom efficiency, functional versatility, and reactivity as the key criteria for the successful valorization of biomass into chemicals.

To guide the choice of renewable chemicals and their production, this review adopts the original *van Krevelen* plots and develops alternative diagrams by introducing a functionality parameter F and a functionality index F:C (rather than O:C). This index is more powerful than the O index to describe the importance of functional groups. Such plots are ideal to assess the effect of several reaction types on the overall functionality in biomass conversion. The atom economy is an additional arbitrator in the evaluation of the reaction types. The assessment is illustrated in detail for the case of carbohydrate resources, and about 25 chemicals, including drop-ins as well as novel chemicals, are selected.

Most of these chemicals would be difficult to synthesize from petrochemicals feeds, and this highlights the unique potential of carbohydrates as feedstocks, but, importantly, the products should have a strong applied dimension in existing or rising markets. Ultimately, the production scales of those markets must be harmonized to the biomass availability and its collection and storage logistics.

Keywords Atom economy · Biomass conversion · Biomass-to-chemicals · Biorefinery · Cellulose · Functionality index · Modified *van Krevelen plot* · Platform chemicals

Contents

1 General Introduction to Biomass Conversion and Biorefineries

The concept of the biorefinery is attracting exponentially growing interest. According to *Chemical Abstracts*, more than 650 publications have appeared to date with the term "biorefinery" in their title, including 170 review papers. If the search is widened to include papers that contain "biorefinery" in either the title or abstract, there are more than 1,800 references, including 436 reviews. The most cited of these is a 2006 review by Ragauskas and coworkers in the journal *Science*, which has a remarkable 1,412 citations as of mid-January 2014 [1].

While the above speaks to the current popularity of the field, the essential concept of the biorefinery predates the term by a long stretch. A historical development of biorefining is complicated by the fact that natural products have been exploited for various industrial purposes since long before petroleum was commercially developed. The use of the terms "biorefinery," "biorefining," or "biomass refining" before the year 2000 was confined to only about 35 papers, the earliest of which appeared in 1981 [2]. Prior to this, however, related terms were used. For example, in 1978, an article entitled "Fuels from biomass: integration with food and materials systems" provided an early insight into the "integrated biorefinery" concept which has become so popular in the modern abstraction of the field [3].

We might suppose a reasonable historical marker of the modern biorefinery could be the point at which the exploitation of cellulose as an alternative to sugars and starches in biotechnology was put forward. However, this is not easy to pin down, since the conversion of cellulose to fermentable sugars was a subject of investigation as far back as 200 years ago [4]. Alternatively, the thermochemical conversion of biomass to secondary products such as fuels might be seen as the dawning of the biorefinery movement. However, here also there is little new under the sun, with studies along these lines having already been described in the eighteenth century [5, 6]. Even chemical conversion methods take root in historical processes. For example, a review entitled "Biorefining of biomass to liquid fuels and organic chemicals" describes the anaerobic digestion of biomass to aliphatic acids, which were then oxidized to alkyl radicals that couple to give alkanes, citing results that hark back to the days of Faraday [2]. Thus, we must consider biorefining a historical, cross-disciplinary movement which has seen a surge in the twenty-first century, rather than a field that has just arisen to confront modern environmental, economic, and political challenges. The latter and arguably most contentious of these challenges was foreseen in a 1982 article by Bungay entitled "Biomass refining," which begins as follows:

> Because there is no shortage of petroleum at present, there is less apparent pressure to develop alternative chemical feedstocks. But the respite may be brief as OPEC (the Organization of Petroleum Exporting Countries) is attempting to cut production and to dissipate oil stockpiles. Having alternatives to petroleum may soon be essential to economic survival [7].

In fact, it is no coincidence that the biorefinery timeline parallels that of historical gasoline prices in the US (Fig. 1) and the consequent launch of a range of federal programs to stimulate research in renewable energy and materials. The inflection just after 2001, the year in which 9–11 occurred, is also significant from a political standpoint.

It is also a fact that the biorefinery movement is by no small measure "scientific" in character. It is the prospect of taking raw biomass components and converting

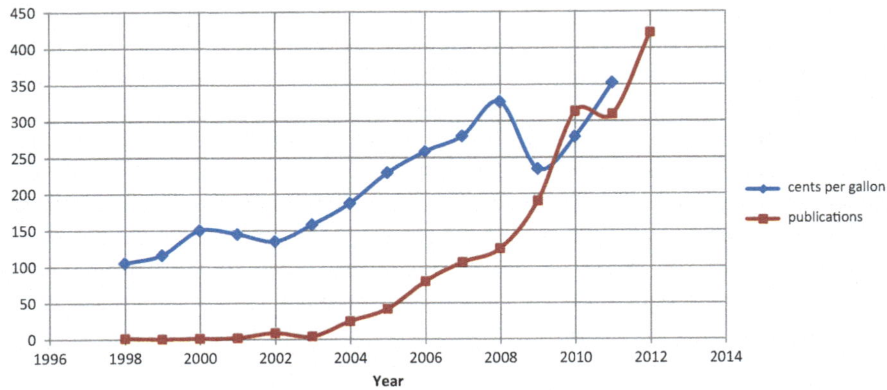

Fig. 1 The average cost of a gallon of gasoline in the US (*Source*: Energy Information Administration, Annual Energy Review 2011, Table 5.24) plotted against the number of hits in a *Chemical Abstracts* search using the index term "biorefinery"

them, using industrially relevant practices, into renewable energy, fuels, materials, and chemicals that challenges the scientist. Generally, this has been approached in three ways, as shown in Fig. 2.

The first and most common of these is microbial. This was the "low hanging fruit," since it is grounded in a knowledge-base with roots in antiquity, i.e., agriculture and brewery/distillery. An additional advantage is that the products (mainly simple alcohols and acids) are "drop-ins" and do not require regulatory approval for most applications. However, fermentation is a relatively slow process, generally involving time scales on the order of days. It also does not usually proceed directly from raw, cellulosic biomass, but requires expensive pretreatment, saccharification, and often post-treatment to remove by-products toxic to microorganisms. Finally, not all the carbon is captured into useful products. For example, glucose is utilized following the equation $C_6H_{12}O_6 \rightarrow 2C_2H_5OH + 2CO_2$. Even assuming 100% theoretical efficiency of the fermentation process, one third of the available carbon is blown out as carbon dioxide, about an equal mass of which is produced per unit mass of ethanol. Also belonging to the above classification is microbial methanogenesis, which proves valuable for certain types of feedstock (generally food and animal wastes) [8]. However, around 40% of the output is again CO_2, and the methane thus produced is for the most part limited in its applications to local power generation. While this constitutes, alongside other power technologies (solar, wind, hydroelectric, geothermal), a useful contribution to the greening up of energy, biomass may be considered to have a more vital role to play in the production of liquid transportation fuels, chemicals and materials. For instance, we will not fly commercial aircraft on solar cells, and neither are heavy transportation vehicles and construction equipment likely to convert to alternative power sources soon. By the same token, a carbon source is needed to make chemicals and materials, such as polymers, coatings, adhesives, solvents, and the like. The chemical diversity of these industrial products and the sheer scale on

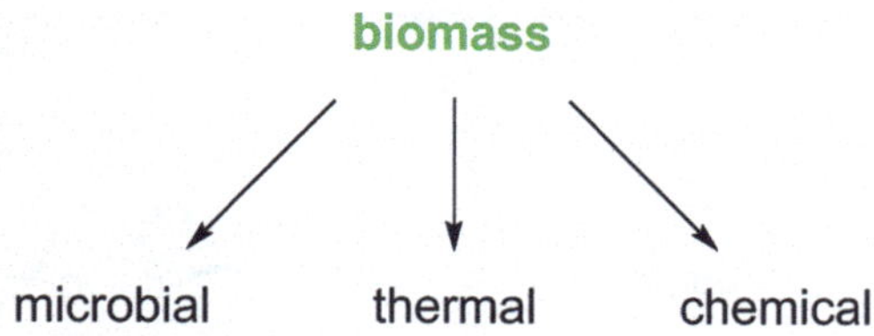

Fig. 2 Approaches to biomass exploitation

which they are consumed indicates they can only be provided in a renewable way via biomass. It follows from the above that the more approaches we develop to valorizing biomass towards these ends, the further along we will be on the way to independence from petroleum.

The second mode of biomass processing, also historically rooted (as noted above), is thermochemical in nature, leading either to bio-gas or bio-oil. This method has also seen a flurry of activity, with several comprehensive reviews having been written in recent years [9]. Nevertheless, this approach also presents limitations. In gasification, biomass is converted into a mixture of combustible gases (principally CO and H_2) by partial oxidation at temperatures from ~800 to 1,000°C. As in fermentation, significant quantities of carbon are diverted into CO_2 and, akin to methanogenic biogas, the product is mainly relevant to direct power generation, although when reconditioned to clean synthesis gas [10], it can be submitted to the (expensive) Fischer–Tropsch process to give liquid fuels [11].

Fast pyrolysis at lower temperatures (~500°C), which can be conducted in the presence of heterogeneous catalysts [12, 13], gives bio-oil as the main product, alongside gases and char. Bio-oil is a highly complex mixture of water, lignin fragments (substituted phenols), furans, carboxylic acids, hydroxyaldehydes and ketones, esters, alcohols, sugars, tar, and other products. This material can be considered "densified biomass" and can also be combusted to generate heat or electricity. Due to its reactive nature, the liquid gradually becomes more viscous and may undergo phase separation, and thus cannot be stored for prolonged periods in its crude state [14]. For use as transportation fuel, extensive catalytic upgrading is required [15]. Materials applications are also limited by its heterogeneous nature.

The third approach to biomass valorization encompasses a group of chemical/catalytic processes, which are usually targeted to a single product or a narrow range of products. These selective catalytic approaches for renewable feedstocks and chemicals are the main subject of this *Topics in Current Chemistry* volume, as introduced in the following section.

Before examining biomass conversion approaches in detail, we first briefly review the chemical composition of plant biomass. The main biomass components [16, 17] are cellulose [18–20], hemicellulose [21], starch, sugars, lignin [22], oils, fats, and waxes [23, 24], proteins [25, 26], and various extractives. Those most relevant to the biorefinery are shown in Fig. 3.

Cellulose, hemicellulose, and starch consist of polymers of simple sugars and thus are known as polysaccharides. Cellulose is the single most abundant organic

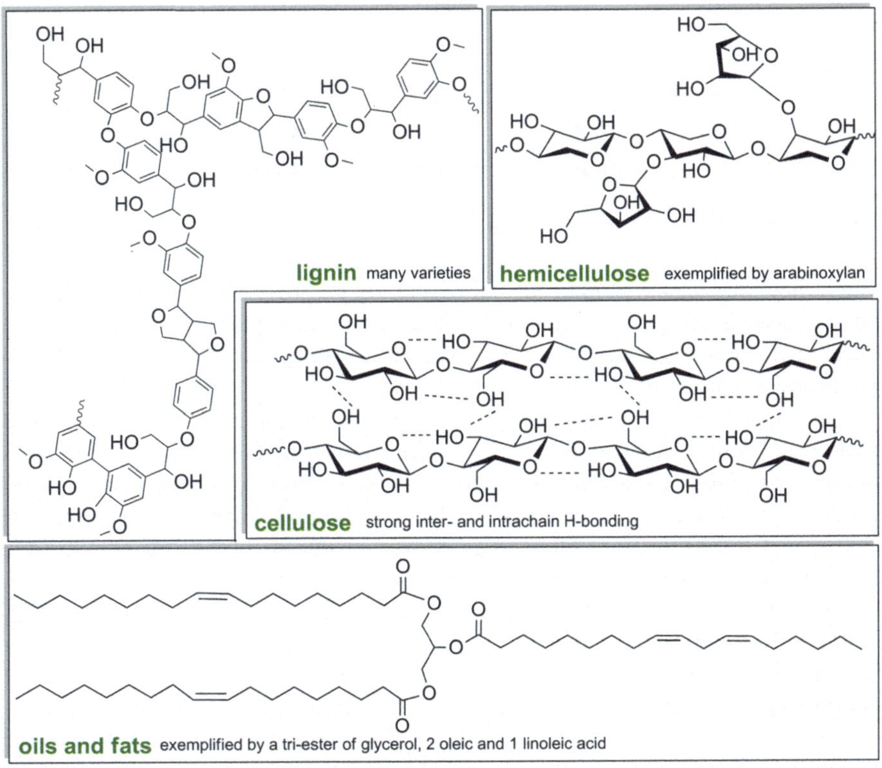

Fig. 3 The general chemical structure of the major biomass components. The *wavy bonds* indicate further attachment in the biopolymer structure

compound on Earth [19, 27]. This biopolymer consists exclusively of glucose, a six-carbon aldose sugar, whereas hemicellulose is primarily a pentosan (e.g., arabinoxylan in Fig. 3) based on five-carbon sugars, of which xylose and arabinose are the main representatives (next to ribose and six-carbon glucose, mannose and, galactose) [19, 20, 28]. The nature of hemicellulose is thus more complex, but, due to side chain branching, this biopolymer is amorphous and amenable to facile hydrolysis. Cellulose, on the other hand, is able to form intra- and inter-chain hydrogen bonds, which leads to the organization of the chains into planes, and the stacking of those plains [18, 29]. This renders cellulose a crystalline and hydrophobic material which is recalcitrant towards hydrolysis [30]. The other biopolymer, starch, is also abundant in nature and is much more easily processed into its glucose monomers than is cellulose due to its more chemically vulnerable α-glycoside linkage and amorphous tertiary structure. Besides the polysaccharides, to a lesser extent, mono- and disaccharides such as glucose, fructose, and sucrose are also encountered in plants, and thus may also be considered as feedstocks for the synthesis of chemicals and fuels. The use of sugars and starch as chemical precursors is, however, somewhat controversial, since these products are a primary source of human nutrition. The controversy about the use of the human edible carbohydrates or oils as replacements for petroleum should be interpreted with care and placed in perspective. It is a case-specific exercise, and highly dependent on issues of scale [25, 31]. Such an exercise would, for instance, be completely different for a low volume chemical target than for a biofuel.

Plant oils and micro-algal oils are generally composed of fatty acid tri-esters of glycerol, although micro-algal oils are often more diverse and can also contain phospholipids [24]. The future for the development of these as biorefinery feeds will depend on the extent to which they can compete with carbohydrates in terms of value as a carbon source for fuels and chemicals. Although they occur in a chemically more reduced state than sugars, they are generally found in much lower abundance in biomass. A co-product of the processing of these oils into fatty acid derivatives is glycerol, which is also considered as a biomass-derived carbon precursor [32–34].

Lignin, of which one variety is shown in Fig. 3, is a highly complex, branched biopolymer of phenolic and allylic alcohols, of which the main representatives are sinapyl, coumaryl, and coniferyl alcohols [22]. This material is amorphous but highly resistant towards catalytic/enzymatic hydrolysis. In Chap. 7 of this volume, the oxidative conversion of lignin is discussed in detail by Bozell and coworkers.

2 Introduction to Chemical Biomass Conversion Strategies

Chemical-catalytic methods of biomass valorization vary widely in their reaction conditions, but are generally fast and have the potential to utilize all of the available carbon in the feedstock. They have, up to now, attracted less commercial attention than fermentative and thermocatalytic or pyrolytic methods, possibly due to the fact that they are new technologies which involve considerable method development and thus invoke greater capital expenses on startup, whereas the classic approaches described above were already well established and available as and when the need arose. All the same, once an economic shakedown of the renewable technology sector comes about, it may well be that purely chemical technologies have the competitive edge due to their potential to process biomass inexpensively into versatile platform molecules in short times and under mild conditions. Catalysis, either in homogeneous or heterogeneous mode, presents a formidable toolbox for this purpose, as will be discussed below.

2.1 Chemical Conversion Routes to Biofuels

The conversion of biomass to automotive fuels has perhaps received the most attention of any chemical biomass conversion process. One of the most visible technologies under this classification is the aqueous phase reforming (APR) process [35], in which the oxygen content of carbohydrate feedstocks is reduced with in-situ generated hydrogen, the ultimate product after final hydrotreating being hydrocarbons, based on work originally described by Huber, Cortright, and Dumesic in 2004 [36, 37]. The main advantage of APR, now being commercially developed by Virent, Inc., is that the product hydrocarbons are already well integrated as fuels in the current automotive infrastructure. The disadvantages are: (1) complex, expensive metal catalysts and relatively high temperatures and pressures are required; (2) the overall yield of useable hydrocarbon products ($\geq C_8$) is often modest, with substantial quantities of carbon being stripped out in the reforming process as CO_2; and (3) the method operates only on sugars or starch, not directly on cellulose or cellulosic biomass. Despite the attention it has received, APR seems unlikely to form the basis of a renewables revolution, being ultimately displaced by technologies that can produce new generations of biofuels at a fraction of the cost. Since the APR process came to light, several other related approaches based on sugar dehydration, condensation reactions, and reductive stripping of oxygen have been described, again by Dumesic [38], but also others [39] in multiple reviews [40–42]. 5-(Hydroxymethyl)furfural (HMF) [37, 43], furfural[44], levulinic acid [45], and its derivatives γ-valerolactone [46, 47] and angelica lactone [48] are other platforms on which either condensation chemistry plus hydrogenation, [49] or conversion to olefins which can be oligomerized by well-described routes, are founded [50]. This subject will be taken up in greater detail in Chap. 2 of this volume Mascal and Dutta.

2.2 *Chemical Conversion Routes to Bio-Derived Chemicals*

As mentioned above, besides the production of low-margin/high-volume biofuels, the integrated lignocellulosic biorefinery should also produce higher value chemicals. However, as Bozell et al. clearly pointed out in 2010, '*the choice of appropriate products for addition to the biorefinery's portfolio is challenged by a lack of broad-based conversion technology coupled with a plethora of potential targets*'[51]. Due to the vast number of potential derivatives of carbohydrates, the development of a set of broad-based catalytic conversion technologies is hampered by the need to catalyze a diverse range of reactions effectively. In practice, the use of tailored catalysts to effect specific transformations within a reaction network to produce a desired set of chemicals is necessary. For instance, cellulose might be hydrolyzed to glucose, which might be dehydrated to HMF, rehydrated to levulinic acid, and hydrogenated to γ-valerolactone, or it might be hydrogenated to sorbitol, then dehydrated to sorbitans and ultimately to isosorbide.

The approach to choosing chemicals and processes will form the subject of the remainder of this chapter. A summary of the most important chemicals from cellulose and hemicellulose is shown in Fig. 4.

To start with, the most abundant five- and six-carbon sugars and some of their isomers, as well as shorter chained sugars, are indicated in green. Chemicals produced by fermentative or biocatalytic processes of carbohydrate feedstocks are

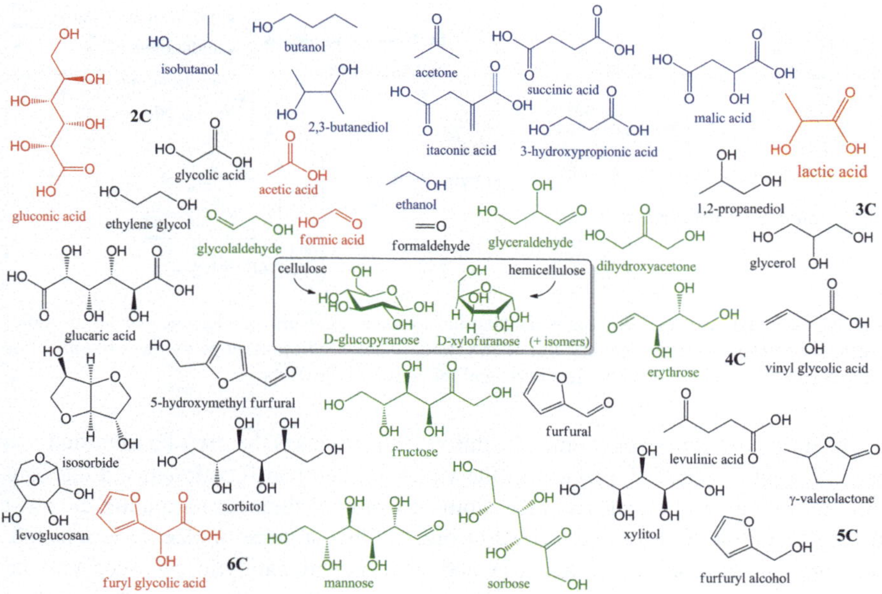

Fig. 4 Prominent conversion products of carbohydrates. *Green*: sugars. *Black*: via chemical catalysis. *Blue*: via bio-catalysis. *Red*: via bio-and/or chemical catalysis

indicated in blue. The chemicals in black are usually formed via catalytic processing. For the chemicals in red, both bio- as well chemocatalytic routes starting from hexoses and pentoses have been reported. Aside from the fermentation products, within more or less similar reaction conditions, different chemicals will be formed from different thermodynamically allowed reaction channels, each by using different catalyst types or by altering the balance of the active sites accordingly. A fundamental understanding of the potential reaction network is therefore crucial in order to tackle the selective conversion of biomass towards targeted molecules and to design the process steps and the catalysts required.

The choice of the most cost-effective approach to producing chemicals from carbohydrates is thus challenging, as it depends on many chemical, technical, and economic factors. The purpose of this introduction is not to provide a comprehensive techno-economic study for each target chemical, but rather to establish a methodology for assessing and selecting the most viable chemicals from biomass, here illustrated for carbohydrate biomass, with respect to the preservation of functionality and high atom economy from a green chemist's perspective. Several reviews have been published summarizing the literature on the catalytic conversion of cellulose to chemicals [17, 19, 20, 52, 53]. The content discussed in the present chapter significantly complements this work, as it proposes to offer a toolbox for the selection of the most relevant chemicals from carbohydrate biomass and to suggest the preferred synthetic routes, in a context that bears in mind the competition with classic petrochemicals derived from oil and natural gas.

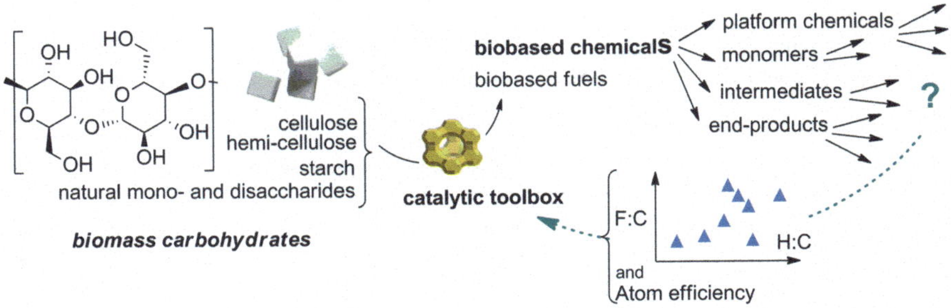

Fig. 5 Conversion of carbohydrates to chemicals requires the informed choice of target molecules. Preservation of functionality and atom economy are two important parameters to guide this choice and to search for the best reactions and the required catalysts

To guide the choice of chemicals, this review proposes the use of a modified *van Krevelen* plot in Sect. 3.1 to assess the value of chemical functionality. Functionality of the target chemicals and the atom economy of their formation are the most decisive guiding tools, as visualized in Fig. 5. Once a target molecule is selected, pursuing an atom-efficient reaction and an efficient catalytic protocol will be imperative. Viable reactions within this context make up the scope of Sect. 3.2. Note that the catalytic conversion of lignin to chemicals is not within the theme of this chapter, but a similar exercise as performed here for carbohydrates may likewise be executed. Though somewhat more difficult than cellulose conversion because of its complex and variable chemical structure, catalytic lignin conversion has become important as well and will receive increasing attention in the future. The reader is referred to a recent review regarding the potential of lignin in the chemical industry [22] and to Chap. 7 in this volume.

3 A Chemist's View of the Selection of Viable Target Molecules and Their Formation in Cellulosic Biomass Conversion to Chemicals

3.1 Modification of van Krevelen Plots to Verify the Selection of Bio-Derived Chemicals

3.1.1 Introduction

In contrast to natural gas, coal, and oil, polysaccharide feedstocks have a very high oxygen content, with an *O*:*C* ratio near unity. In the assessment of valorization paths for the use of carbohydrate biomass to produce chemicals, a US Department of Energy report was published in 2004. The final top 12 chemicals out of more than 300 potential candidates were selected, mainly based on their market potential. The

Table 1 Top chemical opportunities from carbohydrates as given by Bozell and Petersen [51], ordered by total carbon number (#C) with functional group analysis (defined below)

#C	Compound	*F* value	*F*:*C*
2	Ethanol	1	0.5
3	Lactic acid	4	1.33
3	3-Hydroxypropanoic acid	4	1.33
3	3-Hydroxypropionaldehyde	3	1
3	Glycerol	3	1
4	Succinic acid	6	1.5
5	Levulinic acid (LA)	5	1
5	Xylitol	5	1
5	Furfural	6	1.2
5	Isoprene (biohydrocarbon)	2	0.4
6	Sorbitol	6	1
6	5-(Hydroxymethyl)furfural (HMF)	7	1.17
6	Furandicarboxylic acid (FDCA)	10	1.67

report has been used by many researchers as their guideline to identify and develop routes to the most important platform chemicals [54]. Bozell and Petersen re-analyzed this list in 2010, mainly based on literature citations (as a measure for technology development) and multiple product applicability, among seven other criteria. Details of their revised top 13 chemical opportunities from biorefinery carbohydrates are collected together in Table 1 [51]. Clearly, the selected molecules, including carboxylic acids, alcohols, ethers and aldehydes, contain a high degree of chemical functionality, bearing at least one functional group in their molecular structure. Such functionality is beneficial as these selected platform chemicals can serve as entry points for further chemical transformation [16].

Beyond this careful selection of platform chemicals, the production of a number of other polymer building blocks, chemical intermediates, and end products from carbohydrates has been reported. Due to the unique oxygen-rich composition of carbohydrates, their conversion into renewable chemicals that preserve the functional groups is an advantage over the current petroleum and natural gas conversion routes. Biomass conversion with high atom efficiency is a key aspect of the competitive synthesis of chemicals and chemical-based products.

The production strategy for biofuels contrasts with that of renewable chemicals: here, it is not so much the atom economy but rather the heating value of the product that is of ultimate interest. The energy content of molecules is mainly associated with the number of C–H bonds present. Therefore, complete defunctionalization protocols, using a set of reactions such as C–O hydrogenolysis, decarbonylation, decarboxylation, and hydrogenation are followed, resulting in a lowered atomic *O*:*C* ratio, enhancing the energy content of the resulting fuel [13, 41, 43, 55–58]. An inherent trait of defunctionalization is the consumption of (expensive) H_2 and/or CO_2 formation. Biofuel synthesis from cellulosic feedstocks thus requires a cheap source of hydrogen to be economically viable, unless the energy content in the molecule is provided by a loss of CO_2 as in the fermentative production of ethanol, but then a low carbon efficiency is obtained.

As the synthesis of transportation fuels is a high-volume, low-margin business, process economics and feedstock logistics might be problematic with biomass, while such scale-related issues are of less concern for renewable chemical and polymer synthesis. Several countries have programs in place which offer governmental incentives and subsidies to compensate for the economic shortcomings of biofuel production. Only in Brazil has the production of biofuels like ethanol been economically sustainable, but it is a matter of strong debate whether the complete replacement of liquid transportation fuels by biofuels is feasible worldwide. Thorough life cycle assessments and unbiased economic analyses should help to determine the sustainability and viability of biofuel production and use [59–62].

3.1.2 The Original *Van Krevelen* Diagram

The original *van Krevelen* diagram, proposed in 1950 by Dirk van Krevelen [63], is frequently used in geochemistry and coal research. It typically plots the atomic H:C ratio (H index) of a chemical feedstock on the y-axis vs the atomic O:C molar ratio (O index) on the x-axis. A classic plot of this nature is exemplified in Fig. 6.

As can be seen, major biogeochemical classes of compounds like lignin, lignite, peat, anthracite, cellulose, coal, and oils occupy different regions in the diagram. Typically, the *van Krevelen* diagram has been used to assess the evolution and origin of coal and oil samples. In such plots, trends along lines are indicative of structural relationships among groups of compounds, which arise from a number of typical reactions that influence both the H and O index – e.g., dehydration, decarboxylation, demethanation, dehydrogenation, hydrogenation, and oxidation [65]. van Krevelen mainly applied this methodology to monitor coalification series [63]. In a way, the formation of coal or crude oil from plant matter containing lignin, hemicellulose, and cellulose – a process taking thousands of years in the crust of the Earth – is comparable to the transformation of cellulose or lignin into fuels and chemicals in a chemical reactor, albeit on a different time scale. Different conditions are obviously applied and, due to the enormous global demand for chemicals, the use of rate-accelerating catalysts is paramount to achieving fast conversion on a feasible time scale, rendering the processes cost-efficient. The O index in the *van Krevelen* plot is ideal for assessing the production of fuels and deoxygenation series with a focus on increasing the calorific value, e.g., from wood to peat to coal (Fig. 6) or the upgrading of bio-oils [66].

However, when looking at the *van Krevelen* plot for the prominent carbohydrate-based platform chemicals, monomers, and intermediates containing functional groups (Fig. 7), the plot is not very informative for distinguishing between various choices of chemicals. For example, all sugars occupy one point on the diagram ($x, y = 1, 2$), a place shared with other chemicals such as lactic and acetic acids. Deoxygenated chemicals are indeed found in one region of this plot, namely on the left-hand-side, but apart from this trend, evaluation of the

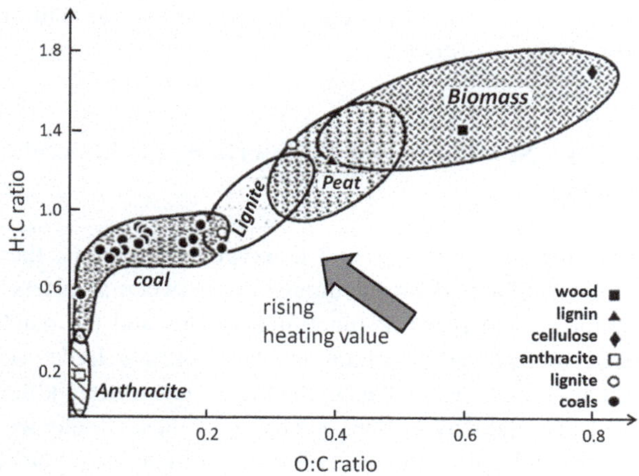

Fig. 6 A typical van Krevelen diagram, loosely based on [64]

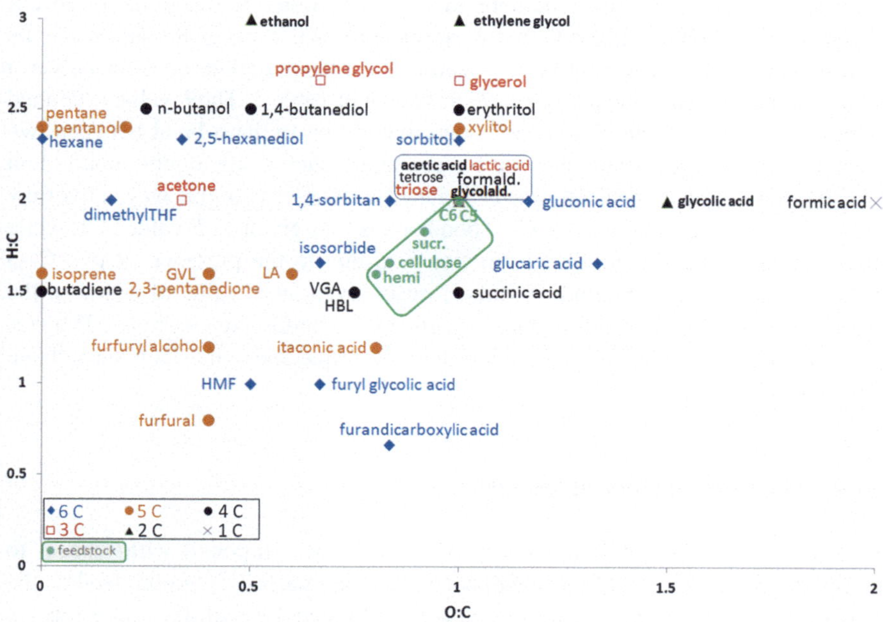

Fig. 7 *van Krevelen* plot for carbohydrate-derived chemicals, colored by carbon number. *VGA*: vinyl glycolic acid, *HBL*: hydroxybutyrolactone, *GVL*: γ-valerolactone, *C5*: pentoses, *C6* = hexoses, *LA*: levulinic acid, *Sucr.*: sucrose, *Hemi.*: hemicellulose

functionality of the chemicals is challenging, and requires one to look diagonally at the plot. The main carbohydrate feedstocks of importance are indicated in a green box on the plot: sucrose, cellulose, hemicellulose, glucose, and pentoses. Functional target molecules are found both on the left (e.g., isosorbide and HMF) and right (e.g., glucaric acid) as well as below (e.g., furandicarboxylic acid or FDCA) and above (e.g., sorbitol) the feedstock box.

3.1.3 Introducing a New Parameter *F*, Describing the Molecular Functionality

While appropriate for fuel purposes and deoxygenation series, the original *van Krevelen* plot is not well suited for distinguishing between a variety of chemicals with different numbers and types of functional groups and the reactivity derived therefrom. Indeed, since all carbohydrates and even formaldehyde occupy the same point, one cannot appreciate, for instance, the higher reactivity and functionality of formaldehyde and glycolaldehyde. The reactivity of these aldehydes is more pronounced than in molecules like glucose, which exist in stable, cyclic acetal forms. Thus, to study the functionality of chemicals derived from original biomass feedstocks like cellulose, hemicellulose, sucrose, and monosaccharides, an adaptation is necessary. Here, we propose the introduction of a new functional group parameter, denoted by the letter *F*. This parameter takes into account the functional groups of chemicals based on C, H, and O, but its application is flexible as it could easily be extended also to heteroatoms like N, S, and P. Each functional group is thus given a relative contribution to the parameter *F*, as shown in Table 2. The *F* value is defined as follows: each C–O bond and each bond in addition to the σ-bond between two carbon atoms or one carbon and one oxygen atom, such as the double bond in an alkene or an aldehyde, respectively, is assigned a value of 1. In this way, a hydroxy group, for instance, contains one C–O bond and is thus given an *F* value of 1, while the *F* value of a carboxylic acid function is 3 due to the presence of two C–O σ-bonds and one C–O π-bond. A furan ring is given an *F* value of 4 due to the presence of two double bonds and the two ether C–O bonds. Accordingly, HMF has a total *F* value of 7, whereas glucose has the same value due to its aldehyde and five OH groups.

3.1.4 The Functionality Index *F*:*C*

To assess the degree of functionality of a chemical compound with respect to its carbohydrate feedstock, its functional group parameter *F* is normalized to the number of carbon atoms, giving the ratio *F*:*C*. In analogy with the *van Krevelen* plot, we call *F*:*C* the functionality index. A plot of the functionality index against the oxygen index is given in Fig. 8. The diagonal presents the chemicals for which both indexes share the same value. Chemicals on the diagonal are the mono-, di-, and poly-alcohols, while the origin represents alkanes. It is interesting to note the large number of chemicals off the diagonal, confirming a substantial refinement in the assessment of functionality using the *F*:*C* index. In contrast to the *O* index, the *F*:*C* index can, for example, differentiate chemicals by their degree of unsaturation. Thus, there is a shift to the right for acetone vs propanol due to the

Table 2 Contribution to F of the most common functional groups in oxidized hydrocarbons

Type of functional group	Contribution to F
Alcohol	1
Aldehyde	2
Ketone	2
Ether	2
Carboxylic acid	3
C=C	1
C≡C	2

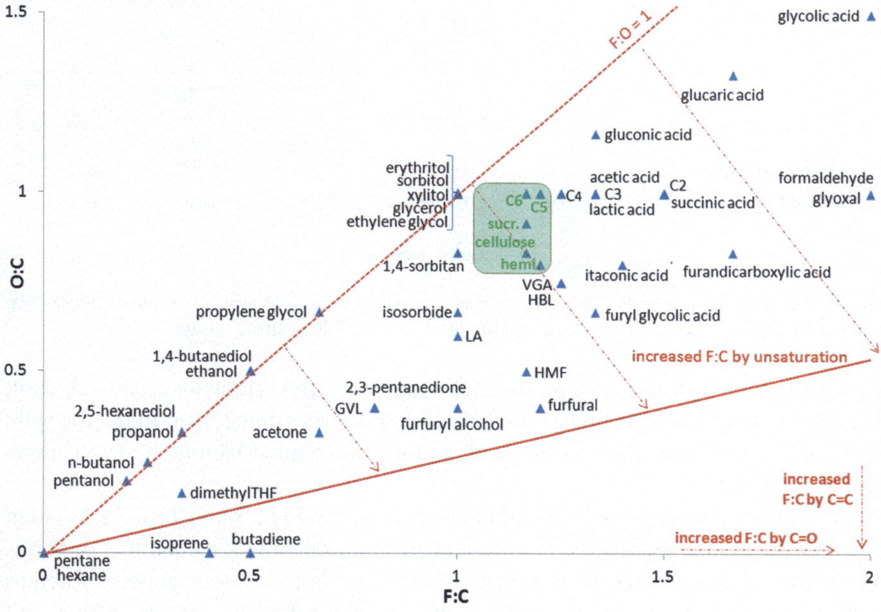

Fig. 8 $O{:}C$ vs $F{:}C$ ratio for a range of carbohydrate-derived chemicals and derivatives thereof. Feedstocks highlighted in *green box*. Cx: resp. x-carbon monosaccharide

increase in functionality, while the position of a chemical drops with the presence of an unsaturation in the carbon skeleton (e.g., olefins and aromatics) relative to oxygen functionality, as seen by comparing 1,4-butanediol and butadiene for instance. In this way, the $F{:}C$ index as such adds useful information about the reactivity of chemicals.

Figure 9 illustrates the $F{:}C$ values for a range of different target molecules from carbohydrate biomass, which are generally representative of the published work in this field, ordered by carbon number. The plot is supplemented with derivatives of these chemicals and some petrochemicals as well for the purposes of comparison. This group of molecules comprises feedstock sugars, sugar alcohols (sorbitol), ethers (isosorbide), aldehydes, ketones, alcohols, diols, glyoxals, and acids. Moreover, a number of compounds containing more than one kind of functional group

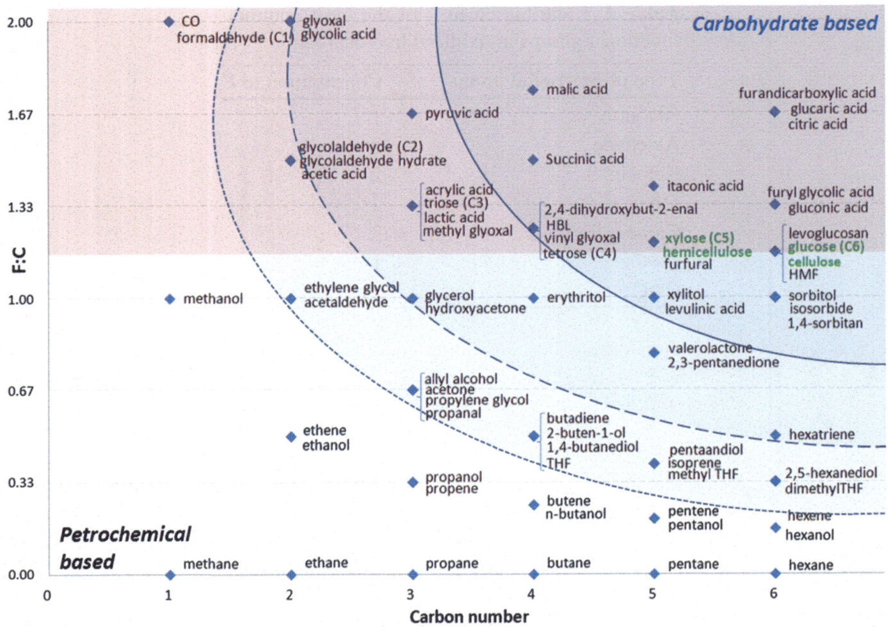

Fig. 9 *F*:*C* ratio for a selection of carbohydrates, carbohydrate derivatives, and other chemicals, ordered by carbon number. Feedstocks highlighted in *green*. *Blue arcs*: see text

are also included: α-hydroxy acids (lactic and vinylglycolic acids), allyl alcohol, furan derivatives such as HMF (furan, alcohol, aldehyde) and 2,5-furandicarboxylic acid, other acids such as levulinic acid, and the substituted lactone hydroxybutyrolactone (HBL).

The chemical structure of most of the compounds in Figs. 8 and 9 can be found in Fig. 4. Table 1 also reports the calculated *F*:*C* values for the top chemical opportunities from carbohydrates as defined by Bozell and Petersen (see above), whereas Table 3 (found under section 3.1.6) provides a list of *F*:*C* values for the majority of commonly reported carbohydrate-based platform chemicals, monomers, and intermediates. The abscissa of the plot in Fig. 9 contains the alkanes, with a zero *F* value; they have no functionality and are considered chemically inert. From a strictly functional point of view, producing alkanes from carbohydrates is not a preferred conversion route.

Carbohydrate feedstocks such as cellulose, sucrose, and starch have an *F*:*C* value of 1.17, whereas hemicellulose has a value of 1.2. A tentative zone has been indicated in red in Fig. 9 for all chemicals with an equal or higher *F*:*C* value than cellulose. A drop below this line is equivalent to a loss of functionality. The biopolymers share their *F*:*C* value with their respective monomers, thereby simplifying hemicellulose to a pure pentosan. While the five- and six-carbon sugar monomers have values near 1.2, shorter-chained sugars have a higher *F*:*C* index; tetroses, trioses, and the smallest sugar, glycolaldehyde, attain *F*:*C* values of 1.2, 1.33, and 1.5, respectively. This rising trend parallels the higher reactivity of the lower sugars.

Most chemicals above the feedstock line are chemicals with a high added-value, among them formaldehyde, glyoxal with its two carbonyl groups, and a number of organic acids such as glycolic, pyruvic, and malic acid, and even furan or sugar-based acids. Formic acid, which can also be sourced from biomass [69, 71, 89, 90], has a value of 3, rendering it unsuitable to plot here for scaling reasons. The highest *F*:*C* of 4 is found in CO_2, but this is an artifact due to over-oxidation and not a reflection of value. The molecules below CO_2 and above the feedstock line are generally obtained by a selectively catalyzed oxidation reaction or fermentation. Such highly functionalized molecules are interesting targets as long as O_2 (or alternatively H_2O_2) can be used as the oxidant. Provided that the selectivity is kinetically under control, such exothermic reactions are favorable. CO may also be formed through endothermic reforming, but such a route is energetically less motivating.

This F:C analysis is based on a chemical's structure and allows comparison to the structure of a possible feedstock, but it does not imply that a product with high F:C is highly valuable or useful. The market for a given chemical will ultimately determine its value, and a (future) market may not per se exist for all chemicals assessed and selected here. The opposite is also true, as a chemical with a low F:C value may still be highly sought after.

It is evident from the plot that the highest degree of functionalization per carbon is obviously achieved in the shortest molecules. However, one should take into account that the shorter the carbon chain, the more competition is expected from petrochemical routes for its formation. Such routes typically use natural gas compounds like methane and ethane and cracked hydrocarbons like ethylene and propylene. The lower left corner is thus occupied by petrochemicals, whereas in the upper right corner highly oxidized six-carbon compounds are found. To highlight the difference between these corners, zones are indicated on the chart by expanding, concentric arcs:

- The innermost zone covers chemicals for which no petrochemical route exists, and therein lies a great opportunity to produce them from carbohydrate feedstocks (in the case that a market for them is present or arises): gluconic [91, 92] and glucaric acids [17], furyl glycolic acid [87], furandicarboxylic acid [85], and fermentation-derived malic, citric, succinic, and itaconic acids [17]. Besides these, the sugar alcohols xylitol [78] and sorbitol [81, 93–95], as well as isosorbide [82, 96, 97], HMF [80, 98], and furfural [44, 99] are also present, along with tetroses and their derivatives obtained by dehydration [100]. For these molecules, it is virtually certain that no petrochemical-based production will ever be developed. Note that levulinic acid [67, 101–103] is at the border of this zone.
- The second zone also contains carbohydrate opportunities, exemplified by, e.g., glycolic [73, 104], pyruvic [105], lactic [74], and acrylic acids as well as glyoxal, glycolaldehyde [72, 106], acetic acid [71], erythritol [70], and γ-valerolactone [47, 77]. A petrochemical route to these chemicals could be derived (e.g., the route to lactic acid via hydrolysis of acrylonitrile [107]), but in general, the carbohydrate route is likely to be cost-competitive. Glyoxal, acetic acid, and

acrylic acid are exceptions, as they are all petrochemicals at this point, although bio-based routes from carbohydrates are receiving more attention, especially for the acids [71, 88, 108].

- The third zone encompasses those chemicals for which there is potential competition between the carbohydrate and petrochemical routes. The ultimate choice will depend on production costs, economies of scale, technological advances, or political directives. This zone includes, for instance, isoprene and butadiene – to which bio-routes, e.g., via ethanol [109] and 1,4-butanediol, are increasingly drawing attention [110, 111], as well as allyl alcohol[112, 113] and ethylene and propylene glycols [75, 114]. The latter three chemicals may be derived from renewable glycerol for instance [32, 115], and for ethylene glycol a direct carbohydrate approach is becoming increasingly viable as an alternative to the petrochemical route [70, 116].

Chemicals with *F*:*C* values comparable to the starting mono-, di-, and polysaccharides are produced without a net loss in functionality. Their formation typically requires a limited number of reagents, and the number of by-products is often low, one generally being water. Examples of such conversions are, for instance, the formation of HMF, sorbitol, and isosorbide from the six-carbon sugars, and furfural, xylitol, furfuryl alcohol, and levulinic acid from the five-carbon sugars (in case of retention of total carbon number; vertical direction on the plot). If the cleavage of C–C bonds is allowed (horizontal direction), the formation of tetroses, vinyl glyoxal, trioses, lactic acid, acrylic acid, methyl glyoxal, ethylene glycol, and glycolaldehyde is possible, among others. Gamma-valerolactone, cyclic ethers like THF and dimethyl furan, hexatriene, diols like the butanediols and propanediols, butadiene, and allylic alcohol, for example, have lost a significant amount of functionality per carbon, compared to a carbohydrate feedstock.

Other chemicals like the one- to six-carbon alkenes and the corresponding alcohols are outside the three zones. According to our assessment, they are better formed by converting natural gas and oil feedstocks, e.g., by thermal cracking chemistry and catalytic oxidation or the Fischer–Tropsch process via syngas [117], ideally from waste, biogas, or biomass resources. They will require a high input of energy and H_2 if generated from cellulosic biomass. Irrespective of the hydrocarbon source used in these cases, H_2 production is always associated with the formation of CO_2, either directly as in the aqueous reforming of biomass [36, 84, 118, 119], or indirectly by compensating for the endothermic nature of the reaction as in case of steam and dry reforming [120–122].

To summarize, the inner circle domain, as presented in Fig. 9, emphasizes the chemicals that are (or will be) exclusively synthesized from cellulosic carbohydrate feedstocks (in case of market demand). The second circle encompasses those chemicals that are preferentially synthesized from carbohydrates, although in some cases alternative petrochemical routes for their formation exist. The competition between both comes down to a question of economic factors. The third circle is a truly competitive region and, for now, petrochemical routes are dominant (e.g.,

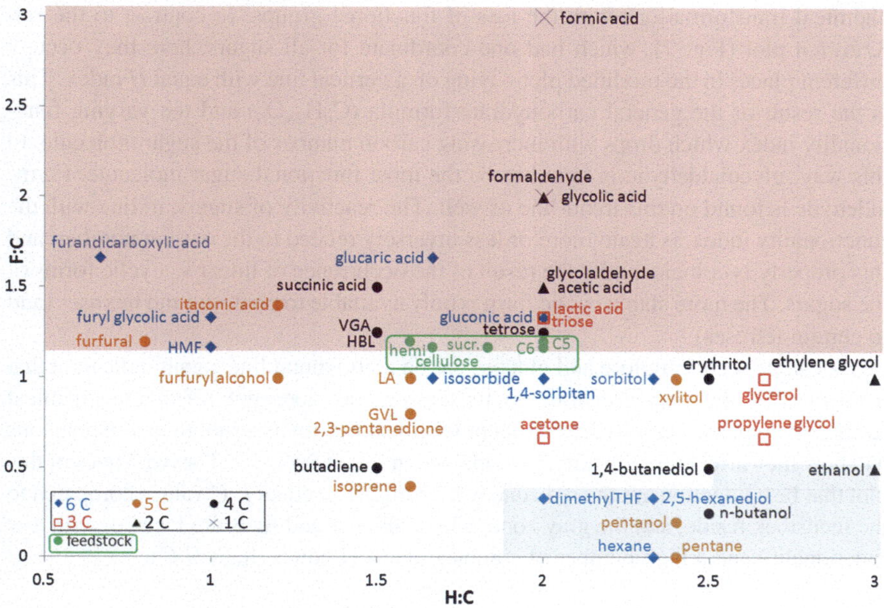

Fig. 10 Modified *van Krevelen* plot for carbohydrates and their derived products, colored by carbon number

butadiene and ethylene glycol). By closer inspection of these chemicals and taking into account their current petrochemical derivation, it is evident that real competition from cellulosic feedstocks can be expected in the event that a cheap source of H_2 gas is identified (possibly solar) and the defunctionalization of carbohydrates thus becomes less expensive. This trend is more pronounced with the size of the molecules, e.g., in the formation of dimethyltetrahydrofuran. Note that the assessment assumes the use of carbohydrate feedstocks, and does not take into account the use of other biomass related feedstocks, such as glycerol. Somewhat different conclusions might be drawn if such feedstocks were included.

3.1.5 Modifying the Original *Van Krevelen* Plot: Functionality vs *H*-Index

Plotting the functionality index F:C as a function of the H:C ratio (the original y-axis in *van Krevelen* plots) also provides an improved tool to assess functionality of bio-derived chemicals and the type of reactions involved in their synthesis. Figure 10 shows such a modified *van Krevelen* plot. The green box marks the cellulosic and hemicellulosic feedstocks, while the other molecules are represented with color depending on their carbon number. The red zone indicates F:C values above the 1.17 line (as in Fig. 9), which represents chemicals with a functionality index higher than or equal to that of the feedstock. Molecules below this line have undergone chemical transformations with the loss of functional groups. In contrast to the *van Krevelen* plot (Fig. 7), which had one coordinate for all sugars, here they occupy different places in the modified plot – lying on a vertical line with equal *H* index. This is the result of the general carbohydrate formula ($C_nH_{2n}O_n$) and the varying functionality index which drops with increasing carbon number of the sugar molecule. In this way, glycolaldehyde is found to be the most functional sugar molecule. Formaldehyde is found on this trend-line as well. The reactivity of sugars, in line with the functionality index, is again more or less inversely related to the carbon number, and this property is coincidentally the result of the occurrence of linear vs cyclic forms of the sugars. The more stable cyclic form is only available to pentoses and hexoses (and to certain tetroses).

As can be seen, cellulose and glucose are on a horizontal line, again indicating that hydrolysis of the polysaccharide to its constituents does not affect the chemical functionality of the carbohydrate. This plot is ideal for use as a tool to assess reactions and how they affect functionality, as explained in full in Sect. 3.2. The two zones of this plot that bear attention are the red zone, where higher or equal *F*:*C* values compared to the feedstock reside, and the gray zone, where alkanes and fuels are located, with low functionality and a high number of combustible C–H bonds (high *H*:*C*).

3.1.6 Guidelines for the Selection of Carbohydrate-Derived Chemicals: Combining Functionality Index and Atom Economy

Atom economy (AE), first defined by Barry Trost in 1991, is a simple but extremely useful tool to describe the conversion efficiency of a chemical process [123]. It is calculated by dividing the molecular weight of the desired product by the total of the molecular weights of the reagents used in the stoichiometric equation for the reaction(s) involved. If no by-product is formed, the atom economy is 100%. While initially developed for the synthesis of fine chemicals, higher costs of raw materials and the increased concern for the environment have made the application of atom economy and other green metrics very popular [124–127]. However, systematic atom-economical approaches to the conversion of biomass feedstocks have not often been reported, in spite of their combined "green" messages. Table 3 lists the atom economy of the chemical reactions used to produce various chemicals. The calculation of atom economy was based on current technologies and the described net reactions in the literature for each chemical. For instance, the fermentative synthesis of ethanol is known to produce two molecules of ethanol from glucose, while also forming two molecules of CO_2, yielding an AE value for ethanol of 51 % as seen in Table 3. In Table 3 in general, the AE was calculated for primary products of pentoses and hexoses (or their polymers), as well as for some interesting downstream derivatives, calculated as if they were produced in a single step from the sugar reagent. An example of this is furandicarboxylic acid via oxidation of HMF. Thus, HMF (formed by triple dehydration of glucose or fructose) is oxidized with 1.5 equiv. O_2 to yield furandicarboxylic acid and 1 equiv. of H_2O. Based on HMF from glucose, this means that the total co-product generation for FDCA from glucose consists of four water molecules.

Table 3 Net reaction stoichiometry

In					Out							
Feed	Mw_r	Co-r	$Mw_{(co-r)}$	Technology/or main reaction	Main product	#	Mw_p	Co-products	$Mw_{(co-p)}$	AE %	*F:C*	References
Cell	342	H_2O	18	Hydrolysis	Glucose	2	180			100	1.17	[30]
Hcell	282	H_2O	18	Hydrolysis	Xylose	2	150			100	1.20	[137]
G	180	$3O_2$	96	Retro-aldol+oxidation	Formic acid	6	46			100	3.00	[69]
G	180	–		Retro-aldol	Formaldehyde	6	30			100	2.00	[69]
G	180	–		Fermentation	Ethanol	2	46	CO_2	88	51	0.50	[17]
G	180	$3H_2$	6	Retro-aldol+H_2	Ethylene glycol	3	62			100	1.00	[70]
G	180	–		Fermentation	Acetic acid	3	60			100	1.50	[71]
G	180	–		Retro-aldol	Glycolaldehyde (C_2)	3	60			100	1.50	[72]
G	180	1.5O2	48	Retro-aldol+oxidation	Glycolic acid	3	76			100	2.00	[73]
G	180	–		Retro-aldol	Triose (C_3)	2	90			100	1.33	[74]
G	180	$2H_2$	4	Retro-aldol+H_2	Glycerol	2	92			100	1.00	[70]
G	180	$4H_2$	8	Retro-aldol+hydrogenolysis	Propylene glycol	2	76	$2H_2O$	36	81	0.67	[75]
G	180	–		Retro-aldol+dehydration	Methyl glyoxal	2	72	$2H_2O$	36	80	1.33	[74]
G	180	–		Retro-aldol+isomerization	Lactic acid	2	90			100	1.33	[74]
2G	360	–		Retro-aldol	Tetrose (C_4)	3	120			100	1.25	[138]
G	180	CO_2	44	Fermentation	Succinic acid	1	118	Acetic+formic acid	106	53	1.33	[17]
2G	360	–		Fermentation	*n*-Butanol	1	74	Acetone+$5CO_2$+$4H_2$	286	21	0.25	[17]
2G	360	–		(Retro-)aldol+isomerization	Vinyl glycolic acid	3	102	$3H_2O$	54	85	1.25	[138]
X	150	–		Fermentation	Isoprene	1	68	Complicated		45	0.40	[–]
X	150	–		Dehydration	Furfural	1	96	$3H_2O$	54	64	1.20	[44]
X	150	H_2	2	Dehydration+H_2	Furfuryl alcohol	1	98	$3H_2O$	54	64	1.00	[77]
X	150	H_2	2	Dehydration+rehydr.+H_2	Pentosic LA	1	116	$2H_2O$	36	76	1.00	[77]
X	150	H_2	2	Hydrogenation	Xylitol	1	152			100	1.00	[78]
G	180	–		Isomerization	Fructose	1	180			100	1.17	[79]
G	180	$1.5O_2$	48	Fermentation	Itaconic acid	1	130	CO_2 + $3H_2O$	98	57	1.75	[17]
G	180	–		Dehydration	HMF	1	126	$3H_2O$	54	70	1.17	[80]

(continued)

Table 3 (continued)

In					Out							
Feed	Mw_r	Co-r	$Mw_{(co-r)}$	Technology/or main reaction	Main product	#	Mw_p	Co-products	$Mw_{(co-p)}$	AE %	*F:C*	References
G	180	–		Dehydration+rehydration	Hexosic LA	1	116	$HCOOH + H_2O$	64	64	1.00	[67]
G	180	H_2	2	Hydrogenation	Sorbitol	1	182			100	1.00	[81]
G	180	H_2	2	Hydrogenation-dehydration	Isosorbide	1	146	$2H_2O$	36	80	1.00	[82]
G	180	$0.5O_2$	16	Oxidation(gold)	Gluconic acid	1	196			100	1.33	[83]
G	180	O_2	32	Oxidation	Glucaric acid	1	212			100	1.67	[17]
G	180	–		Dehydration	Levoglucosan	1	162	H_2O	18	90	1.17	[–]
G	180	H_2	2	Dehydration	Sorbitan	1	164	H_2O	18	90	1.00	[82]
Derivative products via												
G	180	$5H_2$	10	Sorbitol hydrogenolysis	2,5-Hexanediol	1	118	$4H_2O$	72	62	0.33	[41]
G	180	$7H_2$	14	Sorbitol hydrogenolysis	Hexane	1	86	$6H_2O$	108	44	0.00	[84]
X	150	$6H_2$	12	Xylitol hydrogenolysis	Pentane	1	72	$5H_2O$	90	44	0.00	[84]
G	180	$1.5O_2$	48	HMF oxidation	Furandicarboxylic acid	1	156	$4H_2O$	72	68	1.67	[85]
G	180	H_2	2	HMF rehydr.+H_2	Hexosic GVL	1	100	$HCOOH + 2H_2O$	82	55	0.80	[47]
X	150	$2H_2$	4	Furfural rehydr.+H2	Pentosic GVL	1	100	$3H_2O$	54	65	0.80	[77]
G	180	$5H_2$	10	HMF hydrogen-ation/olysis	DimethylTHF	1	100	$5H_2O$	90	53	0.33	[150]
G	180	O_2	32	Cortalcerone dehydration	Furyl glycolic acid	1	142	$2H_2O + H_2O_2$	70	67	1.33	[87]
G	180	–		Lactic acid dehydration	Acrylic acid	2	72	$2H_2O$	36	80	1.33	[88]

AE atom efficiency and *F:C* value for common biomass primary products and some derivatives. Calculated straight from pentose or glucose-based biomass. *Cell*: cellulose, *Hcell*: hemicellulolse, *G*: glucose, *X*: xylose, *Mw*: molecular weight, *co-r*: co-reagents, *co-p*: co-products, *GVL*: γ-valerolactone, *LA*: Levulinic acid, dimethylTHF: dimethyltetrahydrofuran #: stoichiometry of product formation with respect to the feedstock

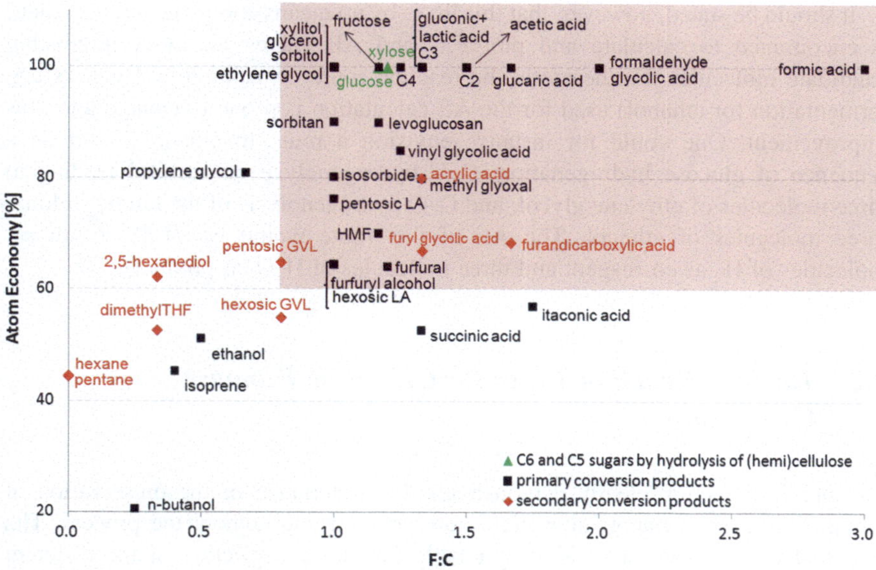

Fig. 11 Atom economy vs *F*:*C* plot for common biomass primary products and some derivatives, calculated directly from the pentose or glucose biomass. Cx: resp. x-carbon monosaccharide Data: see Table 3

Using the data in this table, an atom economy vs *F*:*C* plot was constructed, as shown in Fig. 11. Where high functionality and atom economy are key objectives, Fig. 11 may be used as a guideline to recognize the preferred chemicals derived from the carbohydrate feedstocks (in green). The best examples of an atom economical and functional group efficient conversion are found in the upper right corner. A tentative red zone has been indicated, with a minimum allowed *F*:*C* and AE of 0.8 and 60%, respectively. While propylene glycol, GVL, and succinic and itaconic acids are borderline examples, most preferred chemicals from carbohydrate biomass based on this analysis - which is strictly chemical and ignores whether or not a product has relevant uses in existing or potential markets - are among the following 25 compounds:

- *In the family of carboxylic acids:* formic acid, glycolic acid, glucaric acid, gluconic acid, acetic acid, and lactic acid and, to a lesser extent, in following order: vinyl glycolic acid, acrylic acid, furandicarboxylic acid, furyl glycolic acid, and levulinic acid (the latter, preferably, from five-carbon sugars due to a higher AE).
- *In the family of polyols*: ethylene glycol, glycerol, erythritol, xylitol, and sorbitol.
- *In the family of ethers*: sorbitan, isosorbide, and levoglucosan.
- *In the family of aldehydes*: formaldehyde, glycolaldehyde, and methyl glyoxal (pyruvic aldehyde).
- *In the family of aromatics*: HMF, furfural, and furfuryl alcohol

It should be stated, however, that this list is by no means complete and the reader is encouraged to calculate and plot *F*:*C* and AE values for other interesting candidate molecules. Furthermore, the reaction route or technology (such as e.g. fermentation for ethanol) used for the AE calculation is prone to change and thus improvement. One could for instance envision a route to ethanol based on a sequence of glucose hydrogenation, C–C hydrogenolysis of sorbitol leading to three molecules of ethylene glycol, and C–O hydrogenolysis of the latter, yielding three molecules of ethanol. The AE of this route would be 71.8% (with six molecules of H_2 as co-reagent and three molecules of H_2O co-product).

3.2 *Justified Reaction Types for Cellulosic Biomass Conversion*

The previous section clearly demonstrates the importance of the preservation of functionality, while maintaining high atom economy throughout the process. The modified *van Krevelen* plot is also suitable for closer inspection of the different chemical reactions leading to these molecules in order to identify the most advantageous reaction types for processing cellulosic biomass in an atom economical fashion to functional products. It reveals that a number of reactions are ideal for preserving functionality, among them hydrolysis, retro-aldol, rehydration, and dehydration (retro-Michael, 1,2-elimination, or 1,4-cyclodehydration), isomerization, and partial oxidation. Selective hydrogenolysis and hydrogenation are to some extent useful as well.

3.2.1 Dehydration–Rehydration

In Fig. 12, a modified *van Krevelen* plot is shown, which contains cellulose and common sugars as well as reaction arrows. The latter are drawn to emphasize the impact of dehydration (in blue). As can be seen, a typical dehydration reaction preserves the level of functionality in our assessment while decreasing the *H* index of the compound. In general, dehydration involves a horizontal move to the left-hand-side of the plot.

In catalytic biomass conversion, the dehydration of polyol moieties is a key reaction, forming either an olefin bond, an ether, or a carbonyl group (after tautomerization). This type of reaction usually requires the aid of acid catalysis. Both Brønsted and Lewis acids are known to catalyze dehydration. The most famous dehydration in the context of biomass conversion is the formation of HMF from six-carbon sugars. As can be seen in Fig. 12, HMF from fructose (or glucose) preserves the *F*:*C* value of 1.17, while the *H* index falls from 2 to 1. Similarly, pentoses lead to furfural in aqueous media under acid catalysis. The catalytic formation of HMF and furfural from hexoses and pentoses, respectively, and HMF production directly from cellulose, have been reported frequently over

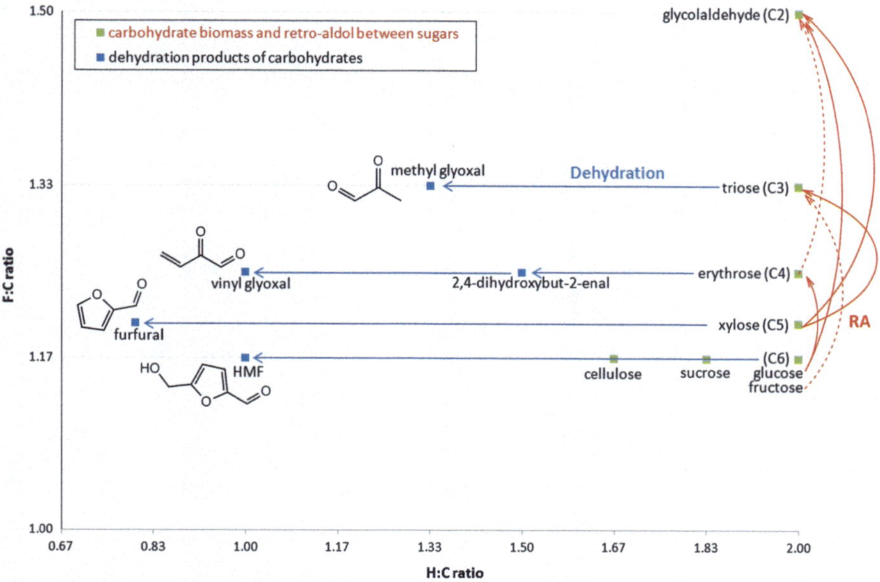

Fig. 12 Modified *van Krevelen* plot for cellulose, sugars, and the dehydration products thereof. *RA* = retro-aldol reaction

the last 10 years and insightful reviews on this conversion can be found for HMF [80, 98] and furfural [44, 128], as well as in Chap. 2 of this volume (Mascal et al.). Both HMF and furfural are considered platform molecules for the synthesis of, e.g., levulinic acid and γ-valerolactone and are present in Table 1 [44, 51].

As an illustration of the importance of selective dehydration, the reaction of tetrose sugars in alcoholic media with soluble Sn halides has recently been reported [100]. This presents a homogeneous catalytic system, which delivers both Brønsted (as HCl) and Lewis (as Sn^{2+} or Sn^{4+}) acids. The final products of this conversion were useful α-hydroxy-acids, such as vinyl glycolic acid as discussed in detail in Chap. 3 of this volume (Dusselier et al.). The first step in the reaction path involves a double dehydration leading to the proposed intermediate vinyl glyoxal, as shown being derived from the tetrose in Fig. 12. The mechanism of the dehydration of tetroses is shown in more detail in steps 1 and 2 in Fig. 13 (tentatively catalyzed by a Sn salt).

Selective dehydration of triose sugars such as dihydroxyacetone and glyceraldehyde leads to another glyoxal, viz. pyruvic aldehyde or methyl glyoxal (Fig. 12). Such glyoxals are notoriously unstable and reactive compounds, mainly due to their conjugated carbonyl groups. They are nevertheless useful intermediates to take into account when searching for unique reaction pathways towards platform chemicals from carbohydrates, such as lactic acid [74, 129] or 1,2-propylene glycol [119]. The dehydration route in Fig. 13 is in principle possible for all carbohydrates, with the exception of glycolaldehyde, due to their structural combination of a carbonyl and a β-hydroxyl group. It proceeds according to a generalized retro-Michael dehydration

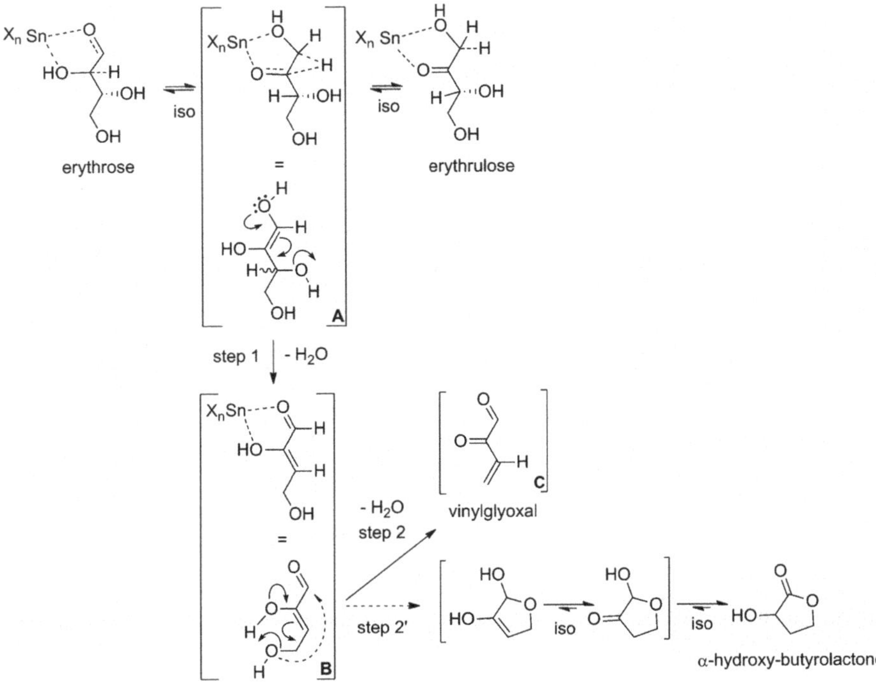

Fig. 13 Dehydration of tetroses (both ketose and aldose) via enolization and retro-Michael reactions (steps 1 and 2). An observed side reaction after the first dehydration is a cyclization (step 2′), followed by isomerizations (= iso) to α-hydroxybutyrolactone HBL [100]

mechanism [16], combined with keto-enol tautomerization. Both mechanisms are encountered for the tetroses in Fig. 13.

Besides sugars, a plethora of hydroxyl group-containing, oxidized hydrocarbons and sugar-derived molecules can undergo selective 1,2-dehydration as well as 1,4-cyclodehydration. Figure 14 displays a few examples of such dehydration series in a modified *van Krevelen* plot. For a start, one may consider the products derived via carbohydrate fermentation processes. These compounds include, for instance, ethanol, mixtures of acetone, butanol, and ethanol (ABE fermentation), and lactic acid (anaerobic fermentation). In recent years, organisms have also been genetically modified to produce atypical products via carbohydrate fermentation. A good example of this is a process based on the metabolic engineering of *Escherichia coli*, which is capable of directly producing 1,4-butanediol by the fermentation of various biomass-derived sugars [111, 130, 131]. Double dehydration of 1,4-butanediol leads to the rubber precursor butadiene with preservation of the *F*:*C* value (0.5), as seen in Fig. 14. The intermediate chemicals in this process are interesting as well: tetrahydrofuran (via a 1,4-cyclodehydration) and the unsaturated alcohol 3-buten-1-ol. The dehydration of lactic acid to acrylic acid [74, 88, 108], and that of *n*-butanol to butene [132, 133], are two other well known examples, as is the dehydration of ethanol to ethylene [16, 134].

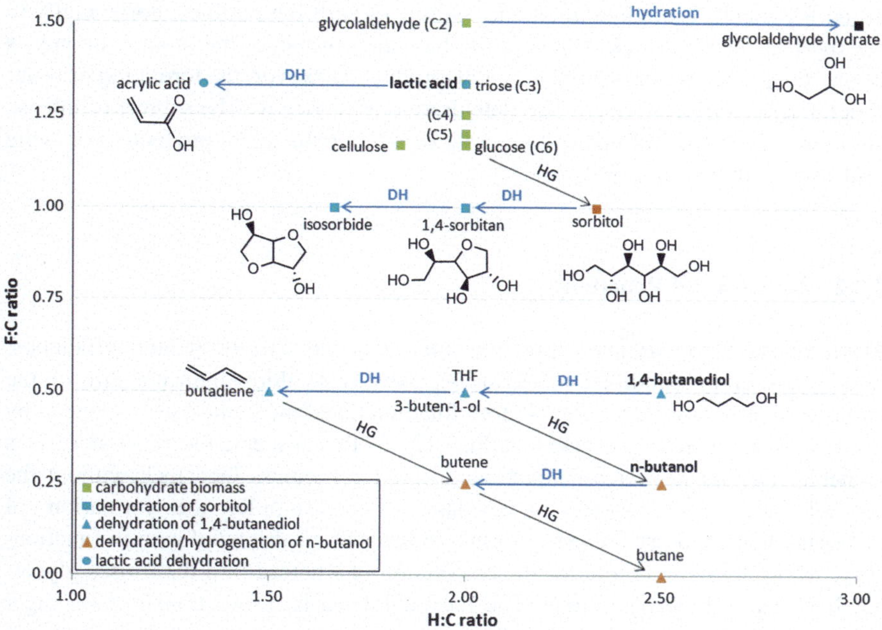

Fig. 14 Modified *van Krevelen plot* for the dehydration of sorbitol (after hydrogenation of glucose), lactic acid, 1,4-butanediol, and *n*-butanol. *DH* = dehydration, *HG* = hydrogenation. The effect of hydration of an aldose (hemi-acetal formation with water) is also shown

Finally, 1,4-cyclodehydration is another important type of reaction for oxygen-containing sugar derivatives, such as, for instance, sugar alcohols. A current topic of high interest is the synthesis of isosorbide from sorbitol, the primary hydrogenation product of glucose. As seen in Fig. 14, it proceeds via two 1,4-cyclodehydrations, via 1,4-sorbitan, culminating in the bicyclic isosorbide ether structure, seen to the left of sorbitol, with equal *F*:*C* on the plot. Recent successful efforts have focused on producing sorbitans and isosorbide in a one-pot approach directly from lignocellulose, using bifunctional catalysis [82, 96, 135, 136]. In this work, acidic sites for hydrolyzing the cellulosic bio-polymer and dehydrating sorbitol are combined with active metal sites for catalyzing the hydrogenation of glucose to produce sorbitol. Interestingly, cellulose and isosorbide are very close to each other with respect to functionality (Fig. 14); only 14% of the original *F*:*C* value of cellulose is lost due to a single hydrogenation step. Isosorbide indeed belongs to the list of preferred chemicals from carbohydrate biomass in this chemist's view, and was found in the innermost zone of Fig. 9, for which no petrochemical competition is expected. Besides undergoing dehydration, sugars (and in general, aldehydes) can be hydrated in water, leading to a hemiacetal with an equal F:C, as shown in Fig. 14 for glycolaldehyde.

In conclusion, dehydration, hydration, and hydrolysis (e.g., of cellulose to glucose or hemicellulose to xylose [137]) reactions fully preserve the degree of functionality of the starting biomass or chemicals derived therefrom. In the modified *van Krevelen* plot, the dehydrated products are located horizontally to the left of the feedstock, while hydration and hydrolysis shift the products horizontally to the right of the feedstock. Dehydration has one downside; the atom economy is never 100% since water is formed as a by-product. However, the formation of water is not a serious disadvantage. The dehydration of trioses will be subject of intense discussion in Chap. 3 of this volume, as a key step in the direct production of lactic acid from carbohydrates [68, 74].

3.2.2 Retro-Aldol Reaction

The retro-aldol is a very interesting reaction as it occurs with 100% atom efficiency. In addition, on the modified *van Krevelen* plots, retro-aldols induce a shift on the curve vertically upwards towards more functionalized molecules, as evidenced by the reactions (in red) presented in Fig. 12. In this respect, the retro-aldol is a valuable reaction to increase functionality of the biomass feedstock without the use of sacrificial reagents or oxidants. For example, the formation of glycolaldehyde and erythrose from glucose leads to a substantial gain in functionality, while the same is true for the conversion of fructose into the trioses glyceraldehyde and dihydroxyacetone. The retro-aldol reaction thus transforms a sugar into two smaller sugars. In line with the aldol addition, this reaction is catalyzed under both Brønsted acidic or basic conditions, as well as by Lewis acidic centers or even amines via imine formation. Similar to the retro-Michael dehydration, a hydroxyl group β to the carbonyl is a prerequisite for the retro-aldol reaction to occur. For a ketose sugar, this implies that dihydroxyacetone (a triose) is formed. Applied to aldoses, a retro-aldol will always lead to at least one C_2 fragment, i.e., glycolaldehyde [138].

In comparison to other common reactions such as dehydration (see above) or hydrogenation (see below), the retro-aldol reaction is more difficult kinetically as a higher activation barrier is encountered [116]. The search for catalysts with high activity and an exclusive preference for the retro-aldol reaction is a very challenging area. Yet control of this reaction type is of major importance in carbohydrate chemistry and lies at the heart of a major energy pathway in living cells, viz. the glycolysis metabolism. In the carbohydrates-to-chemicals field, the retro-aldol reaction may also be useful for instance in the synthesis of ethylene glycol and propylene glycol from glucose and cellulose. Up to now, it has been observed that solutions of tungsten salts [116] and supported tungsten trioxides [75] or carbides [70, 114] in water at temperatures above 200°C will preferentially lead to retro-aldol derived products. Due to the high reaction temperature and the pronounced instability of the smaller sugars, the retro-aldol reaction usually has to be combined with another reaction, such as, for instance, hydrogenation. The selective conversion of cellulose [75, 114] or concentrated glucose syrups (even demonstrated semi-continuously [70, 116]) into ethylene glycol is a good example of this strategy, in which the retro-aldol intermediate glycolaldehyde is immediately hydrogenated upon formation. Another illustrative example is found in the catalytic synthesis of lactic acid and

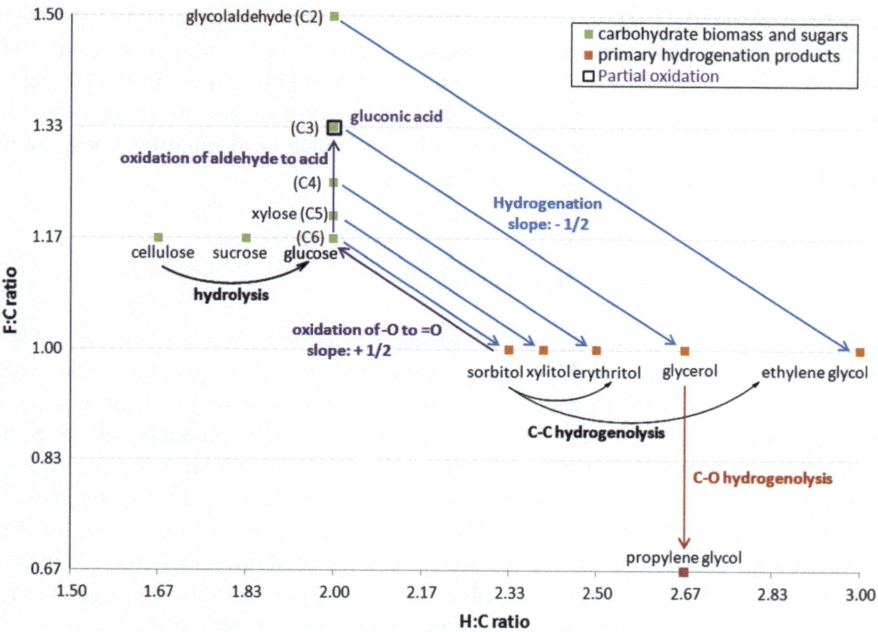

Fig. 15 Modified *van Krevelen* plot for cellulose hydrolysis, carbohydrate hydrogenation and C–C vs C–O hydrogenolysis

methyl lactate from sucrose, glucose and fructose in aqueous or alcoholic media, as discussed in detail in Chap. 3 of this volume.

3.2.3 Partial Oxidation

Partial oxidation of biomass derived sugars or their derivatives is a very useful reaction and a simple way of introducing additional functionality. Typically, on the modified *van Krevelen* plot, the oxidation of an aldehyde to a carboxylic acid involves a vertical shift as no H is removed. The oxidation of an alcohol to an aldehyde, for instance, also results in a shift to higher *F*:*C* values, but according to a +0.5 slope (in fact, the opposite of a hydrogenation). An example of both shifts is given in Fig. 15 for gluconic acid from glucose [76] and glucose from sorbitol (the latter only relevant in theory). Multiple oxidations are, for instance, needed to convert glucose to glucaric acid [17]. The stoichiometry and atom efficiency of these reactions are found in Table 3. Other useful examples of the catalytic oxidation of cellulosic biomass are found in the synthesis of glycolic acid (also found in Chap. 3), formic acid, and acetic acid [69, 71, 89, 139]. Ideally, these oxidations would be performed with simple reagents, preferably with O_2 or H_2O_2. In the case of gluconic acid, the use of both oxidants has been reported [76, 92, 140], and its direct production from cellobiose and even cellulose is known [83, 91]. A simultaneous path to gluconic acid and lactic acid from glucose has also been reported [141], as well as an innovative route from cellobiose without sacrificial agents, based on the simultaneous formation of sorbitol (reduction) and gluconic acid (oxidation) via redox with Ru under argon [142]. When oxidizing sugar molecules or derivatives, the production of CO_2 by over-oxidation should be avoided. Selectivity is usually the biggest challenge in the oxidation of multifunctional compounds such as carbohydrates and their derivatives.

3.2.4 Hydrogenation and Hydrogenolysis

The hydrogenation of carbohydrates into sugar alcohols is a relatively easy aldehyde to alcohol reduction and has been commercially practiced for years in the case of sorbitol [143]. Figure 15 depicts the formation of the five major sugar alcohols. Hydrogenation slightly lowers the functionality index of the products with respect to that of the feedstock, but the atom economy of hydrogenation is always 100%. For instance, 33% of the functionality per carbon is lost by hydrogenating glycolaldehyde to ethylene glycol. The slope of the hydrogenation lines in the modified *van Krevelen* plot is −0.5, indicative of a loss of one functional group per addition of two H atoms. The same observation is apparent in the hydrogenations shown in Fig. 14, e.g., the hydrogenation of butadiene to butene and butane.

To illustrate a good example, the one-pot hydrolytic hydrogenation of cellulose gives nearly quantitative yields of sorbitol, some mannitol, and some sorbitans [19, 52, 81, 93–95, 144, 145].

Besides hydrogenation, the presence of certain (metal) catalysts and H_2 can also promote hydrogenolysis. As far as C–C hydrogenolysis is concerned, the combined F value of both fragments is not affected, while the atom economy remains high (assuming both fragment products are considered relevant and one of them is thus not a side product, which is not often the case). This can be seen in Fig. 15 by following the hydrogenolysis of sorbitol into erythritol and ethylene glycol. If, however, C–O bonds are hydrogenolyzed, there is a drop in F:C index, as exemplified by the formation of propylene glycols (1,2- or 1,3-propanediol) from glycerol (red arrow). Although carbohydrate C–O hydrogenolysis reduces the functionality index, this hydrodeoxygenation reaction is becoming increasingly important these days, especially when fuels and molecules with a high number of C–H bonds are targeted.

3.2.5 Isomerization

Isomerization reactions are of high importance to the exploitation of carbohydrates. These transformations retain the functionality parameter of the feedstock, while maintaining complete atom-efficiency, since the reaction only involves a reshuffle of atoms within the molecule. Keto-enol tautomerization between aldoses and ketoses as encountered in Fig. 13, as well as the furanose or pyranose ring formation via cyclic hemi-acetalization of pentoses and hexoses, are simple examples. The epimerization of glucose into mannose has been described using the Lewis acid molybdenum oxide in a slightly acid environment [146] or with Ca^{2+} under slightly alkaline conditions, and even with Sn-zeolite in the presence of borates [147], while isomerization to fructose has recently been shown to proceed in the presence of Sn-containing catalysts via a 1,2-hydride shift [148, 149]. This can be exploited towards the synthesis of HMF-based chemicals, as fructose more readily dehydrates to HMF [79, 150]. Such 1,2-hydride shifts also occur in the synthesis of carboxylic acids or esters from two to four-carbon sugars (e.g., lactic acid from hydrated methylglyoxal, viz. Chap. 3 of this volume) [74, 86]. Another striking example is the isomerization of glucose to sorbose via a titanium-β-zeolite catalyzed intramolecular 1,5-hydride shift [151].

3.3 Key Lessons of the Assessment

To conclude this section, the key lessons of the assessment are summarized here:

- Cellulosic and hemicellulosic feedstocks may be converted into 100+ chemicals, among them drop-in products but also several novel chemicals. Aside from biofuels and their precursors, they may be classified as end products, platform chemicals, monomers, and chemical intermediates (leading to a specific end-product). Some of these are already produced commercially from carbohydrates whereas others are currently petroleum or natural gas derived. Most of the chemicals described here are not produced commercially to date.
- Limited feedstock supply in biorefineries demands a careful selection of value-added and/or platform chemicals from carbohydrates.
- Functionality (chemical reactivity approach) and atom economy (green chemistry approach) are the driving criteria of this assessment, which is developed from a strictly chemical point of view. Ultimately, the applications of the product need to be considered, next to the existence of drop-in or rising potential markets; as well as technological measures.
- Functionality is defined here by a new functionality parameter F, and its normalized functionality index $F{:}C$, rather than with the original oxygen index ($O{:}C$) used for fuel and deoxygenation series.
- Plotting $F{:}C$ against the calculated atom economy provides a beneficial tool for the selection of a viable set of target molecules from carbohydrate biomass that may compete with classic petrochemical routes for their formation.
- A list of 25 chemicals based on the above analysis was proposed.
- The modified *van Krevelen* diagram, plotting functionality vs H index ($F{:}C$ vs $H{:}C$), is also a useful tool for visualizing the impact of the various reaction types involved in the synthesis of carbohydrate derivatives.
- Preferred reaction types for the conversion of carbohydrates preserve $F{:}C$ or result in an increase in functionality (or minimal decrease) with respect to the feedstock.
- The methodology thus allows evaluation of processes and feedstocks for their maintenance of function in the synthesis of chemicals.

- The analysis also suggests that derivatives with a degree of functionality similar to that of the feedstock might be more efficiently produced than those that require more extensive transformations, and therefore have the potential to become valuable intermediates, if a market exists or arises for them.
- α-Hydroxy acids are excellent candidates following this approach, and lactic acid, vinyl glycolic acid, glycolic acid, and furyl glycolic acid are included in the target chemical list. Some of these feature 100% atom economy from cellulose or glucose, while all of them present a higher $F{:}C$ value than cellulose. Given their demonstrated usefulness in the context of existing and potential markets, selective catalytic approaches for their synthesis would be highly valuable and will be described in full in Chap. 3 in this volume.

4 Summary, Conclusions, Outlook

The future development of bio-derived chemicals will depend on multiple technological, economic, and environmental factors. This chapter has explored new modes of assessment to evaluate the competitiveness and sustainability of the use of carbohydrates, preferably from lignocellulosic feedstocks, to produce chemicals. The assessment is based on the preservation of functionality and a high atom economy in the process, and a toolbox with a novel functionality index and modified *van Krevelen* plots is provided to assess the advantages of processes and feedstocks. In essence, highly functionalized chemicals are very versatile and the most cost-effective derivatives of carbohydrates, while chemical conversions should involve maximum conservation of atoms in the desired products. Based on the criteria, a list of about 25 chemicals was proposed and several preferred reaction types were discussed. High scoring, drop-in chemicals from carbohydrates are formic acid, ethylene glycol, acetic acid, glycolic acid, and acrylic acid, while furfural, furfuryl alcohol, sorbitol, lactic and levulinic acids and isosorbide are already exclusively produced commercially from carbohydrates in the chemical industry. Currently non-commercial chemicals like vinyl and furyl glycolic acid and HMF are projected to be products of high interest according to the two criteria of the assessment. The proposed methodology presented here is easily adapted to the assessment of chemicals and their formation routes from other feedstocks, such as lignin for example, but it does not take into account the usefulness and potential market demand of the products, which in the end determines their real value.

Successful integration of carbohydrate chemistry in a biorefinery will greatly depend on the availability and price of the lignocellulosic feedstock and its fractionation cost. Since the collection of biomass is limited by its volume and density, biorefineries are likely to be smaller and more highly distributed than petroleum refineries. In order to produce meaningful volumes of products, a biorefinery should focus on a specific set of platform chemicals rather than attempting to provide an extended portfolio of products. There are two technological obstacles to the delivery of such a biorefinery, and effort is required to overcome them. First, the sustainable and cheap fractionation of biomass into its components – cellulose, hemicellulose, lignin, proteins, and minerals, is a key issue. Current schemes often focus on the separation and processing of one or two of these fractions, while all the plant components should be integrated. Second, while a chemical equation may predict high efficiency (in terms of thermodynamics and atom economy), in practice, lower performance due to kinetic (selectivity) issues or process/technological challenges are often seen. Although there are already many elegant examples in the literature, further technological development in the area of chemical and bio-catalysis and in some cases their symbiosis is required to make a biorefinery sustainable in competition with petrochemistry.

Selective catalysis is a challenge, as many target chemicals as well as their feedstocks are highly functionalized and chemically reactive, and their production from carbohydrates may involve complicated multistep reaction cascades with parallel pathways and reactive intermediates. The tuning of different catalytic sites with regard to number, activity and location on the catalyst or in the medium, and their stabilization under various reaction conditions, are certainly two of the biggest future challenges in the commercial development of biorefineries.

Acknowledgements M.D. acknowledges FWO Vlaanderen (Research Foundation - Flanders) for a post-doctoral fellowship. B.F.S thanks the Research Council of the KU Leuven (IDO-3E090504) for financial support, as well as the Belgian government for its funding through IAP (Belspo).

References

1. Ragauskas AJ, Williams CK, Davison BH, Britovsek G, Cairney J, Eckert CA, Frederick WJ, Hallett JP, Leak DJ, Liotta CL, Mielenz JR, Murphy R, Templer R, Tschaplinski T (2006) The path forward for biofuels and biomaterials. Science 311:484–489
2. Levy PF, Sanderson JE, Kispert RG, Wise DL (1981) Biorefining of biomass to liquid fuels and organic chemicals. Enzyme Microb Tech 3:207–215
3. Lipinsky ES (1978) Fuels from biomass: integration with food and materials systems. Science 199:644–651
4. Bracannot H (1819) Sur la conversion du corps ligneux en gomme, en sucre, et en un acide d'une nature particuliere, par le moyen de l'acide sulfurique; conversion de la même substance ligneuse en ulmine par la potasse. Ann Chim Phys 12:172–195
5. Vogel O (1908) History of wood distillation II. Dusseldorf Chem Ztg 32:561
6. Fawsitt CA (1885) Wood naphtha. J Soc Chem Ind Lond 4:319–321
7. Bungay HR (1982) Biomass refining. Science 218:643–646
8. Gunaseelan VN (1997) Anaerobic digestion of biomass for methane production: a review. Biomass Bioenerg 13:83–114
9. Bridgwater AV, Peacocke GVC (2000) Fast pyrolysis processes for biomass. Renew Sust Energ Rev 4:1–73
10. Yung MM, Jablonski WS, Magrini-Bair KA (2009) Review of catalytic conditioning of biomass-derived syngas. Energ Fuel 23:1874–1887
11. Baliban RC, Elia JA, Floudas CA (2013) Biomass to liquid transportation fuels (BTL) systems: process synthesis and global optimization framework. Energ Environ Sci 6:267–287

12. Cheng Y-T, Jae J, Shi J, Fan W, Huber GW (2012) Production of renewable aromatic compounds by catalytic fast pyrolysis of lignocellulosic biomass with bifunctional Ga/ZSM-5 catalysts. Angew Chem Int Ed 51(6):1387–1390
13. Carlson TR, Vispute TP, Huber GW (2008) Green gasoline by catalytic fast pyrolysis of solid biomass derived compounds. ChemSusChem 1(5):397–400
14. Elliott DC, Oasmaa A, Meier D, Preto F, Bridgwater AV (2012) Results of the IEA round robin on viscosity and aging of fast pyrolysis bio-oils: long-term tests and repeatability. Energ Fuel 26:7362–7366
15. Mortensen PM, Grunwaldt J-D, Jensen PA, Knudsen KG, Jensen AD (2011) A review of catalytic upgrading of bio-oil to engine fuels. Appl Catal A Gen 407:1–19
16. Centi G, van Santen RA (eds) (2007) Catalysis for renewables: from feedstock to energy production. Wiley-VCH, Weinheim
17. Corma A, Iborra S, Velty A (2007) Chemical routes for the transformation of biomass into chemicals. Chem Rev 107:2411–2502
18. Nishiyama Y, Sugiyama J, Chanzy H, Langan P (2003) Crystal structure and hydrogen bonding system in cellulose Iα from synchrotron X-ray and neutron fiber diffraction. J Am Chem Soc 125(47):14300–14306
19. Van de Vyver S, Geboers J, Jacobs PA, Sels BF (2011) Recent advances in the catalytic conversion of cellulose. ChemCatChem 3(1):82–94
20. Geboers JA, Van de Vyver S, Ooms R, Op de Beeck B, Jacobs PA, Sels BF (2011) Chemocatalytic conversion of cellulose: opportunities, advances and pitfalls. Catal Sci Technol 1(5):714–726
21. Saha B (2003) Hemicellulose bioconversion. J Ind Microbiol Biotechnol 30(5):279–291
22. Zakzeski J, Bruijnincx PCA, Jongerius AL, Weckhuysen BM (2010) The catalytic valorization of lignin for the production of renewable chemicals. Chem Rev 110(6):3552–3599
23. Biermann U, Bornscheuer U, Meier MAR, Metzger JO, Schäfer HJ (2011) Oils and fats as renewable raw materials in chemistry. Angew Chem Int Ed 50(17):3854–3871
24. Foley PM, Beach ES, Zimmerman JB (2011) Algae as a source of renewable chemicals: opportunities and challenges. Green Chem 13(6):1399–1405
25. Tuck CO, Pérez E, Horváth IT, Sheldon RA, Poliakoff M (2012) Valorization of biomass: deriving more value from waste. Science 337(6095):695–699
26. Lammens TM, De Biase D, Franssen MCR, Scott EL, Sanders JPM (2009) The application of glutamic acid alpha-decarboxylase for the valorization of glutamic acid. Green Chem 11 (10):1562–1567
27. Kromus S, Kamm B, Kamm M, Fowler P, Narodoslawsky M (2008) Green biorefineries: the green biorefinery concept – fundamentals and potential. In: Biorefineries-industrial processes and products. Wiley-VCH, Weinheim, pp 253–294
28. Song J, Fan H, Ma J, Han B (2013) Conversion of glucose and cellulose into value-added products in water and ionic liquids. Green Chem 15(10):2619–2635
29. Jarvis M (2003) Chemistry: cellulose stacks up. Nature 426(6967):611–612
30. Rinaldi R, Schüth F (2009) Acid hydrolysis of cellulose as the entry point into biorefinery schemes. ChemSusChem 2(12):1096–1107
31. Vennestrøm PNR, Osmundsen CM, Christensen CH, Taarning E (2011) Beyond petrochemicals: the renewable chemicals industry. Angew Chem Int Ed 50(45):10502–10509
32. ten Dam J, Hanefeld U (2011) Renewable chemicals: dehydroxylation of glycerol and polyols. ChemSusChem 4(8):1017–1034
33. Zhou C-H, Beltramini JN, Fan Y-X, Lu GQ (2008) Chemoselective catalytic conversion of glycerol as a biorenewable source to valuable commodity chemicals. Chem Soc Rev 37:527–549
34. Sels B, D'Hondt E, Jacobs P (2007) Catalytic transformation of glycerol. In: Catalysis for renewables. Wiley-VCH, Weinheim, pp 223–255
35. Virent Inc., http://www.virent.com/technology/bioforming/ Accessed 14 Jan 2014

36. Huber GW, Cortright RD, Dumesic JA (2004) Renewable alkanes by aqueous-phase reforming of biomass-derived oxygenates. Angew Chem Int Ed 43:1549–1551
37. Huber GW, Chheda JN, Barrett CJ, Dumesic JA (2005) Production of liquid alkanes by aqueous-phase processing of biomass-derived carbohydrates. Science 308(5727):1446–1450
38. Kunkes EL, Simonetti DA, West RM, Serrano-Ruiz JC, Gartner CA, Dumesic JA (2008) Catalytic conversion of biomass to monofunctional hydrocarbons and targeted liquid-fuel classes. Science 322:417–421
39. Corma A, de la Torre O, Renz M, Villandier N (2011) Production of high-quality diesel from biomass waste products. Angew Chem Int Ed 50:2375–2378
40. Tompsett GA, Li N, Huber GW (2011) Catalytic conversion of sugars to fuels. In: Thermochemical processing of biomass. John Wiley & Sons, Ltd, pp 232–279
41. Alonso DM, Bond JQ, Dumesic JA (2010) Catalytic conversion of biomass to biofuels. Green Chem 12(9):1493–1513
42. Climent MJ, Corma A, Iborra S (2014) Conversion of biomass platform molecules into fuel additives and liquid hydrocarbon fuels. Green Chem 16(2):516–547
43. Sutton AD, Waldie FD, Wu R, Schlaf M, 'Pete' Silks LA, Gordon JC (2013) The hydrodeoxygenation of bioderived furans into alkanes. Nat Chem 5(5):428–432
44. Lange J-P, van der Heide E, van Buijtenen J, Price R (2012) Furfural—a promising platform for lignocellulosic biofuels. ChemSusChem 5(1):150–166
45. Bozell JJ, Moens L, Elliott DC, Wang Y, Neuenscwander GG, Fitzpatrick SW, Bilski RJ, Jarnefeld JL (2000) Production of levulinic acid and use as a platform chemical for derived products. Resour Conservat Recycl 28(3–4):227–239
46. Lange J-P, Price R, Ayoub PM, Louis J, Petrus L, Clarke L, Gosselink H (2010) Valeric biofuels: a platform of cellulosic transportation fuels. Angew Chem Int Ed 49(26):4479–4483
47. Alonso DM, Wettstein SG, Dumesic JA (2013) Gamma-valerolactone, a sustainable platform molecule derived from lignocellulosic biomass. Green Chem 15(3):584–595
48. Mascal M, Dutta S, Gandarias I (2014) Hydrodeoxygenation of the Angelica lactone dimer, a cellulose-based feedstock: simple, high-yield synthesis of branched C7–C10 gasoline-like hydrocarbons. Angew Chem Int Ed 53:1854–1857
49. Serrano-Ruiz JC, Wang D, Dumesic JA (2010) Catalytic upgrading of levulinic acid to 5-nonanone. Green Chem 12:574–577
50. Bond JQ, Martin-Alonso D, Wang D, West RM, Dumesic JA (2010) Integrated catalytic conversion of γ-valerolactone to liquid alkenes for transportation fuels. Science 327:1110–1114
51. Bozell JJ, Petersen GR (2010) Technology development for the production of biobased products from biorefinery carbohydrates – the US Department of Energy's "Top 10" revisited. Green Chem 12:539–554
52. Kobayashi H, Fukuoka A (2013) Synthesis and utilisation of sugar compounds derived from lignocellulosic biomass. Green Chem 15(7):1740–1763
53. Gallezot P (2012) Conversion of biomass to selected chemical products. Chem Soc Rev 41:1538–1558
54. Werpy T, Petersen G (2004) Top value added chemicals from biomass. DOE/GO-102004-1992 1. http://www.nrel.gov/docs/fy04osti/35523.pdf.
55. Huber GW, Corma A (2007) Synergies between bio- and oil refineries for the production of fuels from biomass. Angew Chem Int Ed 46(38):7184–7201
56. Chheda JN, Huber GW, Dumesic JA (2007) Liquid-phase catalytic processing of biomass-derived oxygenated hydrocarbons to fuels and chemicals. Angew Chem Int Ed 46 (38):7164–7183
57. Carlos Serrano-Ruiz J, Dumesic JA (2009) Catalytic upgrading of lactic acid to fuels and chemicals by dehydration/hydrogenation and C-C coupling reactions. Green Chem 11(8):1101–1104
58. Serrano-Ruiz JC, Dumesic JA (2011) Catalytic routes for the conversion of biomass into liquid hydrocarbon transportation fuels. Energ Environ Sci 4(1):83–99

59. Yan X, Inderwildi OR, King DA (2010) Biofuels and synthetic fuels in the US and China: a review of well-to-wheel energy use and greenhouse gas emissions with the impact of land-use change. Energ Environ Sci 3(2):190–197
60. Inderwildi OR, King DA (2009) Quo vadis biofuels? Energ Environ Sci 2(4):343–346
61. Zinoviev S, Müller-Langer F, Das P, Bertero N, Fornasiero P, Kaltschmitt M, Centi G, Miertus S (2010) Next-generation biofuels: survey of emerging technologies and sustainability issues. ChemSusChem 3(10):1106–1133
62. Azapagic A, Perdan S, Clift R (eds) (2004) Sustainable development in practice: case studies for engineers and scientists. John Wiley & Sons Ltd. Chichester, UK.
63. Van Krevelen D (1950) Graphical-statistical method for the study of structure and reaction processes of coal. Fuel 29(12):269–284
64. IFRF Combustion Handbook - file 23 (2000) International Flame Research Foundation (IFRF). http://www.handbook.ifrf.net/handbook/cf.html?id=23.
65. Kim S, Kramer RW, Hatcher PG (2003) Graphical method for analysis of ultrahigh-resolution broadband mass spectra of natural organic matter, the van Krevelen diagram. Anal Chem 75(20):5336–5344
66. Wildschut J, Iqbal M, Mahfud FH, Cabrera IM, Venderbosch RH, Heeres HJ (2010) Insights in the hydrotreatment of fast pyrolysis oil using a ruthenium on carbon catalyst. Energ Environ Sci 3(7):962–970
67. Van de Vyver S, Thomas J, Geboers J, Keyzer S, Smet M, Dehaen W, Jacobs PA, Sels BF (2011) Catalytic production of levulinic acid from cellulose and other biomass-derived carbohydrates with sulfonated hyperbranched poly(arylene oxindole)s. Energ Environ Sci 4(9):3601–3610
68. de Clippel F, Dusselier M, Van Rompaey R, Vanelderen P, Dijkmans J, Makshina E, Giebeler L, Oswald S, Baron GV, Denayer JFM, Pescarmona PP, Jacobs PA, Sels BF (2012) Fast and selective sugar conversion to alkyl lactate and lactic acid with bifunctional carbon–silica catalysts. J Am Chem Soc 134(24):10089–10101
69. Li J, Ding D-J, Deng L, Guo Q-X, Fu Y (2012) Catalytic air oxidation of biomass-derived carbohydrates to formic acid. ChemSusChem 5(7):1313–1318
70. Ooms R, Dusselier M, Geboers JA, Op de Beeck B, Verhaeven R, Gobechiya E, Martens J, Redl A, Sels BF (2014) Conversion of sugars to ethylene glycol with nickel tungsten carbide in a fed-batch reactor: high productivity and reaction network elucidation. Green Chem 16:695–707
71. Zhang J, Sun M, Liu X, Han Y (2014) Catalytic oxidative conversion of cellulosic biomass to formic acid and acetic acid with exceptionally high yields. Catal Today. doi:10.1016/j.cattod.2013.12.010
72. Sasaki M, Goto K, Tajima K, Adschiri T, Arai K (2002) Rapid and selective retro-aldol condensation of glucose to glycolaldehyde in supercritical water. Green Chem 4(3):285–287
73. Zhang J, Liu X, Sun M, Ma X, Han Y (2012) Direct conversion of cellulose to glycolic acid with a phosphomolybdic acid catalyst in a water medium. ACS Catal 2(8):1698–1702
74. Dusselier M, Van Wouwe P, Dewaele A, Makshina E, Sels BF (2013) Lactic acid as a platform chemical in the biobased economy: the role of chemocatalysis. Energ Environ Sci 6 (5):1415–1442
75. Liu Y, Luo C, Liu H (2012) Tungsten trioxide promoted selective conversion of cellulose into propylene glycol and ethylene glycol on a ruthenium catalyst. Angew Chem Int Ed 51 (13):3249–3253
76. Biella S, Prati L, Rossi M (2002) Selective oxidation of D-glucose on gold catalyst. J Catal 206(2):242–247
77. Bui L, Luo H, Gunther WR, Román-Leshkov Y (2013) Domino reaction catalyzed by zeolites with brønsted and lewis acid sites for the production of γ-valerolactone from furfural. Angew Chem Int Ed 52(31):8022–8025
78. Yi G, Zhang Y (2012) One-pot selective conversion of hemicellulose (xylan) to xylitol under mild conditions. ChemSusChem 5(8):1383–1387

79. Nikolla E, Román-Leshkov Y, Moliner M, Davis ME (2011) "One-pot" synthesis of 5-(hydroxymethyl)furfural from carbohydrates using tin-beta zeolite. ACS Catal 1(4):408–410
80. Wang T, Nolte MW, Shanks BH (2014) Catalytic dehydration of C6 carbohydrates for the production of hydroxymethylfurfural (HMF) as a versatile platform chemical. Green Chem 16(2):548–572
81. Van de Vyver S, Geboers J, Dusselier M, Schepers H, Vosch T, Zhang L, Van Tendeloo G, Jacobs PA, Sels BF (2010) Selective bifunctional catalytic conversion of cellulose over reshaped Ni particles at the tip of carbon nanofibers. ChemSusChem 3(6):698–701
82. Op de Beeck B, Geboers J, Van de Vyver S, Van Lishout J, Snelders J, Huijgen WJJ, Courtin CM, Jacobs PA, Sels BF (2013) Conversion of (ligno)cellulose feeds to isosorbide with heteropoly acids and Ru on carbon. ChemSusChem 6(1):199–208
83. An D, Ye A, Deng W, Zhang Q, Wang Y (2012) Selective conversion of cellobiose and cellulose into gluconic acid in water in the presence of oxygen, catalyzed by polyoxometalate-supported gold nanoparticles. Chem Eur J 18(10):2938–2947
84. Huber GW, Dumesic JA (2006) An overview of aqueous-phase catalytic processes for production of hydrogen and alkanes in a biorefinery. Catal Today 111(1–2):119–132
85. Gorbanev YY, Klitgaard SK, Woodley JM, Christensen CH, Riisager A (2009) Gold-catalyzed aerobic oxidation of 5-hydroxymethylfurfural in water at ambient temperature. ChemSusChem 2(7):672–675
86. Assary RS, Curtiss LA (2011) Theoretical study of 1,2-hydride shift associated with the isomerization of glyceraldehyde to dihydroxyacetone by lewis acid active site models. J Phys Chem A 115:8754–8760
87. Schwartz TJ, Goodman SM, Osmundsen CM, Taarning E, Mozuch MD, Gaskell J, Cullen D, Kersten PJ, Dumesic JA (2013) Integration of chemical and biological catalysis: production of furylglycolic acid from glucose via cortalcerone. ACS Catal 3(12):2689–2693
88. Zhang J, Zhao Y, Pan M, Feng X, Ji W, Au C-T (2011) Efficient acrylic acid production through bio lactic acid dehydration over NaY zeolite modified by alkali phosphates. ACS Catal 1:32–41
89. Wolfel R, Taccardi N, Bosmann A, Wasserscheid P (2011) Selective catalytic conversion of biobased carbohydrates to formic acid using molecular oxygen. Green Chem 13(10):2759–2763
90. Albert J, Wolfel R, Bosmann A, Wasserscheid P (2012) Selective oxidation of complex, water-insoluble biomass to formic acid using additives as reaction accelerators. Energ Environ Sci 5(7):7956–7962
91. Zhang J, Liu X, Hedhili MN, Zhu Y, Han Y (2011) Highly selective and complete conversion of cellobiose to gluconic acid over Au/Cs2HPW12O40 nanocomposite catalyst. ChemCatChem 3(8):1294–1298
92. Wang Y, Van de Vyver S, Sharma KK, Roman-Leshkov Y (2014) Insights into the stability of gold nanoparticles supported on metal oxides for the base-free oxidation of glucose to gluconic acid. Green Chem 16(2):719–726
93. Van de Vyver S, Geboers J, Schutyser W, Dusselier M, Eloy P, Dornez E, Seo JW, Courtin CM, Gaigneaux EM, Jacobs PA, Sels BF (2012) Tuning the acid/metal balance of carbon nanofiber-supported nickel catalysts for hydrolytic hydrogenation of cellulose. ChemSusChem 5(8):1549–1558
94. Geboers J, Van de Vyver S, Carpentier K, Jacobs P, Sels B (2011) Efficient hydrolytic hydrogenation of cellulose in the presence of Ru-loaded zeolites and trace amounts of mineral acid. Chem Commun 47(19):5590–5592
95. Kobayashi H, Ito Y, Komanoya T, Hosaka Y, Dhepe PL, Kasai K, Hara K, Fukuoka A (2011) Synthesis of sugar alcohols by hydrolytic hydrogenation of cellulose over supported metal catalysts. Green Chem 13(2):326–333
96. de Almeida RM, Li J, Nederlof C, O'Connor P, Makkee M, Moulijn JA (2010) Cellulose conversion to isosorbide in molten salt hydrate media. ChemSusChem 3(3):325–328

97. Rose M, Palkovits R (2012) Isosorbide as a renewable platform chemical for versatile applications—Quo Vadis? ChemSusChem 5(1):167–176
98. Rosatella AA, Simeonov SP, Frade RFM, Afonso CAM (2011) 5-Hydroxymethylfurfural (HMF) as a building block platform: biological properties, synthesis and synthetic applications. Green Chem 13(4):754–793
99. Saha B, Mosier N, Abu-Omar M (2014) Catalytic dehydration of lignocellulosic derived xylose to furfural. In: McCann MC, Buckeridge MS, Carpita NC (eds) Plants and bioenergy, vol 4, Advances in plant biology. Springer, New York, pp 267–276
100. Dusselier M, Van Wouwe P, de Clippel F, Dijkmans J, Gammon DW, Sels BF (2013) Mechanistic insight into the conversion of tetrose sugars to novel α-hydroxy acid platform molecules. ChemCatChem 5(2):569–575
101. Lin H, Strull J, Liu Y, Karmiol Z, Plank K, Miller G, Guo Z, Yang L (2012) High yield production of levulinic acid by catalytic partial oxidation of cellulose in aqueous media. Energy Environ Sci 5(12):9773–9777
102. Choudhary V, Mushrif SH, Ho C, Anderko A, Nikolakis V, Marinkovic NS, Frenkel AI, Sandler SI, Vlachos DG (2013) Insights into the interplay of Lewis and Brønsted acid catalysts in glucose and fructose conversion to 5-(hydroxymethyl)furfural and levulinic acid in aqueous media. J Am Chem Soc 135(10):3997–4006
103. Weingarten R, Conner WC, Huber GW (2012) Production of levulinic acid from cellulose by hydrothermal decomposition combined with aqueous phase dehydration with a solid acid catalyst. Energ Environ Sci 5(6):7559–7574
104. Dapsens PY, Mondelli C, Kusema B, Verel R, Perez-Ramirez J (2013) Continuous process for glyoxal valorisation using tailored Lewis-acid zeolite catalysts. Green Chem
105. Xu P, Qiu J, Gao C, Ma C (2008) Biotechnological routes to pyruvate production. J Biosci Bioeng 105(3):169–175
106. Vitasari CR, Meindersma GW, de Haan AB (2012) Laboratory scale conceptual process development for the isolation of renewable glycolaldehyde from pyrolysis oil to produce fermentation feedstock. Green Chem 14:321–325
107. Chahal SP, Starr JN (2000) Lactic acid. In: Ullmann's encyclopedia of industrial chemistry. Wiley, Weinheim
108. Peng J, Li X, Tang C, Bai W (2014) Barium sulphate catalyzed dehydration of lactic acid to acrylic acid. Green Chem 16(1):108–111
109. Makshina EV, Janssens W, Sels BF, Jacobs PA (2012) Catalytic study of the conversion of ethanol into 1,3-butadiene. Catal Today 198(1):338–344
110. Bruijnincx PCA, Weckhuysen BM (2013) Shale gas revolution: an opportunity for the production of biobased chemicals? Angew Chem Int Ed 52(46):11980–11987
111. Yim H, Haselbeck R, Niu W, Pujol-Baxley C, Burgard A, Boldt J, Khandurina J, Trawick JD, Osterhout RE, Stephen R, Estadilla J, Teisan S, Schreyer HB, Andrae S, Yang TH, Lee SY, Burk MJ, Van Dien S (2011) Metabolic engineering of *Escherichia coli* for direct production of 1,4-butanediol. Nat Chem Biol 7(7):445–452
112. Boucher-Jacobs C, Nicholas KM (2013) Catalytic deoxydehydration of glycols with alcohol reductants. ChemSusChem 6(4):597–599
113. Yi J, Liu S, Abu-Omar MM (2012) Rhenium-catalyzed transfer hydrogenation and deoxygenation of biomass-derived polyols to small and useful organics. ChemSusChem 5(8):1401–1404
114. Ji N, Zhang T, Zheng M, Wang A, Wang H, Wang X, Chen JG (2008) Direct catalytic conversion of cellulose into ethylene glycol using nickel-promoted tungsten carbide catalysts. Angew Chem 120(44):8638–8641
115. Wang S, Yin K, Zhang Y, Liu H (2013) Glycerol hydrogenolysis to propylene glycol and ethylene glycol on zirconia supported noble metal catalysts. ACS Catal 3(9):2112–2121
116. Zhao G, Zheng M, Zhang J, Wang A, Zhang T (2013) Catalytic conversion of concentrated glucose to ethylene glycol with semicontinuous reaction system. Ind Eng Chem Res 52 (28):9566–9572

117. Dry ME (2002) The Fischer–Tropsch process: 1950–2000. Catal Today 71(3–4):227–241
118. Cortright RD, Davda RR, Dumesic JA (2002) Hydrogen from catalytic reforming of biomass-derived hydrocarbons in liquid water. Nature 418(6901):964–967
119. D'Hondt E, Van de Vyver S, Sels BF, Jacobs PA (2008) Catalytic glycerol conversion into 1,2-propanediol in absence of added hydrogen. Chem Commun (Cambridge):6011–6012
120. Corthals S, Van Nederkassel J, De Winne H, Geboers J, Jacobs P, Sels B (2011) Design of active and stable NiCeO2ZrO2MgAl2O4 dry reforming catalysts. Appl Catal B 105 (3–4):263–275
121. Corthals S, Van Nederkassel J, Geboers J, De Winne H, Van Noyen J, Moens B, Sels B, Jacobs P (2008) Influence of composition of $MgAl_2O_4$ supported $NiCeO_2ZrO_2$ catalysts on coke formation and catalyst stability for dry reforming of methane. Catal Today 138 (1–2):28–32
122. Palo DR, Dagle RA, Holladay JD (2007) Methanol steam reforming for hydrogen production. Chem Rev 107(10):3992–4021
123. Trost B (1991) The atom economy–a search for synthetic efficiency. Science 254 (5037):1471–1477
124. Sheldon RA (2012) Fundamentals of green chemistry: efficiency in reaction design. Chem Soc Rev 41(4):1437–1451
125. Sheldon RA (2008) E factors, green chemistry and catalysis: an odyssey. Chem Commun 29:3352–3365
126. Sheldon RA (2007) The E factor: fifteen years on. Green Chem 9(12):1273–1283
127. Trost BM (1995) Atom economy—a challenge for organic synthesis: homogeneous catalysis leads the way. Angew Chem Int Ed English 34(3):259–281
128. Danon B, Marcotullio G, de Jong W (2014) Mechanistic and kinetic aspects of pentose dehydration towards furfural in aqueous media employing homogeneous catalysis. Green Chem 16(1):39–54
129. Hayashi Y, Sasaki Y (2005) Tin-catalyzed conversion of trioses to alkyl lactates in alcohol solution. Chem Commun 2716–2718
130. Genomatica (2013) Successful commercial-scale production of BDO. http://www.genomatica.com/products/bdo/.
131. Burk MJ, Van DSJ, Burgard A, Niu W (2008) A synthetic metabolic pathway for the biosynthesis of 1,4-butanediol and a transgenic microorganism for the fermentation of the diol. WO2008115840A2
132. Mascal M (2012) Chemicals from biobutanol: technologies and markets. Biofuel Bioprod Biorefining 6(4):483–493
133. West RM, Braden DJ, Dumesic JA (2009) Dehydration of butanol to butene over solid acid catalysts in high water environments. J Catal 262(1):134–143
134. Le Van MR, Levesque P, McLaughlin G, Dao LH (1987) Ethylene from ethanol over zeolite catalysts. Appl Catal 34:163–179
135. Sun P, Long X, He H, Xia C, Li F (2013) Conversion of cellulose into isosorbide over bifunctional ruthenium nanoparticles supported on niobium phosphate. ChemSusChem 6(11):2190–2197
136. Liang G, Wu C, He L, Ming J, Cheng H, Zhuo L, Zhao F (2011) Selective conversion of concentrated microcrystalline cellulose to isosorbide over Ru/C catalyst. Green Chem 13(4):839–842
137. Chung P-W, Charmot A, Olatunji-Ojo OA, Durkin KA, Katz A (2014) Hydrolysis catalysis of miscanthus xylan to xylose using weak-acid surface sites. ACS Catal 4(1):302–310
138. Dusselier M, Van Wouwe P, De Smet S, De Clercq R, Verbelen L, Van Puyvelde P, Du Prez FE, Sels BF (2013) Toward functional polyester building blocks from renewable glycolaldehyde with Sn cascade catalysis. ACS Catal 3:1786–1800
139. Jin F, Yun J, Li G, Kishita A, Tohji K, Enomoto H (2008) Hydrothermal conversion of carbohydrate biomass into formic acid at mild temperatures. Green Chem 10(6):612–615

140. Comotti M, Della Pina C, Falletta E, Rossi M (2006) Aerobic oxidation of glucose with gold catalyst: hydrogen peroxide as intermediate and reagent. Adv Synth Catal 348(3):313–316
141. Onda A, Ochi T, Kajiyoshi K, Yanagisawa K (2008) A new chemical process for catalytic conversion of D-glucose into lactic acid and gluconic acid. Appl Catal A 343:49–54
142. Komanoya T, Kobayashi H, Hara K, Chun W-J, Fukuoka A (2013) Simultaneous formation of sorbitol and gluconic acid from cellobiose using carbon-supported ruthenium catalysts. J Energ Chem 22(2):290–295
143. Schiweck H, Bär A, Vogel R, Schwarz E, Kunz M, Dusautois C, Clement A, Lefranc C, Lüssem B, Moser M, Peters S (2000) Sugar alcohols. In: Ullmann's encyclopedia of industrial chemistry. Wiley-VCH, Weinheim
144. Geboers J, Van de Vyver S, Carpentier K, de Blochouse K, Jacobs P, Sels B (2010) Efficient catalytic conversion of concentrated cellulose feeds to hexitols with heteropoly acids and Ru on carbon. Chem Commun 46(20):3577–3579
145. Ruppert AM, Weinberg K, Palkovits R (2012) Hydrogenolysis goes bio: from carbohydrates and sugar alcohols to platform chemicals. Angew Chem Int Ed 51(11):2564–2601
146. Bilik V (1972) Reactions of saccharides catalyzed by molybdate ions. II. Epimerization of D-glucose and D-mannose. Chem Zvesti 26:183–186
147. Gunther WR, Wang Y, Ji Y, Michaelis VK, Hunt ST, Griffin RG, Román-Leshkov Y (2012) Sn-beta zeolites with borate salts catalyse the epimerization of carbohydrates via an intramolecular carbon shift. Nat Commun 3:1109
148. Bermejo-Deval R, Assary RS, Nikolla E, Moliner M, Román-Leshkov Y, Hwang S-J, Palsdottir A, Silverman D, Lobo RF, Curtiss LA, Davis ME (2012) Metalloenzyme-like catalyzed isomerizations of sugars by Lewis acid zeolites. Proc Natl Acad Sci U S A 109 (25):9727–9732
149. Roman-Leshkov Y, Moliner M, Labinger JA, Davis ME (2010) Mechanism of glucose isomerization using a solid Lewis acid catalyst in water. Angew Chem Int Ed 49:8954–8957
150. Nakagawa Y, Tamura M, Tomishige K (2013) Catalytic reduction of biomass-derived furanic compounds with hydrogen. ACS Catal 3(12):2655–2668
151. Gounder R, Davis ME (2013) Titanium-beta zeolites catalyze the stereospecific isomerization of D-glucose to L-sorbose via intramolecular C5–C1 hydride shift. ACS Catal 3(7):1469–1476

Top Curr Chem (2014) 353: 41–84
DOI: 10.1007/128_2014_536

Published online: 20 May 2014

Chemical-Catalytic Approaches to the Production of Furfurals and Levulinates from Biomass

Mark Mascal and Saikat Dutta

Abstract The synthesis and chemistry of 5-(hydroxymethyl)furfural (HMF), 5-(chloromethyl)furfural (CMF), and levulinic acid (LA), three carbohydrate-derived platform molecules produced by the chemical-catalytic processing of lignocellulosic biomass, is reviewed. Starting from the historical derivation of these molecules and progressing through modern approaches to their production from biomass feedstocks, this review will then survey their principal derivative chemistries, with particular attention to aspects of commercial relevance, and discuss the relative merits of each molecule in the future of biorefining.

Keywords 5-(Chloromethyl)furfural · 5-(Hydroxymethyl)furfural · Biomass · Biomass derivatives · Biorefinery · Catalysis · CMF · Green chemistry · HMF · Levulinic acid · Platform chemicals · Renewable chemistry

Contents

M. Mascal (✉) and S. Dutta
Department of Chemistry, University of California Davis, 1 Shields Avenue, Davis, CA 95616, USA
e-mail: mjmascal@ucdavis.edu

1 Introduction

As discussed in detail in Chapter 1 [1], the chemical-catalytic approach to biomass valorization is poised to come to the fore of biorefinery operations due to its advantages over microbial and thermochemical processing of lignocellulosic feedstocks. Below, we consider three mainstream platform chemicals, collectively referred to as "furanics," that are derived from the acid-catalyzed dehydration of carbohydrates. The first, 5-(hydroxymethyl)furfural, or HMF **1**, is an icon of the biorefinery movement. With derivatives that branch out over multiple product manifolds, HMF is a recognized commercial opportunity for whoever can manage to produce it economically, and approaches towards the realization of this aim will be discussed.

Obstacles that persist in the path of the economic production of HMF have shifted some attention towards its halogen-substituted congeners (XMFs), i.e., 5-(chloromethyl)furfural (CMF) **2** and 5-(bromomethyl)furfural (BMF) **3**. These molecules have the advantage of being obtainable in high yields directly from cellulosic biomass and also of being easy to isolate from the medium of their production. Additionally, they share all of the derivative chemistry of their forerunner, HMF.

Finally, the HMF rehydration and ring cleavage product levulinic acid (LA) **4**, which can be produced from HMF, XMFs, or directly from biomass, has also entered into the mainstream of renewable chemistry. With its easy availability and a derivative chemistry that rivals that of the furans, LA is currently leading the charge towards commercialization within the furanics movement, and piloted approaches to its production and applications are described.

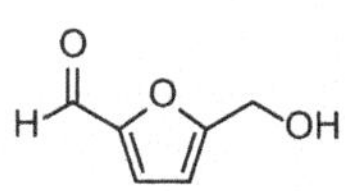

1

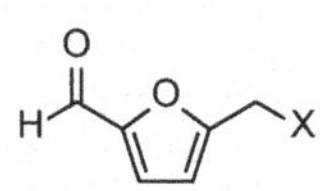

2 X=Cl
3 X=Br

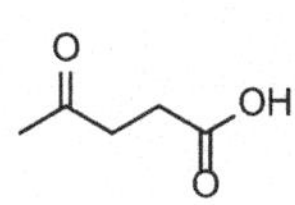

4

2 5-(Hydroxymethyl)furfural (HMF)

2.1 Synthesis: Introduction

The synthesis of HMF was first reported in 1895, where inulin (a β-2,1-fructan) was heated in acidic aqueous solution followed by solvent extraction [2, 3]. The structure was not definitively assigned until 1910 [4]. Since then, thousands of papers have been published that involve HMF in one context or another, and several reviews dedicated to its preparation and chemistry have appeared [5–15]. Further interest in HMF has been generated around the possible health effects of its presence in heat-treated foods and beverages [16].

Since so many detailed reviews are available [5–15], in this contribution we only tabulate data for the best published outcomes under each of the below mentioned representative approaches to HMF synthesis, involving fructose, sugars other than fructose (glucose or sucrose), inulin, cellulose, and finally biomass itself (Table 1). This is followed by a brief discussion of the respective advantages and practical limitations of the feedstocks, catalysts, and media. We then also give highlights of the derivative chemistry of HMF and discuss its future prospects as a renewable platform chemical.

Fructose to HMF in aqueous systems under homogeneous catalysis (entry 1)
Fructose to HMF in aqueous systems under heterogeneous catalysis (entry 2)
Fructose to HMF in organic solvents under homogeneous catalysis (entry 3)
Fructose to HMF in organic solvents under heterogeneous catalysis (entry 4)
Fructose to HMF in biphasic solvent systems under homogeneous catalysis (entry 5)
Fructose to HMF in biphasic solvent systems under heterogeneous catalysis (entry 6)
Fructose to HMF in ionic liquids under homogeneous catalysis (entry 7)
Fructose to HMF in ionic liquids under heterogeneous catalysis (entry 8)
Glucose or sucrose to HMF in aqueous systems under homogeneous catalysis (entry 9)
Glucose or sucrose to HMF in aqueous systems under heterogeneous catalysis (entry 10)
Glucose or sucrose to HMF in organic solvents (entry 11)
Glucose or sucrose to HMF in biphasic solvent systems (entry 12)
Glucose or sucrose to HMF in ionic liquids (entry 13)
Inulin to HMF in aqueous systems under homogeneous catalysis (entry 14)
Inulin to HMF in aqueous systems under heterogeneous catalysis (entry 15)
Inulin to HMF in biphasic solvent systems (entry 16)
Inulin to HMF in ionic liquids (entry 17)
Cellulose to HMF in aqueous systems under homogeneous catalysis (entry 18)
Cellulose to HMF in aqueous systems under heterogeneous catalysis (entry 19)
Cellulose to HMF in organic solvents (entry 20)

Table 1 Representative examples of HMF synthesis from sugars, carbohydrates and biomass feedstocks

Entry	Feedstock	Medium	Catalyst	Reaction conditions	Yield (%)	References
1	Fructose	Water	aq. HCl	95°C, 1.5 h	68	[17]
2	Fructose	Water-acetone	Dowex-50wx8-100	150°C, 15 min	73	[18]
3	Fructose	DMSO	NH_4Cl	100°C, 45 min	100	[19]
4	Fructose	DMSO	Amberlyst-15 powder	120°C, 2 h	100	[20]
5	Fructose	1:1 Water-DMSO/ 7:3 MIBK/ 2-BuOH	HCl	170°C, 4 min	85	[21]
6	Fructose	1:7 DMSO/MIBK	Ion exchange resin (acidic)	76°C	97	[22]
7	Fructose	[BMIm]Cl	NHC/$CrCl_2$	100°C, 6 h	96	[23]
8	Fructose	[HexylMIm]Cl	SO_4^{2-}/ZrO_2	100°C, 30 min	89	[24]
9	Glucose	Water	H_3PO_4/Nb_2O_5	120°C, 3 h	52	[25]
10	Glucose	Water	TiO_2/ZrO_2	250°C, 5 min	29	[26]
11	Sucrose	DMA	$CrCl_3/NH_4Br$	100°C, 1 h	87	[27]
12	Glucose	1:2.25 Water-MIBK	$Ag_3PW_{12}O_{40}$	130°C, 4 h	76	[28]
13	Sucrose	[BMIm]Cl	$CrCl_3$	120°C, 4 h	100	[29]
14	Inulin	Water	CO_2 (6 MPa)	200°C, 45 min	53	[30]
15	Inulin	Water	Cubic ZrP_2O_7	100°C, 1 h	70	[31]
16	Inulin	2:3 Water-2-butanol	Modified hydrated tantalum oxide	160°C, 140 min	87	[32]
17	Inulin	ChoCl	Oxalic acid	80°C, 2 h	64	[33]
18	Cellulose	Water	HCl	RT→300°C, 30 min	21	[34]
19	Cellulose	Water	$Cr[(DS)H_2PW_{12}O_{40}]_3$	150°C, 2 h	53	[35]
20	Cellulose	DMA+10% LiCl	$CrCl_3$/HCl	140°C, 2 h	33	[36]
21	Cellulose	1:5 Water/MIBK	TiO_2 (fixed bed)	270°C, 2 min	30	[37]
22	Cellulose	[EMIm]Cl	$CrCl_2$	120°C, 6 h	89	[38]
23	Oak wood chips	Steam	H_2SO_4	286°C, 90 s	50-80	[39]
24	Cassava waste	Water	Sulfonated carbon	250°C, 1 min	11	[40]
25	tapioca flour	Water/acetone/ DMSO	WO_3-ZrO_2	230°C	22	[41]
26	Pine wood	[BMIm]Cl	$CrCl_3$	200°C, 3 min	35	[38]

Abbreviations: *DMSO* (dimethyl sulfoxide), *DMA* (*N*,*N*-dimethylacetamide), *MIBK* (methylisobutylketone), *[BMIm]Cl* (1-butyl-3-methylimidazolium chloride), *[HexylMIm]Cl* (1-hexyl-3-methylimidazolium chloride), *[EMIm]Cl* (1-ethyl-3-methylimidazolium chloride), *ChoCl* (choline chloride), *NHC* (*N*-heterocyclic carbene), *DS* (dodecyl sulfate)

Cellulose to HMF in biphasic solvent systems (entry 21)
Cellulose to HMF in ionic liquids (entry 22)
Biomass to HMF in aqueous systems under homogeneous catalysis (entry 23)
Biomass to HMF in aqueous systems under heterogeneous catalysis (entry 24)
Biomass to HMF in organic solvents (entry 25)
Biomass to HMF in ionic liquids (entry 26)

2.2 *Synthesis: Discussion*

2.2.1 Feedstocks

Bearing in mind from the outset that processes that employ pure sugars as feedstock are unlikely to be competitive with those that start from raw biomass, the current status of HMF production at scale is tenuous, since no industrially relevant approaches have been developed that do not depend on fructose as the starting material. In effect, no technology is any more scalable than the practical accessibility of its feedstock. The historical dominance of fructose in this chemistry is easily understood by the fact that HMF is simply a dehydration product of fructose (Scheme 1), whereas any other hexose would first have to isomerize to fructose under the reaction conditions. In any case, however, yields of HMF are generally high starting from fructose, in fact reaching up to quantitative (entries 3, 4, 6, 7). A recent review tabulated data on >300 fructose dehydration reactions leading to HMF and its derivatives [6], so clearly this is a well-trodden reaction path.

The other commodity sugars, glucose and sucrose, have also often been targeted as HMF feedstocks, and in some cases, particularly under $CrCl_3$ catalysis (entries 11 and 13), yields are comparable to those of fructose.

While glucose and sucrose are produced in sufficient volumes to supply a commercial HMF market, much interest has centered on the fructan inulin as a feedstock for HMF, the advantage being that, unlike edible sugars, it does not enter into the food vs fuel controversy. Commercial interest in inulin centers almost entirely around the prospect of cultivating the Jerusalem artichoke as an energy crop [42], the tubers of which can yield up to ca. 80% inulin by mass. Since inulin is essentially an easily hydrolyzed polymer of fructose [43], it gives comparable HMF yields (entries 14–17).

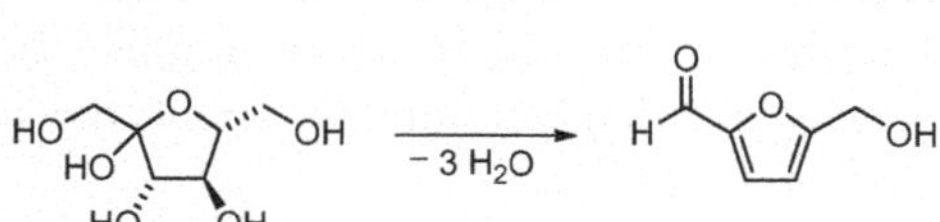

fructose

1

Scheme 1 Dehydration of fructose to HMF

Finally, attempts to produce HMF from cellulose and raw biomass have also been reported. The challenges here are significant, and only under exceptional conditions (e.g., entry 22) are high yields reported.

2.2.2 Catalyst and Medium

An acid catalyst of some description is required in most cases, which may be either homogeneous or heterogeneous in nature. For homogeneous catalysis, either mineral acids (e.g., HCl, H_2SO_4) or organic acids (e.g., pTSA, oxalic acid) can be used, where reactions are generally fast and proceed under mild conditions. However, acid recovery can be problematic. HMF **1** is unstable in acidic media and either rehydrates to levulinic acid **4** or decomposes entirely into humic material, which limits yields. Heterogeneous catalysts tend to perform better in this regard, but may require harsher reaction conditions. Although recovery of the catalyst is facilitated, recycling may be limited by fouling.

HMF synthesis has classically involved either water or polar aprotic organic solvents such as DMSO as the reaction medium, since the sugar feedstocks are soluble in these media. Of all the technical issues that have confronted the industrialization of HMF production, first and foremost has been the high solubility of HMF in these solvents, which complicates product isolation. This problem has been mitigated to some extent by the recent development of biphasic reaction systems that involve continuous extraction of HMF into lower boiling solvents [5]. Salting out strategies can also increase the efficiency of this approach.

A major movement in HMF synthesis came out of the field of ionic liquids, which can in some cases act both as catalyst and solvent. Particularly good outcomes are seen when ionic liquids are coupled with chromium salt catalysis, first introduced by Zhang in 2007 [44]. While good to excellent results in terms of conversion and selectivity have generally been seen in this medium, there are significant drawbacks to large scale applications. Since the carbohydrate to HMF reaction is a dehydration, quantities of water will accumulate in these hydrophilic (and often hygroscopic) solvents, eventually necessitating a potentially expensive drying step. Since ionic liquids are excellent solvents for polar molecules, they also tend to solubilize polar reaction by-products and any miscellaneous compounds that may be introduced via the feedstock, presenting a recyclability/purification challenge. This will be particularly relevant where raw biomass is used. Other issues, such as viscosity, cost, stability, toxicity, product isolation, and reagent recovery from the ionic liquid are together likely to limit the applications of these solvents in large-scale industrial settings [45].

2.3 Synthesis: Perspective

The following key challenges remain to the adoption of HMF as a renewable platform chemical. (1) The continued reliance on fructose as a feedstock for HMF production cannot be sustained. To be competitive in the long term, even glucose and sucrose can only be stopgap solutions on the way to a biomass-based HMF process. While inulin may also present opportunities in the short term, there are questions associated even with marginal land use for planting energy crops that may limit the future prospects of the Jerusalem artichoke. A long horizon view must involve the processing of raw biomass into HMF, and although such approaches have shown promise at the bench, issues involving reaction media and product isolation may prevent scale up. (2) Aqueous and polar organic solvents have commonly been used in HMF synthesis, but ionic liquids now show the most promise with virtually all feedstocks. However, as discussed above, ionic liquids suffer from a range of serious drawbacks that do not present simple solutions for scale up. (3) HMF is a fairly sensitive molecule with well recognized decomposition paths that have long plagued its production under acid catalysis. Thus, quick isolation of HMF from its reaction medium is desirable but practically difficult. Biphasic reaction systems have had a measure of success using pure sugars as feedstocks, but have yet to be proved with biomass.

The ideal HMF production process would use raw biomass as its feedstock, without the necessity for extensive drying or pretreatment (apart from mechanical reduction to particle sizes which do not suffer mass transfer limitations). Reactions would proceed in high yield over short time scales under mild conditions in inexpensive media and use simple, non-foulable catalysts. The HMF product would be isolated without recourse to distillation or protracted solvent extraction, and all materials would be easily recyclable. Except for product yield, none of these objectives has currently been met in such a way as to be reducible to practice on an industrial scale. In the end, it is a matter of economics. When the dust settles, only the most competitive, industrially viable processes will be left standing, and the rest will be consigned to history.

2.4 Derivatives

The great interest around HMF is the result of its recognition as a renewable platform chemical of exceptional promise. HMF has three chemical functionalities; the hydroxymethyl group, the aldehyde, and the furan ring itself. Together, these offer a diverse combination of chemistries for derivative synthesis. Thus, the hydroxymethyl group can be acylated, alkylated, substituted with nucleophiles, or oxidized to the aldehyde or carboxylic acid oxidation state, both of which have multiple derivatives towards which to branch out. The aldehyde can undergo nucleophilic addition or be reduced to the hydroxymethyl group, with attendant opportunities as already noted, or oxidized to the carboxylic acid, which can be

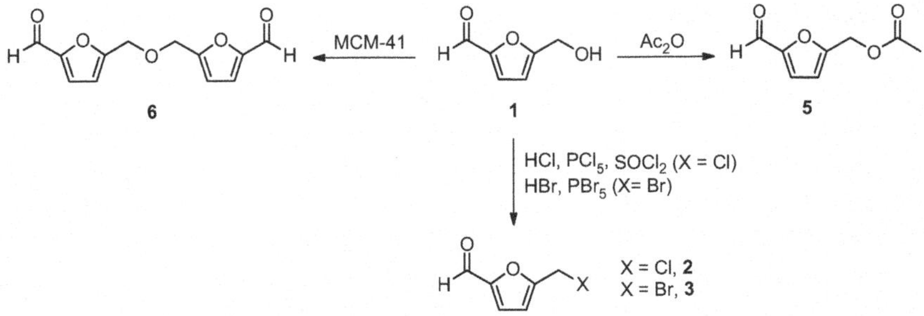

Scheme 2 Functional transformations of the hydroxymethyl group in HMF

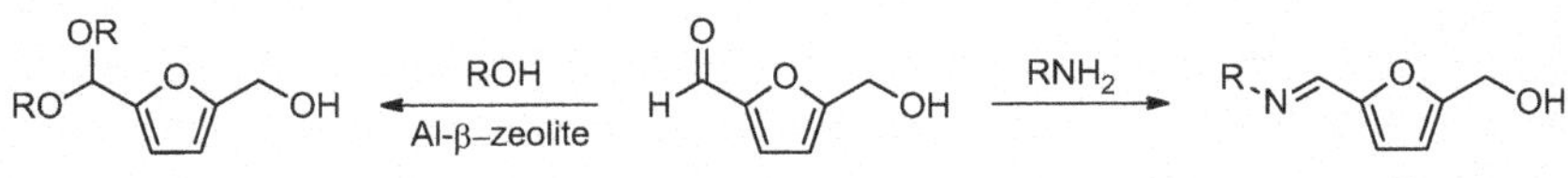

Scheme 3 Functional transformations of the aldehyde group in HMF

variously derivatized. The furan ring can undergo cycloaddition, electrophilic aromatic substitution, ring opening, or hydrogenation to the corresponding tetrahydrofuran. Chosen synthetic transformations can be applied to yield derivatives such as biofuels, monomers for novel polymeric materials, or specialty chemicals (e.g., agrochemicals, pharmaceuticals). Examples of useful applications of this chemistry are discussed below. In each case, selected, high-yielding approaches to specific derivatives will be highlighted, but for full coverage of this field the reader is directed to comprehensive reviews [5–15].

2.4.1 Simple Derivatization of the Hydroxymethyl or Aldehyde Group

The hydroxymethyl group of HMF has been variously derivatized. For example, acetylation of HMF to **5** expedites its isolation from reaction media during its preparation from carbohydrates [46], as well as facilitating the deoxygenation of the hydroxymethyl group (discussed below) [47]. Intermolecular dehydration over the mesoporous aluminosilicate MCM-41 gives the symmetric ether 5,5′(oxybis(methylene))bis(furan-2-carbaldehyde) **6** in high yield, which has been proposed as a potentially useful monomer [48] (Scheme 2). Substitution of the OH group for halogens **2**, **3** can be carried out with the typical reagents (SOX_2, PX_n, HX, etc.) [49] or can be effected in situ during the formation of HMF from various carbohydrate sources [50], and this latter reactivity will be reviewed in Sect. 3 of this chapter.

The carbonyl group of HMF participates in typical reactions of aromatic aldehydes, including the formation of various imines **7** [51] and acetals **8** [52, 53] (Scheme 3), the latter being interesting both as novel surfactants and potential biodiesel fuels.

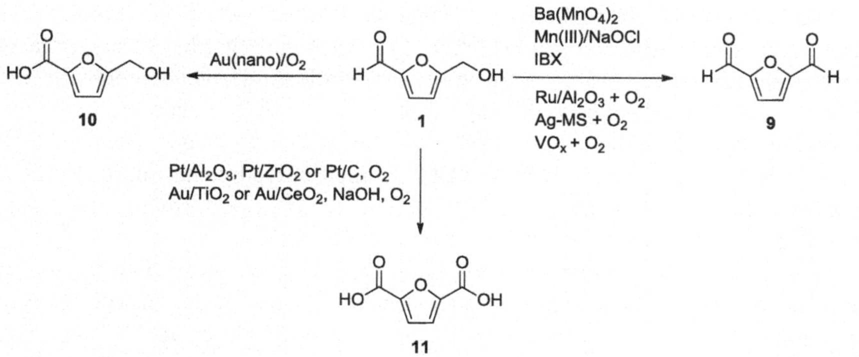

Scheme 4 Oxidized derivatives of HMF

2.4.2 Oxidation of the Hydroxymethyl and/or Aldehyde Group

2,5-Diformylfuran

Oxidation of the hydroxymethyl group of HMF to an aldehyde leads to 2,5-diformylfuran (DFF) **9**, a monomer of considerable interest in the polymer industry [54, 55]. Synthetic approaches to this molecule from HMF were reviewed in 2012 by Hu et al. [56]. Apart from various stoichiometric oxidants, e.g., $Ba(MnO_4)_2$ [57], NaOCl [58], ceric ammonium nitrate [59], IBX [60], and dichromates [61], a number of promising catalytic routes have recently been developed. For example, using air as the oxidant and a silver-impregnated molecular sieve catalyst, an essentially quantitative yield of DFF was observed [62]. Comparable results have also been seen with ruthenium-based catalysts such as Ru/C [63], Ru clusters [64], Ru/Al_2O_3 [65], and a range of vanadium oxides [66] (Scheme 4). Thus, there is much promise here, assuming that an industrially viable route to the HMF starting material can be established. Recognizing this issue, some investigators have sought to produce DFF directly from fructose or glucose, since DFF is more easily extracted into hydrophobic solvents than HMF. Yields, however, have been moderate (25–55%) [67–69].

Furancarboxylic Acids

The rate of oxidation of the HMF aldehyde group to the carboxylic acid is substantially faster than that of the alcohol to the aldehyde, and hence the selective preparation of 5-hydroxymethyl-2-furancarboxylic acid **10** is possible. Casanova et al. reported the synthesis of **10** in quantitative yield using a gold-nanoparticle catalyzed oxidation with molecular oxygen [70]. Davis et al. reported similar results using Au/C and Au/TiO_2 [71], while Pasini et al. used an Au-Cu catalyst [72].

Oxidation of both the hydroxymethyl and the aldehyde group to carboxylic acids gives 2,5-furandicarboxylic acid (FDCA) **11**, a molecule that has of late attracted strong interest as a renewable replacement for petroleum-derived terephthalic acid, which is used to produce polyethylene terephthalate, or PET, a mainstream polymer widely used in the production of synthetic fibers and beverage containers. The equivalent polymer generated from FDCA, called polyethylene furanoate (PEF), has been shown to be competitive with PET in terms of performance, and thus a significant future market can be envisaged for FDCA [73].

Like DFF, synthetic approaches to FDCA generally rely on HMF as the starting material. Recent, efficient approaches all involve catalytic oxidation with O_2 as the stoichiometric oxidant, as described in the reviews of Hu et al. [56] and Tong et al. [74]. The catalysts generally used for this transformation are Pd, Pt, or Au, in various forms and on various supports, and the yields are generally near quantitative. Thus, Lew et al. [75] and Davis et al. [76] reported the application of Au/C, Au/TiO_2, Pd/C, and Pt/C as catalysts in a basic reaction medium. Linga et al. [77] used platinum catalysts on various supports (activated carbon, ZrO_2, Al_2O_3) in basic, neutral, and even acidic media and observed excellent yields of FDCA in all cases. Casanova et al. used gold nanoparticle-based catalysts (e.g., Au/TiO_2, Au/CeO_2) under optimized reaction conditions (10 bar O_2, 130°C, aq. NaOH) to achieve >99% yield of FDCA. The same authors recently reported a base-free, aerobic oxidation of HMF to yield dimethyl furan-2,5-dicarboxylate in methanol using an Au/CeO_2 catalyst [78]. The common application of basic media in this chemistry could be seen as problematic due to the generation of a salt waste stream.

2.4.3 Reduction Chemistry of HMF

Reduction of the aldehyde of HMF to a hydroxymethyl group gives 2,5-di(hydroxymethyl)furan (DHMF) **12**, a useful monomer building block in the production of polymers and polyurethane foams [79]. In the bench scale synthesis, $NaBH_4$ is the obvious reducing agent [80–82]. Catalytic hydrogenation is, however, more industrially relevant to this process, and essentially quantitative yields have been reported by hydrogenation over Ir–ReO_x/SiO_2 [83], gold nanoparticles on Al_2O_3 [84], and Pd/C with formic acid as the hydrogen source [85].

One of the most sought-after furan derivatives of recent times has been 2,5-dimethylfuran (DMF) **13**, the product which results from the reduction of both the hydroxymethyl and aldehyde functions of HMF to methyl groups, and several high-profile papers have been devoted to its renewable production [85–88]. In addition to being a high energy density, high octane biofuel [89], DMF can be converted into *p*-xylene, a high volume chemical intermediate used for the production of drop-in terephthalate polymers [90–93]. Notable efforts towards DMF have included hydrogenation of HMF in the presence of a CuRu/C catalyst (71% yield) [87], a one-pot reaction of fructose with formic acid in the presence of H_2SO_4 and Pd/C catalysts (51% yield) [85], and hydrogen transfer to HMF from supercritical methanol with a Cu-doped porous metal oxide catalyst (48% yield) [86]. Most recently, Zu

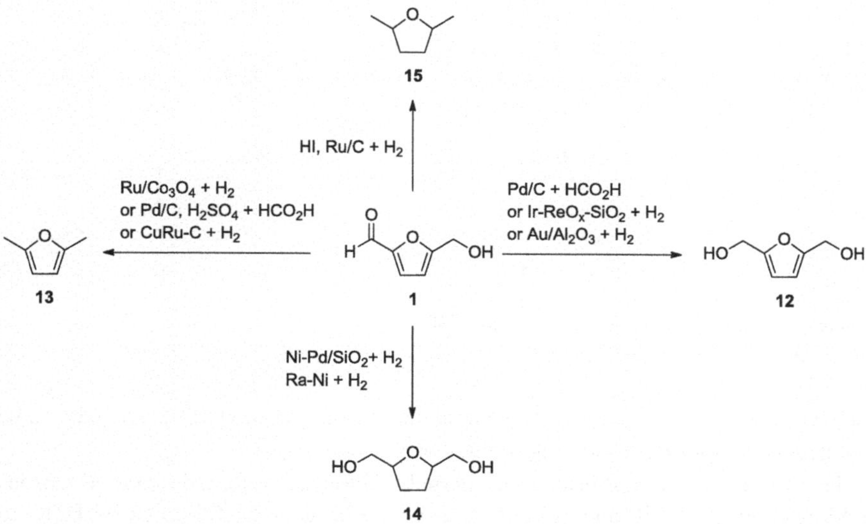

Scheme 5 Catalytic hydrogenations of HMF

and co-workers reported a remarkable 93% yield of DMF by hydrogenation of HMF using a novel Ru/Co_3O_4 catalyst [88] (Scheme 5).

Ring hydrogenation of HMF to tetrahydrofurans is also a highly useful transformation. This can be done while preserving oxygen functionality at the methyl positions, giving 2,5-bis(hydroxymethyl)tetrahydrofuran (BHTHF) **14**. Thus, hydrogenation of HMF on $Ni–Pd/SiO_2$ under mild conditions provided BHTHF in 95% yield, [94] and the use of Ra-Ni [95] or ceria-supported ruthenium catalysts [96] gives similar outcomes.

Complete reduction of HMF to 2,5-dimethyltetrahydrofuran (DMTHF) **15** was studied in detail by Sen and co-workers, where not only HMF but also fructose could be converted to DMTHF by hydrogenation in the presence of HI and a ruthenium catalyst [97, 98]. Despite the high energy content of DMTHF and its potential as a fuel, comparatively few studies have been devoted to its selective production from HMF or biomass in general.

2.4.4 Condensation Chemistry of HMF

In recent years, the production of simple hydrocarbons from biomass has attracted strong interest in the renewables community, due to the fact that the products are considered "drop-in" substitutes for petroleum-derived alkanes, with evident applications to fuels and chemical production. HMF has received much attention in this regard as a platform for extended carbon chain products, the hydrodeoxygenation (HDO) of which gives products which are essentially diesel or aviation fuels, depending on their hydrocarbon distribution. Aldol-type condensation reactions can take place in aqueous solution between HMF (or its derivatives) and biogenic ketones

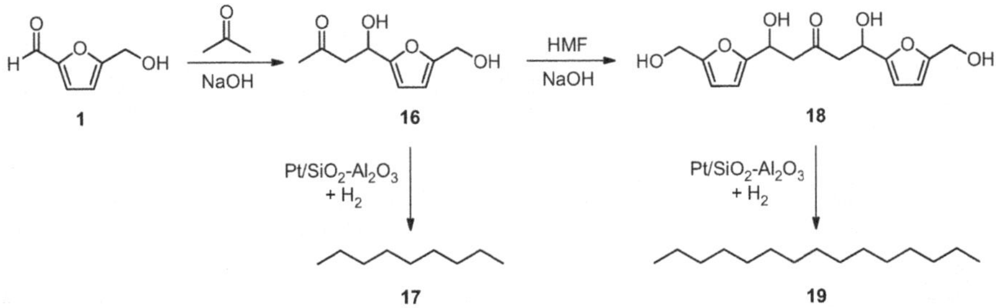

Scheme 6 Synthesis of hydrocarbons from HMF

such as acetone. The method in general has been reviewed [99], so only some representative approaches are highlighted here.

The original work in this area was done by Dumesic et al. and involved various condensations of HMF and related molecules with acetone followed by HDO to give C_1–C_{15} alkanes (for example, **1→17+19**) (Scheme 6) [100]. Liu and co-workers recently described the benzoin condensation of HMF to give a dimer that could be submitted to HDO to give C_{10}–C_{12} alkanes [101]. Sutton and co-workers likewise employed simple aldol chemistry between HMF and acetone to access C_9–C_{15} hydrocarbons [102]. This sugar derivative to alkane chemistry has been successfully piloted by the startup company Virent [103].

2.4.5 Transformations Involving Cleavage of the Furan Ring

Bearing in mind that the HDO chemistry in the preceding section involves first hydrogenation and finally hydrogenolytic cleavage of the THF ring, we now consider other useful derivatives that also take advantage of ring-opening reactions.

Rehydration of HMF in acidic aqueous media leads to hydrolytic ring opening to give equimolar quantities of levulinic acid (LA) **4** and formic acid [104]. As noted in the introduction, LA is a platform chemical in its own right and its production and chemistry will be reviewed in Sect. 4 of this chapter.

Adipic acid **20** is a high-volume commodity chemical used for making nylon polymers. The structural similarity between FDCA and adipic acid provides an obvious route between the two chemicals by hydrogenating FDCA to 2,5-tetrahydrofurandicarboxylic acid followed by reductive cleavage (Scheme 7) [105].

Oxidative cleavage of the HMF ring can be achieved in the presence of singlet oxygen. The reaction is carried out by irradiating an aerated solution of HMF containing a sensitizer (e.g. rose bengal). The reaction proceeds through the endoperoxide intermediate **21**, which undergoes ring opening to a butenolide in aqueous or alcoholic solvent, which subsequently cleaves to 4-oxopent-2-enoic acid or the corresponding ester **22** [106, 107].

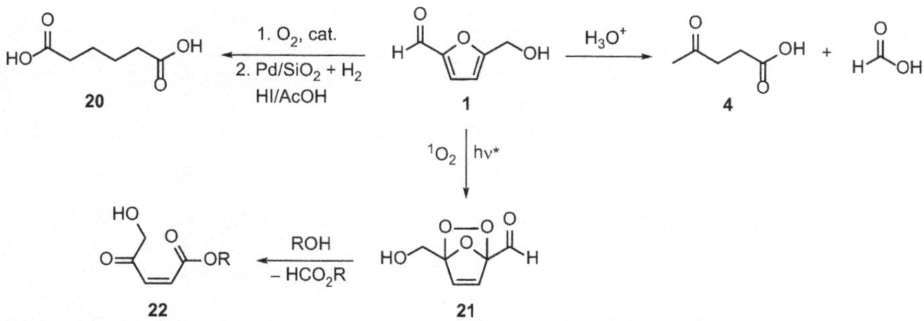

Scheme 7 Transformations of HMF involving furan ring cleavage

3 5-(Halomethyl)furfurals

3.1 Historical Efforts

The first preparation of a 5-(halomethyl)furfural was reported as early as 1899 by Fenton and Gostling [108], the former being well known for the development of "Fenton chemistry." In this work, fructose was treated with a solution of hydrogen bromide in ether, resulting in a modest yield of a bromo-substituted dehydration product which was correctly assigned the 5-(bromomethyl)furfural (BMF) **3** structure. Follow-up studies described yields of up to 28% BMF with cellulose sources (paper, cotton) but rather poor outcomes with starch and sugars other than fructose [109]. 5-(Chloromethyl)furfural (CMF) **2** was first described by the same authors in a 1901 paper, using the same general approach as for BMF (ethereal HX) and reporting a 9% yield from filter paper. A variety of simple derivatives produced by reactions at the halomethyl and aldehyde functional groups were also described [109]. Emil Fischer and co-worker later reported a convenient, scalable synthesis from cane sugar and aqueous HCl, giving yields of up to 17% CMF, which was also used to prepare derivatives [110]. The original work of Fenton and Gostling was re-examined by Hibbert and Hill in 1923, who showed that glucose also gave moderate yields of BMF, refuting the theory that cellulose was composed of ketose subunits, which had been largely based on the former's work [111]. In 1944, Haworth and Jones published a method for making CMF from sucrose that for the first time involved a biphasic acid/solvent reaction, by which CMF could be isolated in 21% yield [112]. Only in a 1978 patent by a Japanese group was the yield of CMF improved to preparative usefulness (up to 77% yield from fructose) when a surfactant was also included in the processing of sugars with aqueous HCl using the biphasic approach, ostensibly the result of a micelle-like state in the reaction mixture [113]. Szmant and Chundury followed up this work in 1981 with a detailed parameter study of the conversion of fructose and high-fructose corn syrup into CMF, ultimately achieving a yield of 95%, although yields from glucose and starch were only 45% and 21%, respectively [114]. Incremental changes to the procedure,

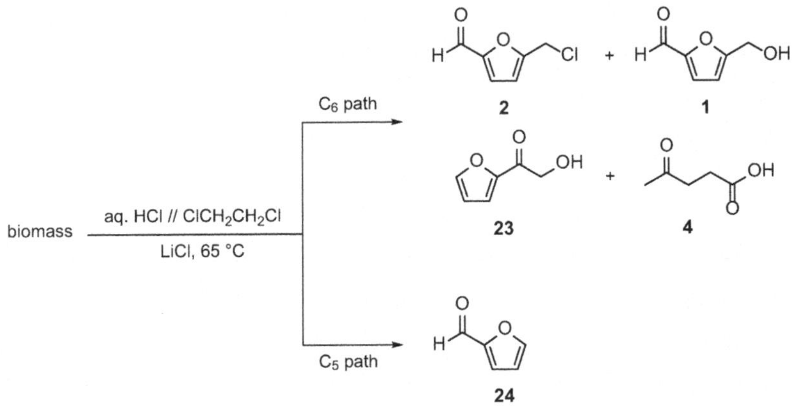

Scheme 8 Production of furanic derivatives by acid-catalyzed dehydration of biomass in a biphasic reactor

for example the inclusion of a magnesium halide as a promoter of the reaction, were later reported [115–117]. As mentioned in Sect. 2, descriptions of the synthesis of CMF by the reaction of HMF **1** with chlorinating agents have also been published [118], but these are not considered practical due to the comparatively poor accessibility of HMF from lignocellulosic feedstocks.

3.2 Modern Synthetic Approaches

In 2008, Mascal and Nikitin reported a method whereby glucose, sucrose, or cellulose was converted into CMF in 71–76% isolated yields, alongside up to 15% of a mixture of 2-(2-hydroxyacetyl)furan (HAF) **23**, HMF **1**, and LA **4** [119] (Scheme 8). The process involved the heating of the feedstock in a biphasic conc. aq. HCl–solvent reactor over several hours with continuous solvent extraction. This was significant in that, up to this point, there had been no report of the conversion of cellulose to CMF in useful yield, apparently due to the harsh conditions required to break cellulose down into glucose units and the extensive decomposition of the furanic products that results. The process was later applied to various forms of raw biomass, which gave CMF yields comparable to that of pure cellulose, based on hexose content. Since plant biomass generally also contains hemicellulose, furfural (**24**) was also isolated in these cases (Scheme 8) [119, 120]. The remarkable net yield of organic products (85–91%, depending on the feedstock) was attributed to the fact that, once formed, the furanics were swept into the organic phase, where they were sheltered from the decomposition pathways that have long plagued the synthesis of HMF (Scheme 9).

Although the CMF process opened a high-yielding pathway from biomass to halomethylfurfurals, the extended reaction times (up to 30 h) could be seen as a

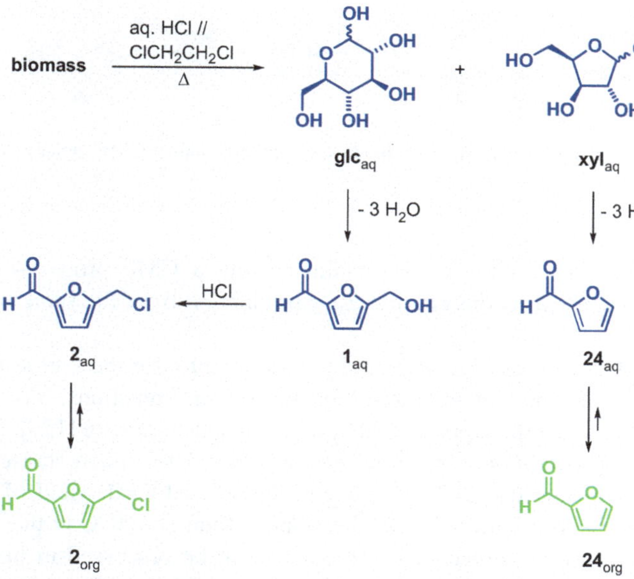

Scheme 9 Schematic representation of the CMF process in an aqueous (*blue*)-organic (*green*) biphasic reactor

disincentive to industrial applications. Working in an open reactor with concentrated HCl meant that the temperature of the process had to be restricted to about 65°C to reduce outgassing of the acid, but this also limited the reaction rate. In 2009, Mascal and Nikitin reported a closed-vessel modification of the method in which the temperature was raised to 100°C, leading to an order of magnitude decrease in reaction time [121]. Instead of carrying out a continuous extraction, the biphasic mixture was extracted hourly. Depending on the feedstock, it was found that the reaction was 85–90% complete after the first hour, with 7–12% of the total product extracted after a second hour, and only about 2–3% after a third. On top of this, the organic extract contained only CMF, with no HAF **23** or HMF **1** observed, and in improved yields of up to 90%. Small quantities of LA (5–8%) could be isolated from the aqueous phase of the reaction, along with lignin as a dark powder in cases where lignocellulosic biomass was used. Raw biomass feedstock loadings of up to 10% w/v in the aqueous solution could be successfully managed in this improved process.

In a 2010 report, the CMF process was applied to oil seed feedstocks for the production of hybrid oleaginous-cellulosic biodiesel [122]. When oil seeds were treated with aqueous HCl in a biphasic aqueous/organic reaction, a mixture of CMF and the triglyceride **25** was obtained in the organic phase which, upon heating with an alcohol, led to a mixture of the levulinate **26** (from the CMF) and fatty acid **27** (from the oil) esters (Scheme 10). As will be noted in the following section on levulinic acid, levulinate esters are promising blendstocks for biodiesel. For a selected feedstock (safflower seeds), a ca. 25% increase in fuel yield (fatty acid

oil seed

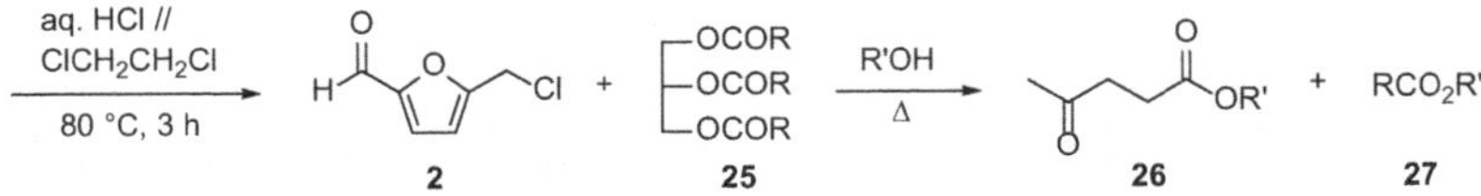

Scheme 10 Levulinate-fatty acid ester biodiesel synthesis using CMF process on oil seed feedstocks

ester/levulinate ester mixture) was obtained using a CMF process+ethanolysis protocol over standard biodiesel ethyl ester production from the same quantity of sample [122].

Recently, other reports have appeared in the literature for the synthesis of CMF using biphasic aqueous hydrochloric/organic solvent reactions. For example, Brasholz et al. adapted the biphasic CMF process to a flow reactor [123]. In contrast to batch reaction conditions, the flow reactor allowed for much shorter reaction times. For example, using DCM as solvent, an isolated CMF yield of 80% was obtained from fructose with only 100 s residence time at 100°C in the reactor. A flow rate of 2 mL/min allowed 10 g of fructose to be processed in just 20 min, giving >6 g of CMF. Glucose and sucrose performed less well, requiring longer residence times and giving CMF in lower yields (50–60%).

Breeden et al. reported a synthesis of CMF from various sugars (e.g., fructose, glucose) and polysaccharides (e.g., cellulose, inulin) using conc. aq. HCl in a biphasic system under microwave irradiation, which allows for selective heating of the aqueous phase, leading to different outcomes compared to conventional heating [124]. Under these conditions, CMF was isolated in 85% yield within 10 min at 70°C using fructose as the feedstock. Glucose, however, gave a lower yield of 39%. The optimal results with cellulose (71% yield) were achieved by pre-treatment in a ball mill prior to the reaction. A concurrent solvent study showed that, while extraction with DCE gave the highest isolated yields of CMF, cyclohexane also performed surprisingly well, and would be the alternative solvent of choice in the event that one wished to avoid the use of halogenated solvents.

In an attempt to produce CMF under milder reaction conditions, Gao et al. described an aqueous–organic biphasic reaction system where a combination of concentrated HCl (37%) and H_3PO_4 (85%) were used in the aqueous phase with chloroform as the extracting solvent at only 45°C [125]. CMF was obtained in 47% isolated yield from fructose, although glucose and cellulose gave poor yields of CMF, 7.3% and 7.8%, respectively. Surprisingly, CMF yields up to 31% were obtained when cellulosic feedstocks (e.g., eucalyptus wood) were used.

High yields (>90%) of both CMF and BMF have been reported by treatment of 3-deoxyglucosone, which is the product of a single dehydration of glucose, with the corresponding HX acids [126].

In a recent patent, Masuno et al. describe a process for making CMF from sugars and biomass using a fluidized bed reactor [91]. Gaseous hydrochloric acid is fed into a reactor containing biomass at a high velocity and temperature. The temperature inside the reactor is ~220°C with pressures up to 15 atm. The hot, pressurized acid causes fluidization of the feedstock, allowing for thorough mixing. After a 2-min residence time, the mixture of gas and solids is separated in a cyclone into gaseous hydrochloric acid and a solid/liquid sludge. The authors observed a mixture of CMF, HMF, and furfural as the major products and levulinic plus formic acids as minor components.

BMF **3**, historically the first halomethylfurfural to be prepared, has recently made a re-appearance in the literature. Thus, Kumari et al. reported the production of BMF by treating sugars, polysaccharides, and cellulosic biomass with a combination of concentrated hydrobromic acid and lithium bromide in a biphasic reactor at 25°C for 48 h [127]. BMF was obtained in up to 82% yield using toluene as the extracting solvent. Interestingly, cellulose gave a better yield of BMF (80%) than did glucose (50%). Use of raw biomass (straw) gave a moderate yield of BMF (68%) when processed at 65°C for 48 h. A related study describes the application of an aqueous HBr/toluene biphasic reactor, resulting in up to 64%, 59%, and 56% yields of BMF from glucose, cellulose, and softwood after 24 h at 65°C with serial solvent extractions [50].

5-(Iodomethyl)furfural (IMF) has been prepared as an intermediate by Finkelstein reaction of CMF with NaI in acetone, although attempts to isolate it have not been successful [114]. In 2012, Sen et al. reported the hydroiodic acid-catalyzed dehydration of sugars during which in situ generated IMF was reduced to 5-methylfurfural (MF) **31** [128].

Finally, 5-(fluoromethyl)furfural has also been made by heating BMF with potassium hydrofluoride or sodium fluoride in acetonitrile solution in the presence of 18-crown-6. Yields of 60–70% are claimed [129].

3.3 *Derivatives of 5-(Halomethyl)furfurals*

CMF is a low-melting solid (m.p. 37–38°C) but its liquid state is easily supercooled and it generally occurs as a colorless or pale yellow liquid at room temperature. It can be isolated by distillation at reduced pressure (b.p. 68–69°C at 0.7 Torr [130] or 137–138°C at 5 Torr [131]), which is predictably lower than that of HMF (110°C at 0.02 Torr) [112]. CMF is indefinitely stable when stored as a 10% solution in an organic solvent at refrigerator temperatures. Pure samples of CMF, however, become increasingly colored on standing, eventually turning from yellow to dark brown, even in the cold. However, stabilization with an epoxy resin such as the commercial DER383 renders even pure samples colorless and stable over long periods of time [130].

Since CMF **2** can be converted into HMF **1** in high yield (see below), all the derivative chemistry of HMF (Sect. 2) applies by proxy to CMF. Like HMF, CMF has two essential reaction manifolds – furanic and levulinic, although there are some subtle differences in reactivity. CMF is more reactive than HMF in substitutions at the CH_2 group, and more soluble in nonpolar solvents, but otherwise, the chemistries of CMF and HMF are largely the same. Here, we give an account of

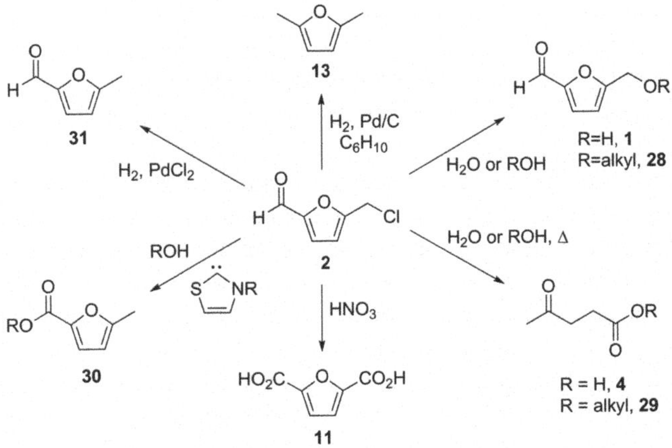

Scheme 11 Preparation of value-added chemicals and intermediates from CMF

a selection of literature reactions involving CMF which establish its place as a new, highly versatile platform chemical.

3.3.1 Conversion to HMF, Alkoxymethylfurfurals, LA, and Levulinic Esters

Mascal and Nikitin reported the conversion of CMF **2** into either HMF **1** or LA **4** with excellent selectivities and yields. HMF was obtained within 30 s in 86% yield by the action of boiling water on CMF. When CMF is hydrolyzed for an extended period (20 min) at 190°C, LA is produced in 91% yield [132]. Similarly, HMF has been obtained in quantitative yield by stirring the more reactive BMF **3** in water at RT [127]. When these same reactions are carried out in alcoholic solution, the corresponding ethers and esters are produced. Thus, both CMF and BMF have been converted into 5-(ethoxymethyl)furfural (EMF) **28** (R=Et), a proposed biofuel, in high yield on treatment with ethanol [50, 120, 127] (Scheme 11). Levulinate esters **29** are likewise produced in high yields from CMF on heating with alcohols at elevated temperatures [132]. In a recent theoretical study, the thermochemistry of the conversion of CMF into HMF, LA, EMF, and ethyl levulinate was calculated using G4 theory and density functional (DFT) methods [133].

3.3.2 Substitution and Reduction Products

The synthesis of the dechlorination product of CMF, 5-methylfurfural (MF) **31**, was first reported by reduction with $SnCl_2$ in low yield (~20%) [134]. In 2001, Hamada et al. reported the palladium-catalyzed hydrogenation of CMF to MF in various organic solvents [115, 116]. Mascal, et al. prepared MF in 88% yield by catalytic

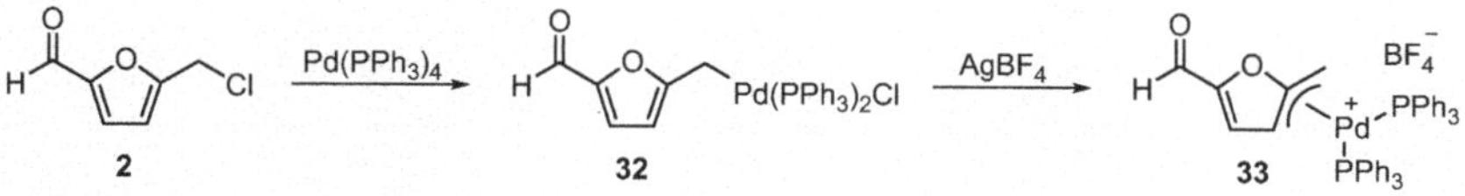

Scheme 12 Preparation of a Pd η^3-complex of CMF

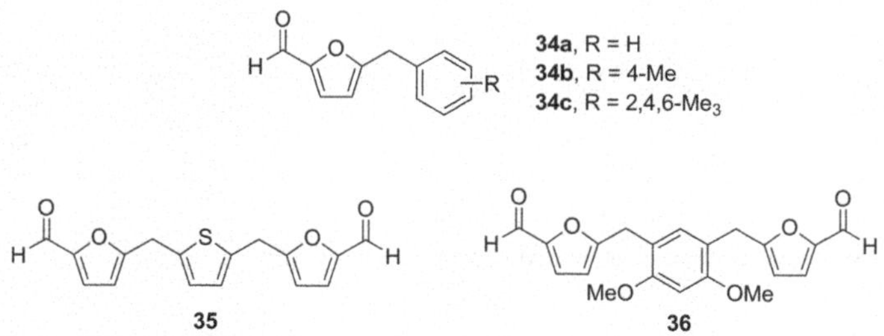

Fig. 1 Friedel-Crafts arylation products of CMF

hydrogenation of CMF using $PdCl_2$ in *N*,*N*-dimethylformamide [119, 120]. Sen et al. produced MF in 68% yield directly from fructose in a biphasic reactor with aqueous hydroiodic acid under hydrogenation conditions via in situ generated IMF [128]. The synthesis of 2,5-dimethylfuran (DMF) **13**, a promising biofuel and a high-profile renewable chemical intermediate (cf. Sect. 2), has also been accomplished from CMF. Hamada et al. reported an 81% yield of DMF by palladium-catalyzed transfer hydrogenation of CMF in refluxing cyclohexene [135]. In a recent patent, the synthesis of 5-methylfuran-2-carboxylic acid esters **30** has been described by the reaction of CMF with an alcohol in the presence of an N-heterocyclic carbene catalyst (Scheme 11) [136]. The resulting products are highly promising fuel additives [137].

A cationic η^3-furfuryl complex of palladium (**33**) has been prepared from CMF (Scheme 12), which has the potential to activate the methyl group towards reactions with various nucleophiles [138].

Friedel–Crafts alkylation of aromatics with CMF to give rise to benzyl derivatives **34a** and **34b** was first demonstrated in 1909 [139] and reprised by Rauchfuss and co-worker in 2013 to provide feedstock **34c** for hydrodeoxygenation to diesel-range hydrocarbons [140]. Similar reactions have been used to produce a range of novel monomers and polymers. Thus, Szmant et al. reported as far back as 1981 the application of CMF to the synthesis of polymeric building blocks **35** and **36**, among others (Fig. 1) [141].

The reaction of CMF with pyrrole, furan, or thiophene gives conjugated, conducting polymers of structure **37** by attack at both the chloromethyl and aldehyde functions (Scheme 13) [142].

The substitution of the chloro group of CMF with triphenylphosphine gives a phosphonium derivative **38** which reacts with base to give annulene 1,4-oxide macrocycles **39** [143] and conjugated poly(2,5-furanylvinylene) polymers **40**

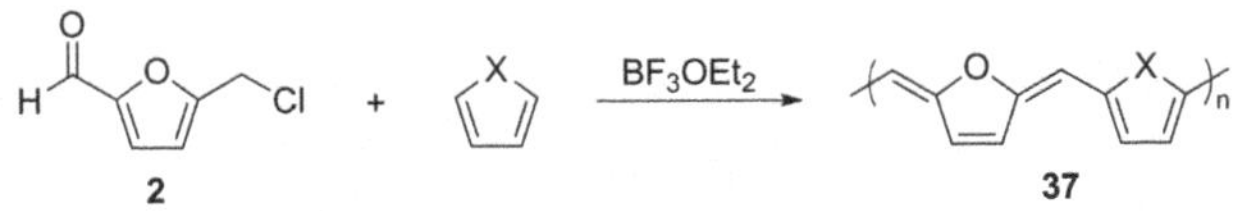

Scheme 13 Synthesis of conducting polyfuranmethine polymers from CMF

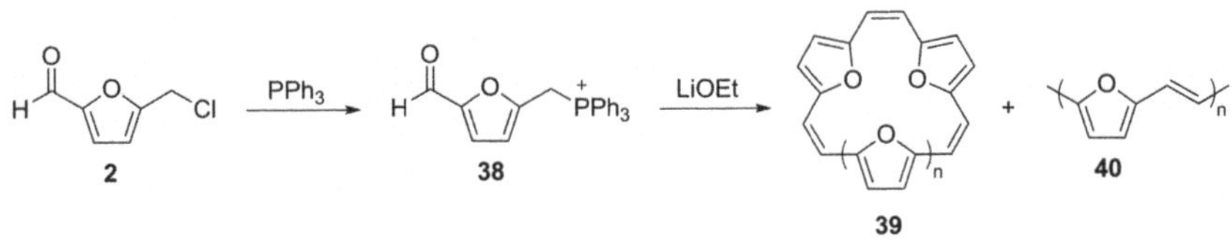

Scheme 14 Synthesis of macrocycles and polymers from CMF

(Scheme 14) [144]. Cram and co-worker also used CMF to incorporate furan 2,5-dimethylene units into crown ethers [145].

3.3.3 Condensation Chemistry of CMF

As was described for HMF in Sect. 2, CMF has likewise been used as a platform for carbon chain extension for the purposes of making higher alkanes. Thus, Silks, et al. describe the direct condensation of CMF with acetone, hydroxyacetone, or dihydroxyacetone in the presence of a zinc-proline complex catalyst [146]. Similarly, Seck recruited CMF as an intermediate in a process that involves various condensations of its derivatives and ultimately hydrotreating to arrive at biofuels [147].

3.3.4 CMF Oxidation

As described in detail in Sect. 2, furan-2,5-dicarboxylic acid (FDCA) **11** has drawn considerable attention in recent years as a renewable alternative to terephthalic acid for the production of phthalate polymers. Synthesis of FDCA from CMF was first reported by Fenton et al. by treatment with nitric acid [141]. Later, Brasholz et al. used the same method to obtain a 59% yield of FDCA [123]. 2,5-Diformylfuran **9** has also been prepared by nitric acid oxidation of CMF [148] (Scheme 15).

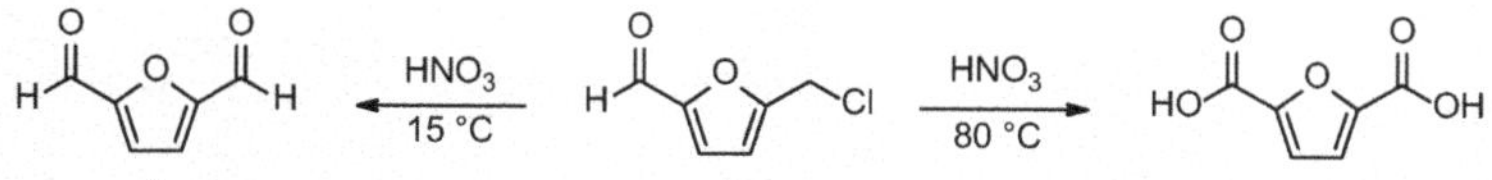

Scheme 15 Oxidation of CMF

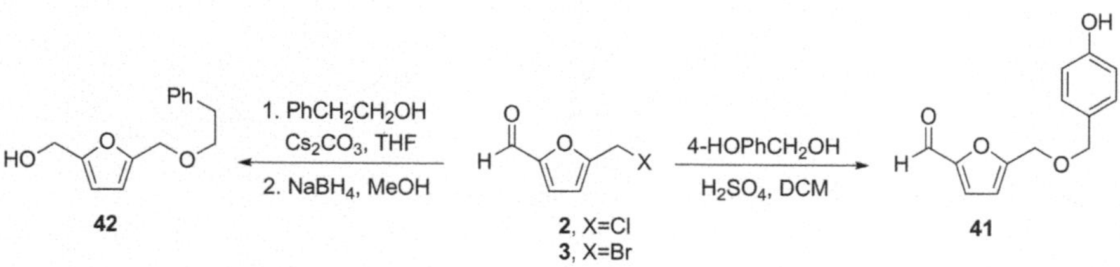

Scheme 16 Synthesis of natural products **41** and **42** from XMFs

3.3.5 Specialty Chemicals from CMF

In recent years, a range of value-added chemicals, including agrochemicals, pharmaceuticals, and natural products, have been synthesized from CMF, as described below.

XMFs as Key Starting Materials for Natural Product Synthesis

Tamariz et al. reported synthesis of the natural products 5-[(4-hydroxybenzyl)oxy] methyl-2-furaldehyde **41** and pichiafuran C **42** from BMF and CMF, respectively. Compound **41** was recently isolated from the rhizome of *Gastrodia elata* Blume and exhibits cytotoxicity against the HT-29 cell line. It was prepared in 81% yield by the reaction of BMF with 4-hydroxymethylphenol in the presence of an acid catalyst. Pichiafuran C **42** was isolated from the yeast *Pichia membranifaciens*, which was derived from the marine sponge *Petrosia* sp. It was synthesized from CMF in two steps in 55% overall yield as shown in Scheme 16 [149].

Klein and Shanklin [150] described the total synthesis of (±)-dimethyl jaconate **46**, a metabolite of the pyrrolizidine alkaloid jacobine, starting from CMF (Scheme 17). Thus, substitution of the chloromethyl group with allyl sulfide followed by reaction with methyllithium gave **43**. Compound **43** then undergoes intramolecular cycloaddition to **44**. Benzylation, ozonolysis, and desulfurization with Raney Ni gives **45**. The hydroxymethyl groups are oxidized with the Jones reagent and esterified with diazomethane, and the benzyl group is finally cleaved to give the dimethyl jaconate product **46** in about 1% overall yield over nine steps.

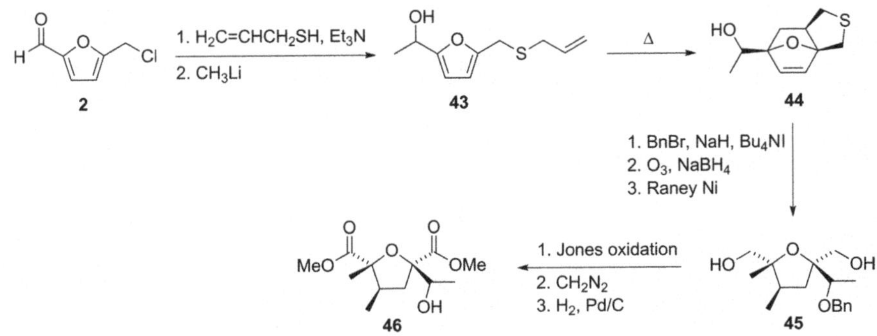

Scheme 17 Synthesis of (±)-dimethyl jaconate **46** from CMF

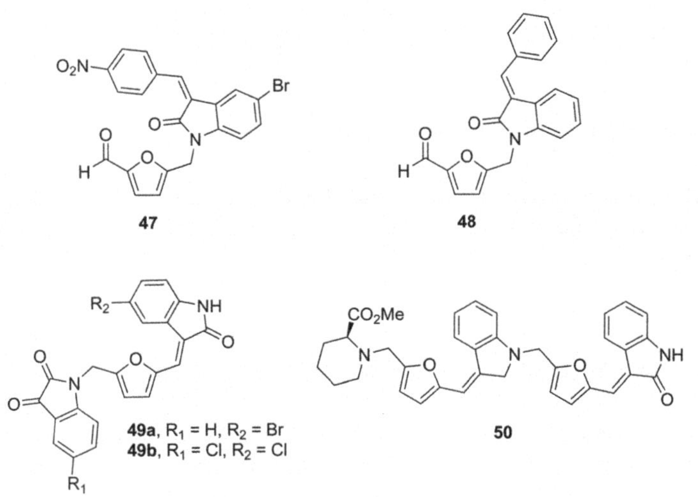

Fig. 2 Biologically active iodolinones and di-indolinones derived from CMF

CMF as a Synthetic Intermediate in Pharmaceutical Lead Generation

Figure 2 gives examples of chemical structures incorporating a furfurylmethyl group which have been synthesized as lead compounds targeting various types of pharmacological action. Thus, indolinone derivatives **47** and **48** [151] and di-indolinones **49** and **50** [152, 153] have been prepared and evaluated for the inhibition of protein tyrosine phosphatases, a family of enzymes that mediate cellular signal transduction, modulators of which are gaining importance as therapeutic agents. The biological activities of **47–50** include anti-tumor and anti-fungal activity.

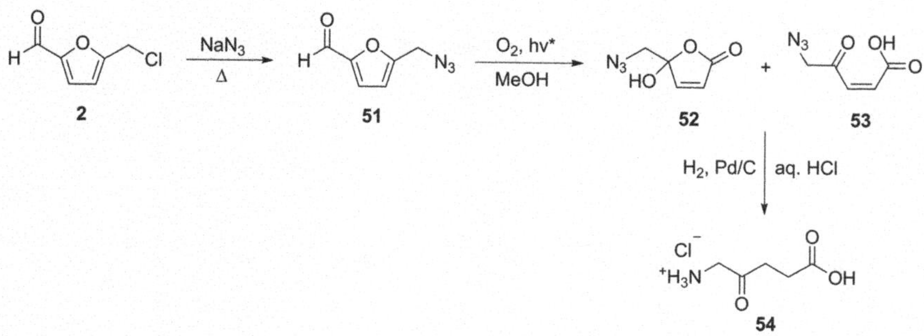

Scheme 18 Synthesis of DALA **54** from CMF

Synthesis of δ-Aminolevulinic Acid from CMF

δ-Aminolevulinc acid (DALA) **54** is a natural, broad spectrum herbicide and pesticide that has attracted much attention due to its effectiveness, low toxicity, and biodegradability. It is also employed as a photodynamic therapy drug. Mascal and Dutta have recently published a high-yielding, three step synthesis of DALA from CMF (Scheme 18) [154]. The route starts by substituting the chloromethyl group with azide anion to give 5-(azidomethyl)furfural (AZF) **51** in 92% isolated yield. Photocatalyzed addition of singlet oxygen to the furan ring of **51** gives a mixture of butenolide **52** and 4-oxopent-2-enoic acid **53**, which is hydrogenated over Pd/C in the presence of aqueous HCl to give DALA **54** as the hydrochloride salt in 68% overall yield from CMF.

Synthesis of Ranitidine from CMF

Ranitidine **58**, popularly known as Zantac™, is a histamine H_2-receptor antagonist used for the treatment of gastric and duodenal ulcers. Introduced by Glaxo in 1981, ranitidine was the first drug to achieve $1 billion in sales. It has recently been reformulated as an over-the-counter general antacid preparation. The original synthesis of ranitidine employed furfural as the starting material and introduced the *N,N*-dimethylaminomethyl functionality on the furan ring later in the synthesis [155]. Mascal and Dutta described a synthetic strategy whereby ranitidine was prepared from CMF in four steps (Scheme 19) [156]. In the first step, commercially available 3-mercapto-*N*-methylpropanamide reacts with CMF in presence of NaH to give **55** in 92% isolated yield. Reductive amination with dimethylamine and $NaBH_4$ in methanol gives **56** in 90% yield. The acetyl protecting group is removed by heating in KOH solution to give **57** in 94% yield. Finally, **57** reacts with commercial 1-methylthio-1-methylamino-2-nitroethylene to give ranitidine **58** in 88% isolated yield. This renewable approach uses inexpensive reagents and requires no chromatographic purification until the final step. The overall yield from CMF to ranitidine was 68%.

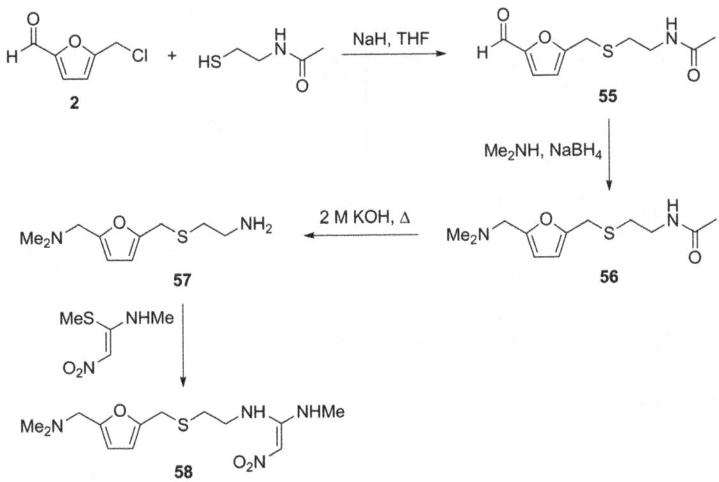

Scheme 19 Synthesis of ranitidine **58** from CMF

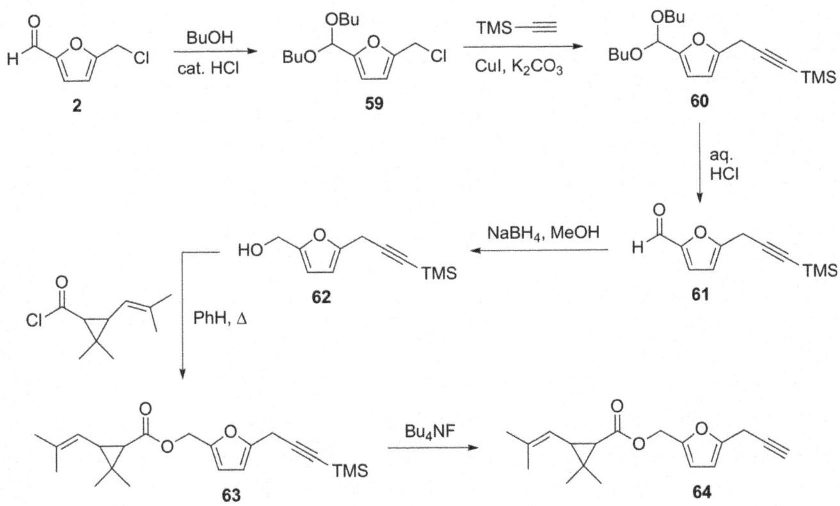

Scheme 20 Synthesis of prothrin **64**, a furan-based pyrethroid, from CMF

Synthesis of Furan-Based Pyrethroids from CMF

Pyrethroids are synthetic analogues of the naturally occurring pyrethrins (isolated from *Chrysanthemum* sp.) and are popular insecticides for agricultural and domestic uses, with favorable properties like biodegradability and low mammalian toxicity. Prothrin **64** is a member of the furan-based class of pyrethroids, which also includes the high-volume commercial insecticide resmethrin. Chang et al. recently

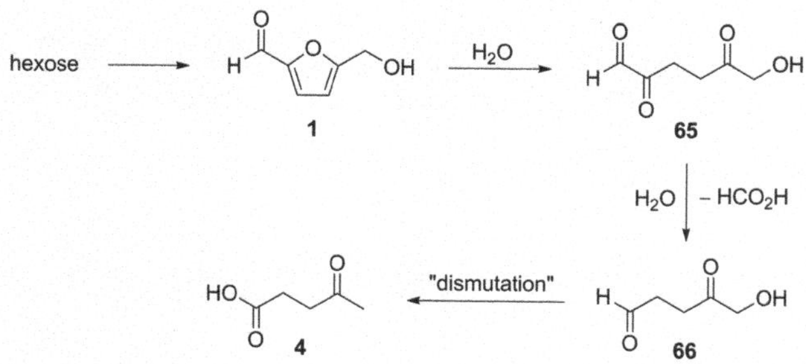

Scheme 21 Early mechanistic speculation on the formation of LA **4** from carbohydrates

published a convenient, high-yielding synthetic route to prothrin via CMF (Scheme 20) [157]. Thus, the aldehyde moiety of CMF was first protected in the form of a dibutyl acetal **59** and coupled with TMS-acetylene in presence of a copper catalyst to give **60**. The acetal was then cleaved and the resulting aldehyde **61** was reduced with $NaBH_4$. Esterification of alcohol **62** with the acid chloride of commercial chrysanthemic acid gave **63**, which was deprotected with Bu_4NF to give prothrin **64** in 65% yield over six steps.

4 Levulinic Acid

4.1 Introduction

The first preparation of levulinic acid (LA) **4** was credited by von Grote and Tollens [158] to Malaguti in 1836 [159] and Mulder in 1840 [160], who heated sucrose with dilute aq. H_2SO_4 and, after removal of the humic material, isolated crude LA as a viscous oil. The structure was unambiguously assigned in 1878 [161]. An *Organic Syntheses* procedure for LA was published in 1929, whereby a 22% yield was obtained from either sucrose or starch using HCl as the catalyst [162]. The yield could be improved to 42% by heating carbohydrates in dilute aq. HCl under pressure [163]. Dahlmann in 1968 described a simpler procedure involving the heating of sugars, starches, or even cellulose at reflux in 20% aq. HCl (azeotropic, b.p. 108°C) to give LA in yields of up to 60% [164]. Even better results were obtained using HBr as the acid, whereby 75%, 64%, and 69% yields were reported with glucose, starch, and sucrose, respectively [165].

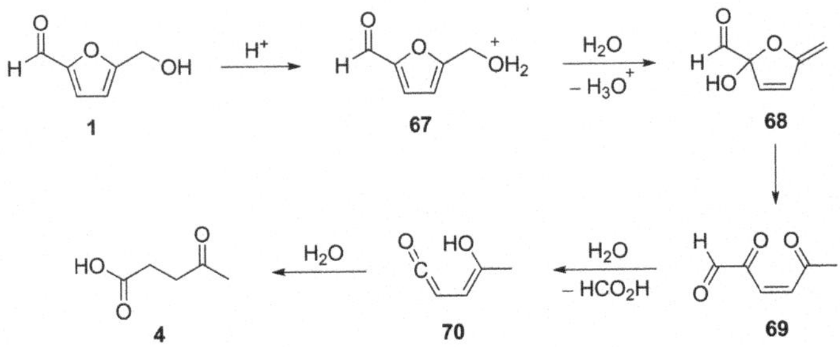

Scheme 22 Mechanism for the formation of LA **4** via HMF **1**

4.2 Mechanism

The mechanism of formation of levulinic acid from carbohydrates bears discussion. The subject was first taken up by Pummerer and co-workers in 1935 [166], who first recognized that HMF **1** was an intermediate which underwent rehydration to open chained products, but otherwise gave an oversimplified interpretation that involved the "dismutation" of oxygen from the CH_2OH group in **66** to the terminal aldehyde (Scheme 21).

Isbell put forward a more intuitive analysis of the process as shown in Scheme 22 [167]. Evidence for intermediate **69** was later obtained by NMR analysis of the hydrolysis of **1** [168]. A somewhat different, but no more reasonable, route between **69** and **4** is generally proposed which involves the hydration of the aldehyde group in **69** and cleavage of formic acid without invoking the ketene intermediate **70**.

4.3 Synthesis

A number of reviews of the synthesis of LA from various carbohydrate sources have been published [169–173]. Approaches vary considerably in terms of selectivity and overall yield, and the reader is directed to these accounts if a detailed treatment is sought. However, given that the only competitive process for the production of LA would use biomass as feedstock, we provide in Table 2 a survey of the highest yielding approaches of this description.

The production of LA from biomass invariably involves chemical processing under relatively forcing conditions. A balance must be found between good conversion and extent of decomposition of the feedstock into humic material. As can be seen, temperatures around 200°C are typical, although heating as low as 150°C over extended periods is also effective in some cases. Reactions are exclusively carried out in aqueous media under homogeneous acid catalysis. Feedstocks include cellulose itself, various energy crops, and industrial, agricultural and forestry

Table 2 Preparation of LA **4** by acid-catalyzed hydrolysis of cellulosic feedstocks

Entry	Feedstock	Catalyst	Reaction conditions	Yield (%)	References
1	Cellulose	H_2SO_4 (3%)	230°C, 4 h	54	[174]
2	Cellulose	HCl (3%)	250°C, 2 h	40	[175]
3	Newspaper	H_2SO_4 (10%)	150°C, 8 h	59	[176]
4	Paper sludge	H_2SO_4 (3.5%)	196–232°C, 20 min	76	[177]
5	Wheat straw	H_2SO_4 (3.5%)	209°C, 37 min	69	[178]
6	Paddy straw	HCl (4.45%)	220°C, 45 min	79.5	[179]
7	Grain sorghum	H_2SO_4	200°C, 40 min	45	[180]
8	Water hyacinth	H_2SO_4 (9.5%)	175°C, 30 min	34	[181]
9	Sugarcane bagasse	H_2SO_4 (1.3%)	165°C, 1 h	61[a]	[182]
10	Water oak	HCl (6%)	160°C, 30 min	45[b]	[183]
11	Sawdust	HCl (1.5%)	190°C, 30 min	36	[184]
12	Bagasse	HCl (4.45%)	220°C, 45 min	82.5	[179]
13	Corn stover	HCl (37%)	100°C, 3 h then 190°C, 20 min	81	[121]
14	Marine algae (*gelidium amansii*)	H_2SO_4 (3%)	160°C, 43 min	19.5	[185]

[a]Calculated based on 30% cellulose content in sugarcane baggase (18.4 wt% compared to starting biomass)
[b]Calculated based on 40.8% cellulose content in water oak

wastes. Two entries (4, 13) based on piloted technologies are singled out for discussion below.

Of all the methods for producing LA, the "Biofine Process" (entry 3) has attracted the most attention. It has won a Presidential Green Chemistry Challenge Award and has been piloted at more than one facility, with plans to go commercial in the near future. Biofine comprises a two-step catalytic process. In the first reactor, cellulosic biomass (0.5 to 1 cm particle size) is mixed with 1.5–3% sulfuric acid at high temperature (210–220°C) and pressure (25 bar) in a plug flow reactor where rapid hydrolysis of cellulose (residence time 12 s) leads first to sugar monomers followed by dehydration to intermediates like HMF (**1**). In the second step, this material is fed into a backmix reactor under less severe conditions (190–200°C, 14 bar, 20 min residence time) during which LA and furfural are produced, with formic acid as a co-product. The crude reaction mixture is then dehydrated and vacuum distilled, leaving behind only a dry "char." The acid is recovered and recycled. LA yields ranging from 59% to 83% of the theoretical have been reported [172].

Entry 13 describes another high yielding approach to LA that involves the intermediacy of CMF **2**, the preparation of which was covered in Sect. 3 of this chapter. Using corn stover as an example, submission to the CMF process (biphasic aq. HCl/solvent reaction, 80°C, 3 h) gives CMF in 80% yield, during which levulinic acid is co-produced in 8% yield. Hydrolysis of the CMF at 190°C for 20 min provides a 91% yield of LA, for a net conversion from biomass of 81% over two steps. The Biofine and CMF methods are co-rendered graphically in Scheme 23.

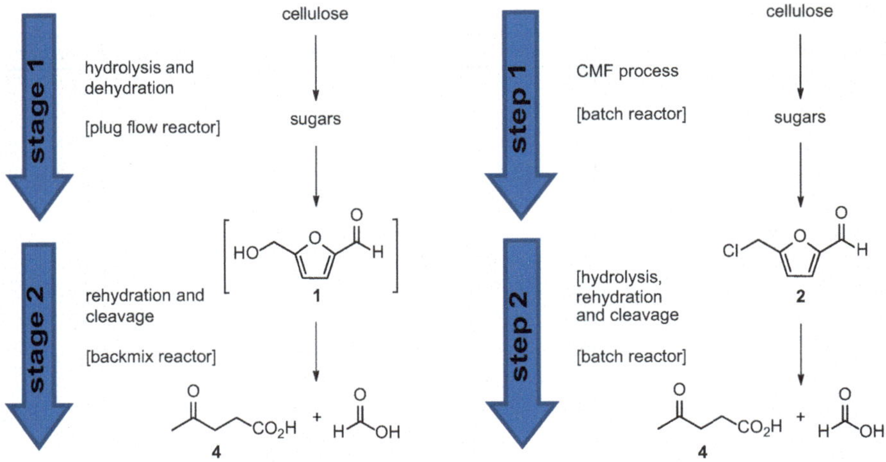

Scheme 23 Production of LA by the Biofine process (*left*) and via CMF (*right*)

4.4 Derivatives

Like HMF **1**, LA **4** is a recognized platform chemical for the generation of a wide array of derivatives with applications across a range of markets. LA has the distinction of appearing on the list of Top Value-Added Chemicals from Biomass published by the US Department of Energy's National Renewable Energy Laboratory (NREL). This list of 12 mainly sugar-derived products was assembled in order to identify major opportunities for "the production of value-added chemicals from biomass that would economically and technically support the production of fuels and power in an integrated biorefinery, and identify the common challenges and barriers of associated production technologies" [186]. In its candidate summary biography, LA is referred to as "one of the more recognized building blocks available from carbohydrates," the derivatives of which "address a number of large volume chemical markets." LA also appears in Bozell's updated review of top biorefinery carbohydrate derivatives [187].

While not as functionally versatile as HMF and CMF, LA nevertheless presents multiple opportunities for derivatization. There are two functional groups – the carboxylic acid and the ketone. In some cases only one of these groups is manipulated, but in many others both are involved. Without purporting to cover the entire breadth of LA chemistry, we discuss below some of the more attractive derivatives with the potential to unlock important industrial markets.

4.4.1 Esters, Amides, Ketals, Alcohols, and Ethers

Alkyl levulinate esters **72** can be prepared by homogeneous acid-catalyzed Fischer esterification of LA [188], under solid acid catalysis [189, 190], by acid-catalyzed

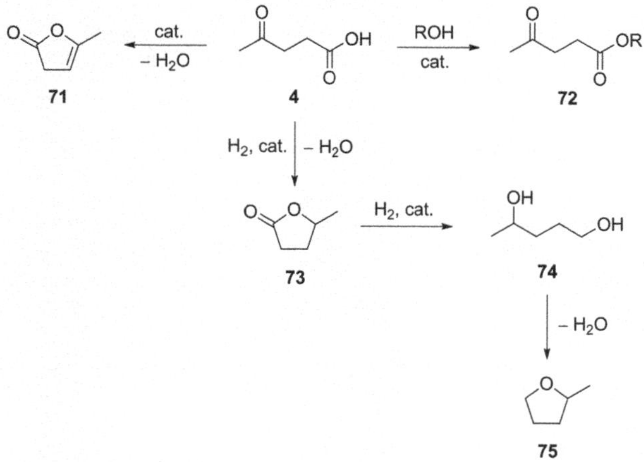

Scheme 24 Synthetic transformations of LA: esters, lactones, alcohols, and ethers

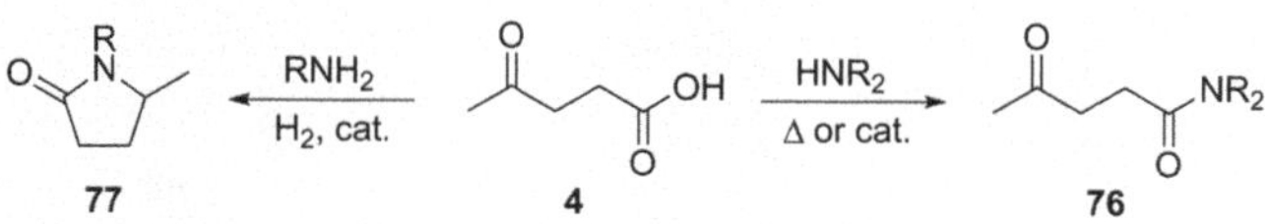

Scheme 25 Synthetic transformations of LA: amides and lactams

reaction with LA with olefins [191], or by reaction of alcohols with the cyclic ester angelica lactone **71** and a catalyst [192] (Scheme 24). Various levulinate esters have been tested as blends with diesel fuel and biodiesel, where they not only act as oxygenates but in the latter case also improve the cold-flow properties of the fuel [193, 194].

Dehydration of LA gives angelica lactone **71**, and reduction of the keto group of LA to the alcohol followed by cyclization gives γ-valerolactone **73** (GVL) (Scheme 24) [195]. While angelica lactone **71** has received relatively little attention in the renewables field, GVL **73** has found wide application as a green solvent and precursor to polymers, chemicals, and a range of biofuels [196]. Catalytic hydrogenation of GVL gives pentane-1,4-diol (1,4-PDO) **74** [197–199]. Cyclodehydration of 1,4-DPO **74** provides 2-methyltetrahydrofuran **75** which, in addition to being a useful solvent, is also a component of EPA-approved P-series flex fuels [200, 201].

Reaction of either LA or angelica lactone with secondary amines gives the corresponding amides **76** (Scheme 25) [202, 203]. Reductive amination of LA with primary amines gives 5-methyl-2-pyrrolidones (MPDs) **77**, which are versatile solvents with a range of industrial applications. Reduction is routinely carried out by hydrogenation over a heterogeneous noble metal-based catalyst (Pd, Pt, Ru, or Au) [180, 204–207]. Recently, Xia et al. reported the reductive amination of LA with various primary alkyl and aromatic amines with a cyclometallated iridium complex catalyst and formic acid as the hydrogen donor [208]. The same authors

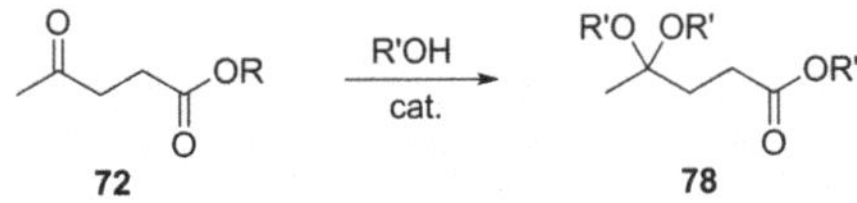

Scheme 26 Synthetic transformations of LA: ketals

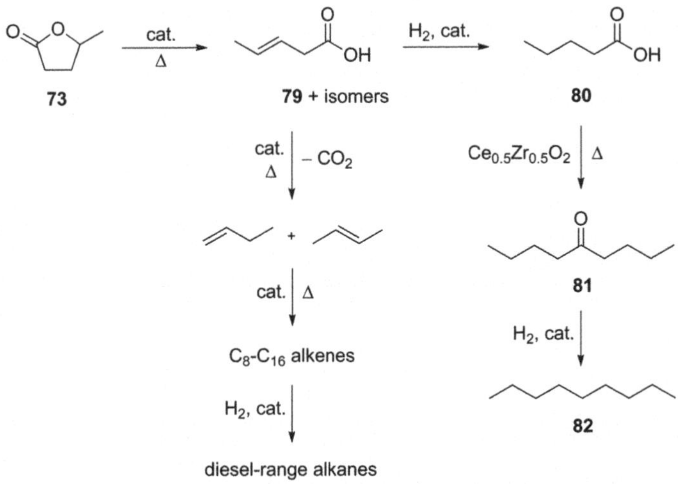

Scheme 27 GVL to hydrocarbon routes

have now also reported a catalyst-free synthesis of MPDs using a combination of formic acid and triethylamine for transfer hydrogenation [209].

Levulinic esters **72** can form acetals **78** in an acid-catalyzed reaction with alcohols, the products of which are variously useful as green plasticizers, solvents, and monomers for renewable polymers (Scheme 26) [210, 211]. Segetis Inc. is commercializing a broad portfolio of levulinic acetals across a range of applications [212].

4.4.2 Transformation into Fuels

GVL **73**, itself a potential biofuel, is also a starting material for pentanoate esters ("valeric biofuels") that have been shown to have outstanding fuel properties [213]. GVL-derived pentanoic acid **80** can be catalytically upgraded by decarboxylative ketonization to 5-nonanone **81**, which can then be variously processed to nonane **82** and other hydrocarbons [214, 215]. GVL has also been converted to butenes over a silica-alumina catalyst, which are then oligomerized on H-ZSM-5 or Amberlyst-70 to give a mixture of C_8–C_{16} alkenes that can be hydrogenated to drop-in fuels (Scheme 27) [216].

Recently, Mascal et al. described a synthetic approach to "cellulosic gasoline," i.e., exclusively branched, C_7–C_{10} hydrocarbons, using LA as the starting material

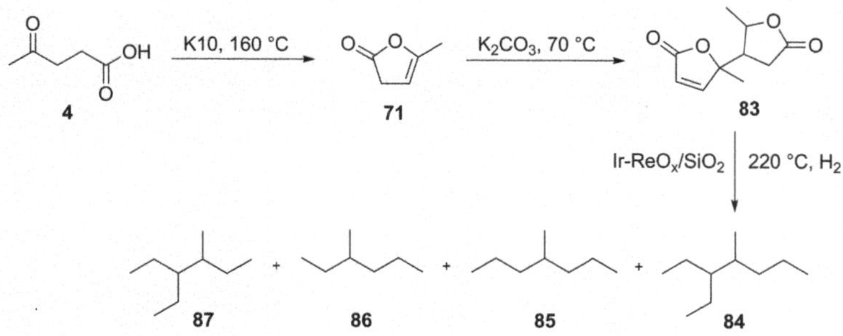

Scheme 28 LA to "cellulosic gasoline"

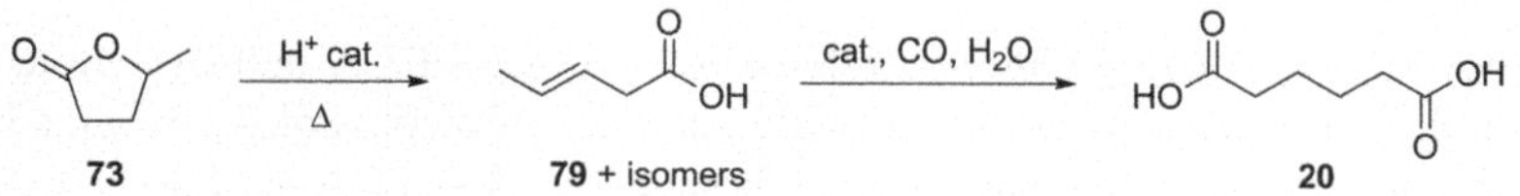

Scheme 29 Adipic acid from LA

[217]. Thus, LA was dehydrated to angelica lactone (AL, **71**) in 92% yield using montmorillonite K10 as an inexpensive, heterogeneous, recyclable catalyst. AL was then dimerized over solid K_2CO_3 to give the angelica lactone dimer (ALD, **83**) in 94% yield. ALD has a C_{10} backbone and, when subjected to hydrodeoxygenation (HDO), gives the branched C_{10} hydrocarbon 3-ethyl-4-methylheptane **84** as the major product, alongside other branched C_7-C_9 products (**85–87**), as shown in Scheme 28. Ir-ReO_x/SiO_2 and Pt-ReO_x/C HDO catalysts performed best, both giving 88% total yield of hydrocarbons from ALD. Considering that LA is available in >80% conversion from biomass, field-to-tank yields of drop-in, gasoline-range hydrocarbons of >60% are achievable by this approach.

Finally, Wheeler has reported high yields of deoxygenated hydrocarbons via simple thermal decomposition of mixtures of LA and formic acid. In contrast to the pyrolysis oils obtained from raw biomass, this product is low in oxygen, non-viscous, of neutral pH, and has a high energy content [218].

4.4.3 Miscellaneous Transformations Leading to Renewable Monomers, Solvents, and Specialty Chemicals

Adipic acid **20**, a component of nylon polyamides, has been prepared by the catalytic hydroformylation of the LA-derived pentenoic acid **79** (Scheme 29) [219], which in turn can be made by ring opening and dehydration of GVL **73** over heterogeneous acid catalysts like SiO_2-Al_2O_3 or ZSM-5 [219, 220]. Yields of up to 67% **20** by carbonylation of **79** have been reported by Phillipe et al. using iridium, ruthenium, and rhodium catalysts [221].

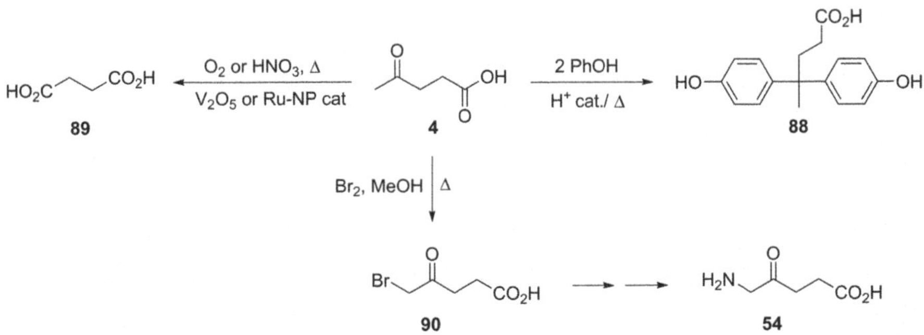

Scheme 30 Preparation of diphenolic acid, succinic acid, and 5-bromolevulinic acid from LA

Diphenolic acid (DPA) **88**, a condensation product of LA and two molecules of phenol, is considered to be a renewable analogue of bisphenol A (BPA), a high-volume chemical used to make polycarbonate plastics. BPA is known to have estrogenic activity [222], which has led to its withdrawal from some markets. Thus, DPA has the potential to displace BPA across the range of its polymer applications [223].

The oxidation of LA can also lead to useful derivatives. High-temperature (365–390°C) oxidation of LA with O_2 in the presence of a V_2O_5 catalyst gives succinic acid **89** in high yield [224] (Scheme 30). This oxidation has also been studied using nitric acid as the oxidant under milder reaction conditions (40°C) with succinic acid yields up to 52% [225]. A ruthenium nanoparticle catalyst also gave good results at 150°C under 14 bar O_2 pressure [226]. Succinic acid is a platform for important derivatives including γ-butyrolactone, 1,4-butanediol, and tetrahydrofuran, which have large-scale applications in the solvent, specialty chemical, and polymer markets.

The synthesis of the natural herbicide δ-aminolevulinic acid (DALA) **54** from CMF was described in Sect. 3 of this chapter, but has also been approached directly from LA. The methods have generally relied on the halogenation of the methyl group of LA, followed by substitution with a nitrogen nucleophile which is then transformed into the primary amine. 5-Bromolevulinic **90** can be obtained by the direct bromination of LA in refluxing methanol, but the reaction suffers from poor regioselectivity, low yield, and difficult product isolation [227, 228].

5 Conclusion and Future Prospects

Just as the chemical-catalytic approach to biomass processing is poised to come to the fore of biorefining, the furanic family of carbohydrate derivatives, HMF **1**, CMF **2**, and LA **4**, are the platforms from which this disruptive innovation in green chemistry will be launched. The primary, secondary, tertiary, etc., generations of their derivatives are like the branches of a tree, fanning out over virtually every aspect of chemistry currently represented by petrochemicals and fermentation products. While the former gradually become more expensive as fossil fuel reserves decline, the latter become less and less competitive in the face of faster, cheaper chemical processes that are more omnivorous with respect to feedstock input. In a sense, because condensed carbohydrate derivatives are emerging as favored feedstocks for hydrodeoxygenation to medium-chain alkanes, every refinery product that is derived from petroleum naphthas (α-olefins, aromatics, etc.) will remain available even in the long-horizon scenario that crude oil raffinates are no longer available as feeds. This, along with syngas from biomass gasifiers, will ensure that the petrochemical industry, on which so many other markets depend, need not go out of existence. In the end, it will simply come down to competition for waste biomass and energy crop feedstocks, and the technologies that can most advantageously supply the petrochemical "drop-in" chain (olefins, alcohols, carbonyl compounds, aromatics) as well as providing innovative products that serve markets otherwise supplied by petroleum (novel fuels and materials, green chemical products, etc.) will win the day, while others, like so many industries that have come and gone, fade into history.

Where will HMF, CMF, and LA fit into this new, post-petroleum economy, and what relative roles will they play? Despite HMF's long history in the literature and the volume of work devoted to its advancement as a cellulosic platform chemical, practical issues, as described in Sect. 2, threaten to hinder its further progress. The greater accessibility of CMF and LA, and their feedstock-agnostic nature, are set to leverage them relative to HMF, provided that issues associated with strong acid catalysis and acid recycle involved in their production can be suitably dealt with. In effect, since HMF and CMF are interconvertible, CMF becomes the new HMF, serving every derivative class and market that can be accessed from HMF, and beyond. As noted in Chap. 1, the biorefinery as such is not a new concept, but the furious pace of its development in the past decade has inevitably brought the issue of commercialization to prominence, which is akin to the natural selection process in the species, and it will be interesting to observe which approaches gain the most traction in the decade to come.

References

1. Dusselier M, Mascal M, Sels BF (2014) Top Curr Chem doi: 10.1007/128_2014_544
2. Düll G (1895) Action of oxalic acid on inulin. Chem Zeit 19:216–217
3. Düll G (1895) A derivative of furfuraldehyde from laevulose. Chem Zeit 19:1003–1005
4. vonEkenstein WA, Blanksma JJ (1910) ω-Hydroxymethylfurfuraldehyde as the cause of certain color reactions of the hexoses. Berichte 43:2355–2361
5. Saha B, Abu-Omar MM (2014) Advances in 5-hydroxymethylfurfural production from biomass in biphasic solvents. Green Chem 16:24–38
6. van Putten R-J, van der Waal JC, de Jong E, Rasrendra CB, Heeres HJ, de Vries JG (2013) Hydroxymethylfurfural, a versatile platform chemical made from renewable resources. Chem Rev 113:1499–1597

7. Li H, Chang F, Zhang Y, Hu D, Jin L, Song B, Yang S (2012) Recent progress towards transition metal-catalyzed direct conversion of cellulose to 5-hydroxymethylfurfural. Curr Catal 1:221–232
8. Tahvildari K, Taghvaei S, Nozari M (2011) The study of hydroxymethylfurfural as a basic reagent for liquid alkanes fuel manufacture from agricultural wastes. Int J Chem Environ Eng 2:62–68
9. Dutta S, De S, Saha B (2012) A brief summary of the synthesis of polyester building-block chemicals and biofuels from 5-hydroxymethylfurfural. ChemPlusChem 77:259–272
10. Amarasekara AS (2011) 5-Hydroxymethylfurfural based polymers. In: Mittal V (ed) Renewable polymers: synthesis, processing and technology. Wiley-Scrivener, Hoboken, pp 381–428, Chap 9
11. Karinen R, Vilonen K, Niemelä M (2011) Biorefining: heterogeneously catalyzed reactions of carbohydrates for the production of furfural and hydroxymethylfurfural. ChemSusChem 4:1002–1016
12. Rosatella AA, Simeonov SP, Frade RFM, Afonso CAM (2011) 5-Hydroxymethylfurfural (HMF) as a building block platform: biological properties, synthesis and synthetic applications. Green Chem 13:754–793
13. Zakrzewska ME, Bogel-Lukasik E, Bogel-Lukasik R (2011) Ionic liquid-mediated formation of 5-hydroxymethylfurfural - a promising biomass-derived building block. Chem Rev 111:397–417
14. Lewkowski J (2001) Synthesis, chemistry and applications of 5-hydroxymethylfurfural and its derivatives. Arkivoc i:17–54
15. Kuster BFM (1990) 5-Hydroxymethylfurfural (HMF): a review focusing on its manufacture. Starch 42:314–321
16. Anese M, Manzocco L, Calligaris S, Nicoli MC (2013) Industrially applicable strategies for mitigating acrylamide, furan, and 5-hydroxymethylfurfural in food. J Agric Food Chem 61:10209–10214
17. Kuster BFM (1977) The influence of water concentration on the dehydration of D-fructose. Carbohydr Res 54:177–183
18. Qi X, Watanabe M, Aida TM, Smith RLJ (2008) Catalytic dehydration of fructose into 5-hydroxymethylfurfural by ion-exchange resin in mixed-aqueous system by microwave heating. Green Chem 10:799–805
19. Brown DW, Floyd AJ, Kinsman RG, Roshan-Ali Y (1982) Dehydration reactions of fructose in non-aqueous media. J Chem Technol Biotechnol 32:920–924
20. Shimizu K, Uozumi R, Satsuma A (2009) Enhanced production of hydroxymethylfurfural from fructose with solid acid catalysts by simple water removal methods. Catal Commun 10:1849–1853
21. Chheda JN, Román-Leshkov Y, Dumesic JA (2007) Production of 5-hydroxymethylfurfural and furfural by dehydration of biomass-derived mono- and poly-saccharides. Green Chem 9:342–350
22. Gaset A, Rigal L, Paillassa G, Salome J-P, Flèche GRF (1986) Process for manufacturing 5-hydroxymethylfurfural. US 4,590,283 A
23. Yong G, Zhang Y, Ying JY (2008) Efficient catalytic system for the selective production of 5-hydroxymethylfurfural from glucose and fructose. Angew Chem Int Ed 47:9345–9348
24. Qi X, Guo H, Li L (2011) Efficient conversion of fructose to 5-hydroxymethylfurfural catalyzed by sulfated zirconia in ionic liquids. Ind Eng Chem Res 50:7985–7989
25. Nakajima K, Baba Y, Noma R, Kitano M, Kondo JN, Hayashi S, Hara M (2011) $Nb_2O_5{\cdot}nH_2O$ as a heterogeneous catalyst with water-tolerant Lewis acid sites. J Am Chem Soc 133:4224–4227
26. Chareonlimkun A, Champreda V, Shotipruk A, Laosiripojana N (2010) Catalytic conversion of sugarcane bagasse, rice husk and corncob in the presence of TiO_2–ZrO_2 and mixed-oxide TiO_2–ZrO_2 under hot compressed water (HCW) condition. Bioresour Technol 101:4179–4186

27. Wang C, Fu L, Tong X, Yang Q, Zhang W (2012) Efficient and selective conversion of sucrose to 5-hydroxymethylfurfural promoted by ammonium halides under mild conditions. Carbohydr Res 347:182–185
28. Fan C, Guan H, Zhang H, Wang J, Wang S, Wang X (2011) Conversion of fructose and glucose into 5-hydroxymethylfurfural catalyzed by a solid heteropolyacid salt. Biomass Bioenerg 35:2659–2665
29. Lima S, Neves P, Antunes MM, Pillinger M, Ignatyev N, Valente AA (2009) Conversion of mono/di/polysaccharides into furan compounds using 1-alkyl-3-methylimidazolium ionic liquids. Appl Catal A Gen 363:93–99
30. Wu S, Fan H, Xie Y, Cheng Y, Wang Q, Zhang Z, Han B (2010) Effect of CO_2 on conversion of inulin to 5-hydroxymethylfurfural and propylene oxide to 1,2-propanediol in water. Green Chem 12:1215–1219
31. Benvenuti F, Carlini C, Patrono P, Raspolli Galletti AM, Sbrana G, Massucci MA, Galli P (2000) Heterogeneous zirconium and titanium catalysts for the selective synthesis of 5-hydroxymethyl-2-furaldehyde from carbohydrates. Appl Catal A Gen 193:147–153
32. Yang F, Liu Q, Yue M, Bai X, Du Y (2011) Tantalum compounds as heterogeneous catalysts for saccharide dehydration to 5-hydroxymethylfurfural. Chem Commun 47:4469–4471
33. Hu S, Zhang Z, Zhou Y, Song J, Fan H, Han B (2009) Direct conversion of inulin to 5-hydroxymethylfurfural in biorenewable ionic liquids. Green Chem 11:873–877
34. Yin S, Pan Y, Tan Z (2011) Hydrothermal conversion of cellulose to 5-hydroxymethylfurfural. Int J Green Energy 8:234–247
35. Zhao S, Cheng M, Li J, Tian J, Wang X (2011) One pot production of 5-hydroxymethylfurfural with high yield from cellulose by a Brønsted-Lewis-surfactant-combined heteropolyacid catalyst. Chem Commun 47:2176–2178
36. Binder JB, Raines RT (2009) Simple chemical transformation of lignocellulosic biomass into furans for fuels and chemicals. J Am Chem Soc 131:1979–1985
37. McNeff CV, Nowlan DT, McNeff LC, Yan B, Fedie RL (2010) Continuous production of 5-hydroxymethylfurfural from simple and complex carbohydrates. Appl Catal A Gen 384:65–69
38. Zhang Y, Du H, Qian X, Chen EY-X (2010) Ionic liquid—water mixtures: enhanced K_w for efficient cellulosic biomass conversion. Energy Fuels 24:2410–2417
39. Snyder FH (1958) Preparation of hydroxymethylfurfural from cellulosic materials. US 2,851,468 A
40. Daengprasert W, Boonnoun P, Laosiripojana N, Goto M, Shotipruk A (2011) Application of sulfonated carbon-based catalyst for solvothermal conversion of cassava waste to hydroxymethylfurfural and furfural. Ind Eng Chem Res 50:7903–7910
41. Dedsuksophon W, Faungnawakij K, Champreda V, Laosiripojana N (2011) Hydrolysis/dehydration/aldol-condensation/hydrogenation of lignocellulosic biomass and biomass-derived carbohydrates in the presence of Pd/WO_3–ZrO_2 in a single reactor. Bioresour Technol 102:2040–2046
42. Li L, Li L, Wang Y, Du Y, Qin S (2013) Biorefinery products from the inulin-containing crop Jerusalem artichoke. Biotechnol Lett 35:471–477
43. Nasab EE, Habibi-Rezaei M, Khaki A, Balvardi M (2009) Investigation on acid hydrolysis of inulin: a response surface methodology approach. Int J Food Eng 5: Article 12
44. Zhao H, Holladay JE, Brown H, Zhang ZC (2007) Metal chlorides in ionic liquid solvents convert sugars to 5-hydroxymethylfurfural. Science 316:1597–1600
45. Gericke M, Fardim P, Heinze T (2012) Ionic liquids – promising but challenging solvents for homogeneous derivatization of cellulose. Molecules 17:7458–7502
46. Hu C, Yang Y, Yan H, Xiang X, Tong D, Zhu L, Li G (2009) Preparation of 5-acetoxymethylfurfural from carbohydrates. Faming Zhuanli Shenqing Gongkai Shuomingshu: CN 10163331 A
47. Rauchfuss TB, Thananatthanachon T (2011) Efficient method for preparing 2,5-dimethylfuran. US Pat Appl 20110263880 A1

48. Casanova O, Iborra S, Corma A (2010) Chemicals from biomass: etherification of 5-hydroxymethyl-2-furfural (HMF) into 5,5'(oxy-bis(methylene))bis-2-furfural (OBMF) with solid catalysts. J Catal 275:236–242
49. Sanda K, Rigal L, Gaset A (1989) Synthesis of 5-(bromomethyl)- and of 5-(chloromethyl)-2-furancarboxaldehyde. Carbohydr Res 187:15–23
50. Bredihhin A, Maeorg U, Vares L (2013) Evaluation of carbohydrates and lignocellulosic biomass from different wood species as raw material for the synthesis of 5-bromomethyfurfural. Carbohydr Res 375:63–67
51. Cukalovic A, Stevens CV (2010) Production of biobased HMF derivatives by reductive amination. Green Chem 12:1201–1206
52. Arias KS, Al-Resayes SI, Climent MJ, Corma A, Iborra S (2013) From biomass to chemicals: synthesis of precursors of biodegradable surfactants from 5-hydroxymethylfurfural. ChemSusChem 6:123–131
53. Balakrishnan M, Sacia ER, Bell AT (2012) Etherification and reductive etherification of 5-(hydroxymethyl)furfural: 5-(alkoxymethyl)furfurals and 2,5-bis(alkoxymethyl)furans as potential bio-diesel candidates. Green Chem 14:1626–1634
54. Gandini A, Belgacem MN (1997) Furans in polymer chemistry. Prog Polym Sci 22:1203–1379
55. Gandini A, Belgacem NM (1998) Recent advances in the elaboration of polymeric materials derived from biomass components. Polym Int 47:267–276
56. Hu L, Zhao G, Hao W, Tang X, Sun Y, Lin L, Liu S (2012) Catalytic conversion of biomass-derived carbohydrates into fuels and chemicals via furanic aldehydes. RSC Advances 2:11184–11206
57. Elhajj T, Masroua A, Martin JC, Descotes G (1987) Synthese de l'hydroxymethyl-5-furanne carboxaldehyde-2 et de ses derives par traitement acide de sucres sur resines echangeuses d'ions. Bull Soc Chim Fr 5:855–860
58. Amarasekara AS, Green D, McMillan E (2008) Efficient oxidation of 5-hydroxymethylfurfural to 2,5-diformylfuran using Mn(III)–salen catalysts. Catal Commun 9:286–288
59. Mehdi H, Bodor A, Lantos D, Horvath IT, DeVos DE, Binnemans K (2007) Imidazolium ionic liquids as solvents for cerium(IV)-mediated oxidation reactions. J Org Chem 72: 517–524
60. Yoon H-J, Choi J-W, Jang, H-S, Cho JK, Byun J-W, Chung W-J, Lee S-M, Lee Y-S (2011) Selective oxidation of 5-hydroxymethylfurfural to 2,5-diformylfuran by polymer-supported IBX amide. SynLett 165–168
61. Cottier L, Descotes G, Lewkowski J, Skowronski R (1995) Ultrasonically accelerated syntheses of furan-2,5-dicarbaldehyde from 5-hydroxymethyl-2-furfural. Org Prep Proc Int 27:564–566
62. Yadav GD, Sharma RV (2014) Biomass derived chemicals: environmentally benign process for oxidation of 5-hydroxymethylfurfural to 2,5-diformylfuran by using nano-fibrous Ag-OMS-2-catalyst. Appl Catal B Environ 147:293–301
63. Nie J, Xie J, Liu H (2013) Activated carbon-supported ruthenium as an efficient catalyst for selective aerobic oxidation of 5-hydroxymethylfurfural to 2,5-diformylfuran. Chinese J Catal 34:871–875
64. Nie J, Xie J, Liu H (2013) Efficient aerobic oxidation of 5-hydroxymethylfurfural to 2,5-diformylfuran on supported Ru catalysts. J Catal 301:83–91
65. Antonyraj CA, Jeong J, Kim B, Shin S, Kim S, Lee K-Y, Cho JK (2013) Selective oxidation of HMF to DFF using Ru/γ-alumina catalyst in moderate boiling solvents toward industrial production. J Ind Eng Chem 19:1056–1059
66. Nie J, Liu H (2012) Aerobic oxidation of 5-hydroxymethylfurfural to 2,5-diformylfuran on supported vanadium oxide catalysts: structural effect and reaction mechanism. Pure Appl Chem 84:765–777

67. Halliday GA, Young RJ, Grushin VV (2003) One-pot, two-step, practical catalytic synthesis of 2,5-diformylfuran from fructose. Org Lett 5:2003–2005
68. Takagaki A, Takahashi M, Nishimura S, Ebitani K (2011) One-pot synthesis of 2,5-diformylfuran from carbohydrate derivatives by sulfonated resin and hydrotalcite-supported ruthenium catalysts. ACS Catal 1:1562–1565
69. Xiang X, He L, Yang Y, Guo B, Tong D, Hu C (2011) A one-pot two-step approach for the catalytic conversion of glucose into 2,5-diformylfuran. Catal Lett 141:735–741
70. Casanova O, Iborra I, Corma A (2009) Biomass into chemicals: aerobic oxidation of 5-hydroxymethyl-2-furfural into 2,5-furandicarboxylic acid with gold nanoparticle catalysts. ChemSusChem 2:1138–1144
71. Davis SE, Houk LR, Tamargo EC, Datye AK, Davis RJ (2011) Oxidation of 5-hydroxymethylfurfural over supported Pt, Pd and Au catalysts. Catal Today 160:55–60
72. Pasini T, Piccinini M, Blosi M, Bonelli R, Albonetti S, Dimitratos N, Lopez-Sanchez JA, Sankar M, He Q, Kiely CJ, Hutchings GJ, Cavani F (2011) Selective oxidation of 5-hydroxymethyl-2-furfural using supported gold-copper nanoparticles. Green Chem 13:2091–2099
73. de Jong E, Dam MA, Sipos L, Gruter G-JM (2012) Furandicarboxylic acid (FDCA), a versatile building block for a very interesting class of polyesters. In: Smith PB, Gross RA (eds) Biobased monomers, polymers, and materials, ACS Symp Ser, Vol. 1105, Chapter 1, pp. 1–13
74. Tong X, Ma Y, Li Y (2010) Biomass into chemicals: conversion of sugars to furan derivatives by catalytic processes. Appl Catal A Gen 385:1–13
75. Lew BW (1967) Method of producing dehydromucic acid. US 3,326,944 A
76. Zope BN, Davis SE, Davis RJ (2012) Influence of reaction conditions on diacid formation during Au-catalyzed oxidation of glycerol and hydroxymethylfurfural. Top Catal 55:24–32
77. Lilga MA, Hallen RT, Gray M (2010) Production of oxidized derivatives of 5-hydroxymethylfurfural (HMF). Top Catal 53:1264–1269
78. Casanova O, Iborra S, Corma A (2009) Biomass into chemicals: one pot-base free oxidative esterification of 5-hydroxymethyl-2-furfural into 2,5-dimethylfuroate with gold on nanoparticulated ceria. J Catal 265:109–116
79. Moreau C, Belgacem MN, Gandini A (2004) Recent catalytic advances in the chemistry of substituted furans from carbohydrates and in the ensuing polymers. Top Catal 27:11–30
80. Cottier L, Descotes G, Soro Y (2003) Heteromacrocycles from ring-closing metathesis of unsaturated furanic ethers. Synth Commun 33:4285–4295
81. Lichtenthaler FW, Brust A, Cuny E (2001) Sugar-derived building blocks. Part 26. Hydrophilic pyrroles, pyridazines and diazepinones from D-fructose and isomaltulose. Green Chem 3:201–209
82. Goswami S, Dey S, Jana S (2008) Design and synthesis of a unique ditopic macrocyclic fluorescent receptor containing furan ring as a spacer for the recognition of dicarboxylic acids. Tetrahedron 64:6358–6363
83. Tamura M, Tokonami K, Nakagawa Y, Tomishige K (2013) Rapid synthesis of unsaturated alcohols under mild conditions by highly selective hydrogenation. Chem Commun 49:7034–7036
84. Ohyama J, Esaki A, Yamamoto Y, Arai S, Satsuma A (2013) Selective hydrogenation of 2-hydroxymethyl-5-furfural to 2,5-bis(hydroxymethyl)furan over gold sub-nano clusters RSC Adv 3:1033–1036
85. Thananatthanachon T, Rauchfuss TB (2010) Efficient production of the liquid fuel 2,5-dimethylfuran from fructose using formic acid as a reagent. Angew Chem Int Ed 49:6616–6618
86. Hansen TS, Barta K, Anastas PT, Ford PC, Riisager A (2012) One-pot reduction of 5-hydroxymethylfurfural via hydrogen transfer from supercritical methanol. Green Chem 14:2457–2461

87. Roman-Leshkov Y, Barrett CJ, Liu ZY, Dumesic JA (2007) Production of dimethylfuran for liquid fuels from biomass-derived carbohydrates. Nature 447:982–985
88. Zu Y, Yang P, Wang J, Liu X, Ren J, Lu G, Wang Y (2014) Efficient production of the liquid fuel 2,5-dimethylfuran from 5-hydroxymethylfurfural over Ru/Co_3O_4 catalyst. Appl Catal B Environ 146:244–248
89. Chen G, Shen Y, Zhang Q, Yao M, Zheng Z, Liu H (2013) Experimental study on combustion and emission characteristics of a diesel engine fueled with 2,5-dimethylfuran–diesel, n-butanol–diesel and gasoline–diesel blends. Energy 54:333–342
90. Brandvold TA (2010) Carbohydrate route to para-xylene and terephthalic acid. US 20100331568 A1
91. Masuno MN, Bissell J, Smith RL, Higgins B, Wood AB, Foster M (2012) Utilizing a multiphase reactor for the conversion of biomass to produce substituted furans. WO 2012170520 A1
92. Shiramizu M, Toste FD (2011) On the Diels–Alder approach to solely biomass-derived polyethylene terephthalate (PET): conversion of 2,5-dimethylfuran and acrolein into p-xylene. Chem Eur J 17:12452–12457
93. Williams CL, Chang C-C, Do P, Nikbin N, Caratzoulas S, Vlachos DG, Lobo RF, Fan W, Dauenhauer PJ (2012) Cycloaddition of biomass-derived furans for catalytic production of renewable p-xylene. ACS Catal 2:935–939
94. Nakagawa Y, Tomishige K (2010) Total hydrogenation of furan derivatives over silica-supported Ni-Pd alloy catalyst. Catal Commun 12:154–156
95. Yao S, Wang X, Jiang Y, Wu F, Chen X, Mu X (2014) One-step conversion of biomass-derived 5-hydroxymethylfurfural to 1,2,6-hexanetriol over Ni-Co-Al mixed oxide catalysts under mild conditions. ACS Sustainable Chem Eng 2:173–180
96. Alamillo R, Tucker M, Chia M, Pagan-Torres Y, Dumesic J (2012) The selective hydrogenation of biomass-derived 5-hydroxymethylfurfural using heterogeneous catalysts. Green Chem 14:1413–1419
97. Grochowski MR, Yang W, Sen A (2012) Mechanistic study of a one-step catalytic conversion of fructose to 2,5-dimethyltetrahydrofuran. Chem Eur J 18:12363–12371
98. Yang W, Sen A (2010) One-step catalytic transformation of carbohydrates and cellulosic biomass to 2,5-dimethyltetrahydrofuran for liquid fuel. ChemSusChem 3:597–603
99. Chheda JN, Dumesic JA (2007) An overview of dehydration, aldol-condensation and hydrogenation processes for production of liquid alkanes from biomass-derived carbohydrates. Catal Today 123:59–70
100. Huber GW, Chheda JN, Barrett CJ, Dumesic JA (2005) Production of liquid alkanes by aqueous-phase processing of biomass-derived carbohydrates. Science 308:1446–1450
101. Liu D, Chen EY-X (2013) Diesel and alkane fuels from biomass by organocatalysis and metal-acid tandem catalysis. ChemSusChem 6:2236–2239
102. Sutton AD, Waldie FD, Wu R, Schlaf M, 'Pete' Silks (III) LA, Gordon JC (2013) The hydrodeoxygenation of bioderived furans into alkanes. Nature Chem 5:428–432
103. Virent, Inc. http://www.virent.com. Accsessed Jan 14, 2014
104. Girisuta B, Janssen LPBM, Heeres HJ (2006) A kinetic study on the decomposition of 5-hydroxymethylfurfural into levulinic acid. Green Chem 8:701–709
105. Boussie TR, Dias EL, Fresco ZM, Murphy VJ, Shoemaker J, Archer R, Jiang H (2010) Production of adipic acid and derivatives from carbohydrate-containing materials. US 20,100,317,823 A1
106. Cottier L, Descotes G, Eymard L, Rapp K (1995) Syntheses of γ-oxo acids or γ-oxo esters by photooxygenation of furanic compounds and reduction under ultrasound: application to the synthesis of 5-aminolevulinic acid hydrochloride. Synthesis 303–306
107. Marisa C, Ilaria D, Marotta R, Roberto A, Vincenzo C (2010) Production of 5-hydroxy-4-keto-2-pentenoic acid by photo-oxidation of 5-hydroxymethylfurfural with singlet oxygen: a kinetic investigation. J Photochem Photobiol A 210:69–76
108. Fenton HJH, Gostling M (1899) Bromomethylfurfuraldehyde. J Chem Soc Trans 75:423–433

109. Fenton HJH, Gostling M (1901) Derivatives of methylfurfural. J Chem Soc Trans 79:807–816
110. Fischer E, von Neyman H (1914) Notiz über ω-chlormethyl- und athoxymethyl-furfurol. Chem Ber 47:973–977
111. Hibbert H, Hill HS (1923) Studies on cellulose chemistry II. The action of dry hydrogen bromide on carbohydrates and polysaccharides. J Am Chem Soc 45:176–182
112. Haworth WN, Jones WGM (1944) The conversion of sucrose into furan compounds. Part 1. 5-Hydroxymethylfurfuraldehyde and some derivatives. J Chem Soc 667–670
113. Hamada K, Suzukamo G, Nagase T (1978) Furaldehydes. Ger Offen DE 2745743
114. Szmant HH, Chundury DD (1981) The preparation of 5-chloromethylfurfuraldehyde from high fructose corn syrup and other carbohydrates. J Chem Technol Biotechnol 31:205–212
115. Hamada K, Suzukamo G, Fujisawa K (1982) 5-Methylfurfural. EP44186A119820120
116. Hamada K, Yoshihara H, Suzukamo G (1982) An improved method for the conversion of saccharides into furfural derivatives. Chem Lett 617–618
117. Hamada K, Yoshihara H, Suzukamo G (1983) 5-Halomethylfurfural. EP 79206A1 19830578
118. Sanda K, Rigal L, Gaset A (1992) Optimisation of the synthesis of 5-chloromethyl-2-furancarboxaldehyde from D-fructose dehydration and in-situ chlorination of 5-hydroxymethyl-2-furancarboxaldehyde. J Chem Technol Biotechnol 55:139–145
119. Mascal M, Nikitin EB (2008) Direct, high-yield conversion of cellulose into biofuel. Angew Chem Int Ed 47:7924–7926
120. Mascal M (2009) High-yield conversion of cellulosic biomass into furanic biofuels and value-added products. US 7,829,732
121. Mascal M, Nikitin EB (2009) Dramatic advancements in the saccharide to 5-(chloromethyl) furfural conversion reaction. ChemSusChem 2:859–861
122. Mascal M, Nikitin EB (2010) Co-processing of carbohydrates and lipids in oil crops to produce a hybrid biodiesel. Energy Fuels 24:2170–2171
123. Brasholz M, von Känel K, Hornung CH, Saubern S, Tsanaktsidis J (2011) Highly efficient dehydration of carbohydrates to 5-(chloromethyl)furfural (CMF), 5-(hydroxymethyl)furfural (HMF) and levulinic acid by biphasic continuous flow processing. Green Chem 13:1114–1117
124. Breeden SW, Clark JH, Farmer TJ, Macquarrie DJ, Meimoun JS, Nonne Y, Reid JESJ (2013) Microwave heating for rapid conversion of sugars and polysaccharides to 5-chloromethyl furfural. Green Chem 15:72–75
125. Gao W, Li Y, Xiang Z, Chen K, Yang R, Argyropoulos DS (2013) Efficient one-pot synthesis of 5-chloromethylfurfural (CMF) from carbohydrates in mild biphasic systems. Molecules 18:7675–7685
126. Jadhav H, Pedersen CM, Solling T, Bols M (2011) 3-Deoxyglucosone is an intermediate in the formation of furfurals from D-glucose. ChemSusChem 4:1049–1051
127. Kumari N, Olesen JK, Pedersen CM, Bols M (2011) Synthesis of 5-bromomethylfurfural from cellulose as a potential intermediate for biofuel. Eur J Org Chem 1266–1270
128. Yang W, Grochowski MR, Sen A (2012) Selective reduction of biomass by hydriodic acid and its in situ regeneration from iodine by metal/hydrogen. ChemSusChem 5:1218–1222
129. Tarabanko VE, Chernyak MY, Morozov AA, Kaigorodov KL (2013) Method of producing 5-fluoromethyl furfural. RU 2,478,097
130. Gilpin JA (1984) Inhibitors for furfurals. US 4433155 A
131. Kawai S, Tanaka S, Terai K, Tezuka M, Nishiwaki T (1960) Synthesis of 1,4,7-cyclononanetrione. Bull Chem Soc Jpn 33:669–674
132. Mascal M, Nikitin EB (2010) High-yield conversion of plant biomass into the key value-added feedstocks 5-(hydroxymethyl)furfural, levulinic acid, and levulinic esters via 5-(chloromethyl)furfural. Green Chem 12:370–373
133. Liu G, Wu J, Zhang IY, Chen Z-N, Li Y-W, Xu X (2011) Theoretical studies on thermochemistry for conversion of 5-chloromethylfurfural into valuable chemicals. J Phys Chem A 115:13628–13641
134. Rinke IJ (1934) 5-Methylfurfural. Org Synth 14:62

135. Hamada K, Yoshihara H, Suzukamo G (2001) Novel synthetic route to 2,5-disubstituted furan derivatives through surface active agent-catalyzed dehydration of D(−)-fructose. J Oleo Sci 50:533–536
136. Mikochik P, Cahana A (2012) Conversion of 5-(chloromethyl)-2-furaldehyde into 5-methyl-2-furoic acid and derivatives thereof. EP 2,606,039 A1
137. xftechnologies.com/technology/products/(Accessed Jan17, 2014)
138. Shi Y, Brenner P, Bertsch S, Radacki K, Dewhurst RD (2012) η3-Furfuryl and η3-thienyl complexes of palladium and platinum of relevance to the functionalization of biomass-derived furans. Organometallics 31:5599–5605
139. Fenton HJH, Robinson F (1909) Homologues of furfuraldehyde. J Chem Soc Trans 95:1334–1340
140. Zhou X, Rauchfuss TB (2013) Production of hybrid diesel fuel precursors from carbohydrates and petrochemicals using formic acid as a reactive solvent. ChemSusChem 6:383–388
141. Szmant HH, Chundury D (1981) Preparation of polymeric building blocks from 5-hydroxymethyl- and 5-chloromethylfurfuraldehyde. Ind Eng Chem Prod Res Dev 20:158–163
142. Jira R, Bräunling H (1987) Synthesis of polyarenemethines, a new class of conducting polymers. Synth Met 17:691–696
143. Elix JA (1969) Synthesis and properties of annulene polyoxides. Aus J Chem 22:1951–1962
144. Nickl J, Naarmann H, Moehwald H (1985) Use of polyheterocyclic compounds of a certain structure as electrode material. Ger Patent DE 3409655 A1
145. Timko JM, Cram DJ (1974) Furanyl unit in host compounds. J Am Chem Soc 96:7159–7160
146. Silks LA, Gordon JC, Wu R, Hanson SK (2011) Method of carbon chain extension using novel aldol reaction. US Pat Appl 20110040109 A1
147. Seck KA (2013) Biorefinery for conversion of carbohydrates and lignocellulosics via primary hydrolysate CMF to liquid fuels. WO 2013122686 A2
148. Cooper WF, Nuttall WH (1912) Furane-2,5-dialdehyde. J Chem Soc Trans 101:1074–1081
149. Florentino HQ, Hernandez-Benitez RI, Avina JA, Burgueno-Tapia E, Tamariz J (2011) Total synthesis of naturally occurring furan compounds 5-{[(4-hydroxybenzyl)oxy]methyl}-2-furaldehyde and pichiafuran C. Synthesis 1106–1112
150. Klein LL, Shanklin MS (1988) Total synthesis of dimethyl jaconate. J Org Chem 53:5202–5209
151. Zhou F, Zheng J, Dong X, Zhang Z, Zhao L, Sha X, Li L, Wen R (2007) Synthesis and antitumor activities of 3-substituted 1- (5-formylfurfuryl) indolin-2-one derivatives. Lett Org Chem 4:601–605
152. Dai H-L, Gao L-X, Yang Y, Li J-Y, Cheng J-G, Li J, Wen R, Peng YQ, Zhang, J-B (2012) Discovery of di-indolinone as a novel scaffold for protein tyrosine phosphatase 1B inhibitors. Bioorg Med Chem Lett 22:7440–7443
153. Dai H-L, Shen Q, Zheng J-B, Li J-Y, Wen R, Li J (2011) Synthesis and biological evaluation of novel indolin-2-one derivatives as protein tyrosine phosphatase 1B inhibitors. Lett Org Chem 8:526–530
154. Mascal M, Dutta S (2011) Synthesis of the natural herbicide δ-aminolevulinic acid from cellulose-derived 5-(chloromethyl)furfural. Green Chem 13:40–41
155. Price BJ, Clitherow JW, Bradshaw J (1978) Aminoalkyl furan derivatives. US 4,128,658
156. Mascal M, Dutta S (2011) Synthesis of ranitidine (Zantac) from cellulose-derived 5-(chloromethyl)furfural. Green Chem 13:3101–3102
157. Chang F, Dutta S, Becnel JJ, Estep AS, Mascal M (2014) Synthesis of the insecticide prothrin and its analogues from biomass-derived 5-(chloromethyl)furfural. J Agric Food Chem 62:476–480
158. von Grote AF, Tollens B (1875) Untersuchungen uber kohlenhydrate. I. ueber die bei einwirkung von schwefelsäure auf zucker entstehende säure (levulinsäure). Liebigs Ann Chem 175:181–204

159. Malaguti (1836) Ueber die einwirkung der verdünnten säuren aus den gemeinen zucker. Liebigs Annalen 17:52–67
160. Mulder GJ (1840) Untersuchungen über die humussubstanzen. J Prakt Chem 21:321–370
161. Conrad M (1878) Ueber acetopropionsäure und ihre identität mit levulinsäure. Berichte 11:2177–2179
162. McKenzie BF (1929) Levulinic acid. Org Synth 9:50
163. Thomas RW, Schuette HA (1931) Studies on levulinic acid. I. Its preparation from carbohydrates by digestion with hydrochloric acid under pressure. J Am Chem Soc 53:2324–2328
164. Dahlmann J (1968) Preparation of levulinic acid. Chem Ber 101:4251–4253
165. Ploetz T (1941) The formation of levulinic acid from carbohydrates. Naturwissenschaften 29:707–708
166. Pummerer R, Guyot O, Birkofer L (1935) Mechanism of levulinic acid formation from hexoses. II. A hydroxyl-free glucosan-like substance. Berichte 68B:480–493
167. Isbell HS (1944) Interpretation of some reactions in the carbohydrate field in terms of consecutive electron displacement. J Res Nat Bur Stand 32:45–59
168. Horvat J, Klaic B, Metelko B, Sunjic V (1985) Mechanism of levulinic acid formation. Tetrahedron Lett 26:2111–2114
169. Galletti AMR, Antonetti C, De Luise V, Valentini G (2011) Conversion of biomass to levulinic acid, a new feedstock for the chemical industry. Chimica e l'Industria 93:112–117
170. Rackemann DW, Doherty WOS (2011) The conversion of lignocellulosics to levulinic acid. Biofuels Bioprod Biorefin 5:198–214
171. Saladino R, Pagliaccia T, Argyropoulos DS, Crestini C (2007) Production of chemicals from cellulose and biomass-derived compounds: advances in the oxidative functionalization of levulinic acid. ACS Symp Ser 954:262–279
172. Hayes DJ, Fitzpatrick S, Hayes MHB, Ross JRH (2006) The Biofine process - production of levulinic acid, furfural, and formic acid from lignocellulosic feedstocks. In: Kamm B, Gruber PR, Kamm M (eds) Biorefineries - industrial processes and products. Wiley-VCH, Weinheim, pp 139–164
173. Timokhin BV, Baransky VA, Eliseeva GD (1999) Levulinic acid in organic synthesis. Russ Chem Rev 68:73–84
174. Efremov AA, Pervyshina GG, Kuznetsov BN (1998) Production of levulinic acid from wood raw material in the presence of sulfuric acid and its salts. Chem Nat Compd 34:182–185
175. Efremov AA, Pervyshina GG, Kuznetsov BN (1997) Thermocatalytic transformations of wood and cellulose in the presence of HCl, HBr, and H_2SO_4. Chem Nat Compd 33:84–88
176. Farone WA, Cuzens JE (2000) Method for the production of levulinic acid and its derivatives. US 6,054,611 A
177. Fitzpatrick SW (1997) Production of levulinic acid from carbohydrate-containing materials. US 5,608,105 A
178. Chang C, Cen P, Ma X (2007) Levulinic acid production from wheat straw. Bioresour Technol 98:1448–1453
179. Yan L, Yang N, Pang H, Liao B (2008) Production of levulinic acid from bagasse and paddy straw by liquefaction in the presence of hydrochloride acid. Clean 36:158–163
180. Fang Q, Hanna MA (2002) Experimental studies for levulinic acid production from whole kernel grain sorghum. Bioresour Technol 81:187–192
181. Girisuta B, Danon B, Manurung R, Janssen LPBM, Heeres HJ (2008) Experimental and kinetic modelling studies on the acid-catalysed hydrolysis of the water hyacinth plant to levulinic acid. Bioresour Technol 99:8367–8375
182. Ramos-Rodriguez E (1972) Process for jointly producing furfural and levulinic acid from bagasse and other lignocellulosic materials. US 3701789 A
183. Carlson LJ (1962) Process for the manufacture of levulinic acid. US 3,065,263 A
184. Sassenrath CP, Shilling WL (1966) Preparation of levulinic acid from hexose-containing material. US 3258481 A

185. Jeong G-T, Park D-H (2010) Production of sugars and levulinic acid from marine biomass Gelidium amansii. Appl Biochem Biotechnol 161:41–52
186. Top value added chemicals from biomass. Vol I, PNNL and the National Renewable Energy Laboratory (http://www1.eere.energy.gov/biomass/pdfs/35523.pdf)
187. Bozell JJ, Petersen GR (2010) Technology development for the production of biobased products from biorefinery carbohydrates-the US Department of Energy's "top 10" revisited. Green Chem 12:539–554
188. Bart HJ, Reidetschlager J, Schatka K, Lehmann A (1994) Kinetics of esterification of levulinic acid with n-butanol by homogeneous catalysis. Ind Eng Chem Res 33:21–25
189. Maheria KC, Kozinski J, Dalai A (2013) Esterification of levulinic acid to n-butyl levulinate over various acidic zeolites. Catal Lett 143:1220–1225
190. Fernandes DR, Rocha AS, Mai EF, Mota CJA, Teixeira da Silva V (2012) Levulinic acid esterification with ethanol to ethyl levulinate production over solid acid catalysts. Appl Catal A Gen 425–426:199–204
191. Fagan PJ, Korovessi E, Manzer LE, Mehta R, Thomas SM (2003) Preparation of levulinic acid esters and formic acid esters from biomass and olefins. WO 2,003,085,071 A1
192. Manzer LE (2005) Preparation of levulinic acid esters from alpha-angelica lactone and alcohols. WO 2,005,097,724 A1
193. Christensen E, Williams A, Paul S, Burton S, McCormick RL (2011) Properties and performance of levulinate esters as diesel blend components. Energy Fuels 25:5422–5428
194. Windom BC, Lovestead TM, Mascal M, Nikitin EB, Bruno TJ (2011) Advanced distillation curve analysis on ethyl levulinate as a diesel fuel oxygenate and a hybrid biodiesel fuel. Energy Fuels 25:1878–1890
195. Zhang J, Wu S, Li B, Zhang H (2012) Advances in the catalytic production of valuable levulinic acid derivatives. ChemCatChem 4:1230–1237
196. Alonso DM, Wettstein SG, Dumesic JA (2013) Gamma-valerolactone, a sustainable platform molecule derived from lignocellulosic biomass. Green Chem 15:584–595
197. Corbel-Demailly L, Ly B-K, Minh D-P, Tapin B, Especel C, Epron F, Cabiac A, Guillon E, Besson M, Pinel C (2013) Heterogeneous catalytic hydrogenation of biobased levulinic and succinic acids in aqueous solutions. ChemSusChem 6:2388–2395
198. Geilen FMA, Engendahl B, Harwardt A, Marquardt W, Klankermayer J, Leitner W (2010) Selective and flexible transformation of biomass-derived platform chemicals by a multifunctional catalytic system. Angew Chem Int Ed 49:5510–5514
199. Geilen FMA, Engendahl B, Holscher M, Klankermayer J, Leitner W (2011) Selective homogeneous hydrogenation of biogenic carboxylic acids with $[Ru(TriPhos)H]^+$: a mechanistic study. J Am Chem Soc 133:14349–14358
200. Pace V, Hoyos P, Castoldi L, Dominguez de Maria P, Alcantara AR (2012) 2-Methyltetrahydrofuran (2-MeTHF): a biomass-derived solvent with broad application in organic chemistry. ChemSusChem 5:1369–1379
201. Du X-L, Bi Q-Y, Liu Y-M, Cao Y, He H-Y, Fan K-N (2012) Tunable copper-catalyzed chemoselective hydrogenolysis of biomass-derived γ-valerolactone into 1,4-pentanediol or 2-methyltetrahydrofuran. Green Chem 14:935–939
202. Haskelberg L (1948) Some derivatives of levulinic acid. J Am Chem Soc 70:2830–2831
203. Lukes R, Koblicova Z, Blaha K (1963) Reaction of angelica lactones with amines. Collect Czech Chem Commun 28:2182–2198
204. Celmer WD, Solomons IA (1963) 1,5-Dimethyl-2-oxo-3-pyrrolidineglyoxylic acid. J Org Chem 28:3221–3222
205. Frank RL, Schmitz WR, Zeidman B (1947) 1,5-Dimethyl-2-pyrrolidone. Org Synth 27:28
206. Manzer LE (2005) Production of 5-methyl-N-(methylaryl)-2-pyrrolidone, 5-methyl-N-(methylcycloalkyl)-2-pyrrolidone and 5-methyl-N-alkyl-2-pyrrolidone by reductive amination of levulinic acid with cyano compounds. WO 2,004,085,048 A3
207. Shilling WL (1966) Making lactams by the vapor phase reductive amination of oxo carboxylic acid compounds US 3235562 A

208. Wei Y, Wang C, Jiang X, Xue D, Li J, Xiao J (2013) Highly efficient transformation of levulinic acid into pyrrolidinones by iridium catalysed transfer hydrogenation. Chem Commun 49:5408–5410
209. Wei Y, Wang C, Jiang X, Xue D, Liu Z-T, Xiao J (2014) Catalyst-free transformation of levulinic acid into pyrrolidinones with formic acid. Green Chem 16:1093–1096
210. Leibig C, Mullen B, Mullen T, Rieth L, Badarinarayana V (2011) Cellulosic-derived levulinic ketal esters: a new building block. ACS Symp Ser 1063:111–116
211. Desai S (2010) Building blocks for a greener industry. Chem Ind London 21–23
212. www.segetis.com (Accessed Jan17, 2014)
213. Lange J-P, Price R, Ayoub PM, Louis J, Petrus L, Clarke L, Grosselink H (2010) Valeric biofuels: a platform of cellulosic transportation fuels. Angew Chem Int Ed 49:4479–4483
214. Serrano-Ruiz JC, Wang D, Dumesic JA (2010) Catalytic upgrading of levulinic acid to 5-nonanone. Green Chem 12:574–577
215. West RM, Liu ZY, Peter M, Dumesic JA (2008) Liquid alkanes with targeted molecular weights from biomass-derived carbohydrates. ChemSusChem 1:417–424
216. Bond JQ, Alonso DM, Wang D, West RM, Dumesic JA (2010) Integrated catalytic conversion of γ-valerolactone to liquid alkenes for transportation fuels. Science 327:1110–1114
217. Mascal M, Dutta S, Gandarias I (2014) The angelica lactone dimer as a renewable feedstock for hydrodeoxygenation: simple, high-yield synthesis of branched C_7–C_{10} gasoline-like hydrocarbons. Angew Chem Int Ed 53:1854–1857
218. Case PA, van Heiningen ARP, Wheeler MC (2012) Liquid hydrocarbon fuels from cellulosic feedstocks via thermal deoxygenation of levulinic acid and formic acid salt mixtures. Green Chem 14:85–89
219. Wong PK, Li C, Stubbs L, Vanmeurs M, AnakKumbang DG, Lim CY, Drent E (2012) Synthesis of diacids. WO 2,012,134,397 A1
220. Bond JQ, Alonso DM, West RM, Dumesic JA (2010) γ-Valerolactone ring-opening and decarboxylation over SiO_2/Al_2O_3 in the presence of water. Langmuir 26:16291–16298
221. Grosselin J-M, Denis P, Metz F, Delis P (1992) Process for preparing adipic acid by hydrocarboxylation of pentenoic acids. EP 0,493,273 B1
222. Vandenberg LN, Maffini MV, Sonnenschein C, Rubin BS, Soto AM (2009) Bisphenol-A and the great divide: a review of controversies in the field of endocrine disruption. Endocr Rev 30:75–95
223. Bozell JJ, Moens L, Elliott DC, Wang Y, Neuenscwander GG, Fitzpatrick SW, Bilski RJ, Jarnefeld JL (2000) Production of levulinic acid and use as a platform chemical for derived products. Resour Conserv Recycl 28:227–239
224. Dunlop AP, Shelbert S (1954) Preparation of succinic acid. US 2,676,186 A
225. Van Es DS, Van der Klis F, Van Haveren J (2012) Succinic acid from biomass. WO 2,012,044,168 A1
226. Podolean I, Kuncser V, Gheorghe N, Macovei D, Parvulescu VI, Coman SM (2013) Ru-based magnetic nanoparticles (MNP) for succinic acid synthesis from levulinic acid. Green Chem 15:3077–3082
227. Ha H-J, Lee S-K, Ha Y-J, Park J-W (1994) Selective bromination of ketones. A convenient synthesis of 5-aminolevulinic acid. Synth Commun 24:2557–2562
228. Manny AJ, Kjelleberg S, Kumar N, de Nys R, Read RW, Steinberg P (1997) Reinvestigation of the sulfuric acid-catalysed cyclisation of brominated 2-alkyllevulinic acids to 3-alkyl-5-methylene-2(5H)-furanones. Tetrahedron 53:15813–15826

Top Curr Chem (2014) 353: 85–126
DOI: 10.1007/128_2014_540

Published online: 14 May 2014

Selective Catalysis for Cellulose Conversion to Lactic Acid and Other α-Hydroxy Acids

Michiel Dusselier and Bert F. Sels

Abstract This review discusses topical chemical routes and their catalysis for the conversion of cellulose, hexoses, and smaller carbohydrates to lactic acid and other useful α-hydroxy acids. Lactic acid is a top chemical opportunity from carbohydrate biomass as it not only features tremendous potential as a chemical platform molecule; it is also a common building block for commercially employed green solvents and near-commodity bio-plastics. Its current scale fermentative synthesis is sufficient, but it could be considered a bottleneck for a million ton scale breakthrough. Alternative chemical routes are therefore investigated using multifunctional, often heterogeneous, catalysis. Rather than summarizing yields and conditions, this review attempts to guide the reader through the complex reaction networks encountered when synthetic lactates from carbohydrate biomass are targeted. Detailed inspection of the cascade of reactions emphasizes the need for a selective retro-aldol activity in the catalyst. Recently unveiled catalytic routes towards other promising α-hydroxy acids such as glycolic acid, and vinyl and furyl glycolic acids are highlighted as well.

Keywords Biomass-to-chemicals · Catalysis · Cellulose · Renewables · Lactic acid · Vinyl glycolic acid · Biodegradable polymers

Contents

M. Dusselier (✉) and B.F. Sels (✉)
Center for Surface Chemistry and Catalysis, KU Leuven, Kasteelpark Arenberg 23, Leuven 3000, Belgium
e-mail: michiel.dusselier@biw.kuleuven.be; bert.sels@biw.kuleuven.be

1 Introduction

The search for oil-independent and renewable alternatives for the production of chemicals, fuels, and materials has led to an increased interest in the use of biomass as a feedstock [1, 2]. The catalytic conversion of biomass, and in particular that of the most abundant and non-edible lignocellulosic feedstock, has led to new pathways for the synthesis of chemicals, biofuels, and polymer building blocks [3–17]. Cellulose is the single most abundant organic compound on Earth [8, 18]. As this biopolymer exclusively consists of glucose, a 6-carbon aldose, as seen in Fig. 1, its selective conversion to desired chemical compounds is feasible [8, 9, 19].

Chemical catalysis, either in homogeneous or heterogeneous mode, is a formidable strategy to convert cellulose and sugar feedstock selectively into chemicals, and, since the turn of the century, publication numbers on catalytic carbohydrate conversion have been rising progressively. In a way, they all contribute to the concept of installing biorefineries in analogy with the existing petrochemical refineries [18]. The ideal biorefinery is thought to supply a select portfolio of both low value biofuels and high value bio-based chemicals and monomers [18, 20–22]. The choice of the desired chemicals from carbohydrate biomass is crucial and, therefore, we have recently developed a chemical selection tool from a chemist's point of view, based on a newly defined functionality index (F:C) of a potential target molecule and the atom economy of its formation. This analysis is found in the first chapter of this volume [23]. When applied to lactic acid, a commercial bio-derived chemical of great interest, a high functionality index of 1.33 is calculated due to its carboxyl and α-hydroxyl group on its 3-carbon backbone. Moreover, as it is a structural isomer of triose sugars, its formation from cellulose, hexoses, or trioses implies a transformation with an atom economy of 100%.

Lactic acid is thus a perfectly suited chemical target to produce from biomass carbohydrates and no petrochemical routes for its formation are likely to take over. Once a proper chemical target molecule is selected, a fundamental understanding of the reaction network is crucial in order to tackle its selective formation from

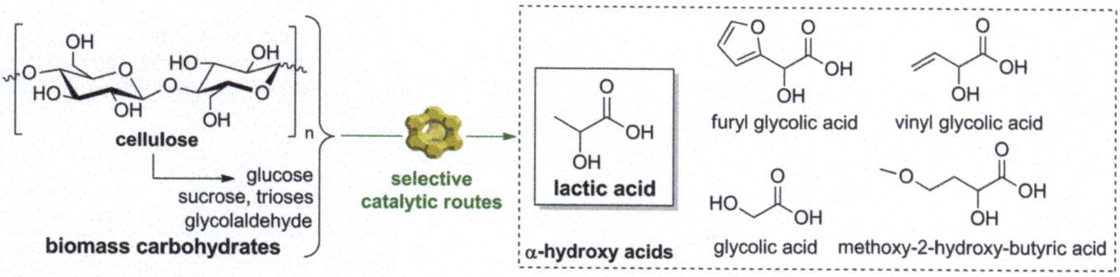

Fig. 1 Catalytic conversion of cellulose and carbohydrates to lactic acid and AHAs

cellulosic biomass and to design the process steps and the catalytic activity required. The transformation of cellulose into lactic acid or other α-hydroxy acids (AHA), as seen in Fig. 1, is a key illustration of the proposed assessment [23] and its novel catalytic formation routes will be discussed in depth in this review chapter.

The next two sections of this review chapter will introduce the reader to the world of lactic acid. The acid is both a key platform chemical of the biorefinery concept, from which other interesting molecules may be formed (Sect. 2), and a monomer for commercial bioplastic polylactic acid (PLA) (Sect. 3). In the platform approach, the assessment from Chap. 1 in this volume [23] proves its value, as it is an equally useful tool to seek out the most desired routes for transforming a biomass-derived platform molecule as it is to select the most relevant carbohydrate-based chemicals from a chemist's point of view. In what follows, the desired catalytic cascade from cellulose to lactic acid will be described (Sect. 4) as well as the specific catalytic data reported for different feedstock (Sects. 5 and 6). Section 7 will introduce the reader to recent synthesis routes for other useful AHA compounds such as furyl and vinyl glycolic acid, as well as others shown in Fig. 1. Before concluding this chapter, Sect. 8 will provide a note on the stereochemistry of the chemically produced AHAs.

2 Lactic Acid as a Platform Molecule

Lactic acid (LA) is one of the top carbohydrate-derived chemicals and it was recently included in Bozell and Petersen's revised selection of the top ten sugar-based chemicals [10, 24]. The conversion of carbohydrates into LA via anaerobic fermentation has been known for ages [25]. The first industrial fermentation was developed by A. Boehringer in 1895 and at the present time the global installed production capacity is estimated at 0.5 Mton year^{-1} [10, 26, 27]. The current fermentation process and its issues will be critically discussed in Sect. 3 in light of the major application of LA today, i.e., as monomer for commercial bioplastic PLA [28]. Besides being used for polyester synthesis, LA is seriously considered as a platform chemical for the synthesis of a diverse range of chemicals such as pyruvic acid, 2,3-pentanedione, and acrylic acid [10, 29].

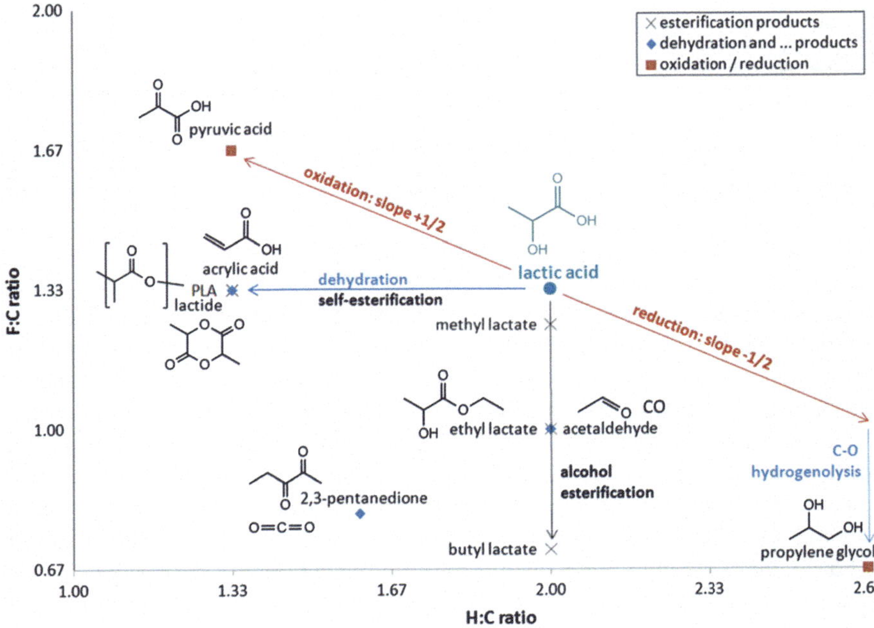

Fig. 2 Modified *van Krevelen* plot demonstrating the conversion of LA into various chemicals. PLA = polylactic acid. F:C = functionality index as defined in Chap. 1 [23]

In the light of preserving functional reactivity and pursuing high atom economy, the synthesis of LA is an excellent example (F:C = 1.33; atom economy AE = 100% [23]) and the original configuration of functional groups in LA provides a high chemical reactivity, allowing for a multitude of different conversion routes. Most of these conversions have been reported feasible in high product selectivity, given the presence of a suitable chemocatalyst (mostly heterogeneous) and the appropriate process conditions. A lactic acid-product family tree was developed in our recent review with detailed emphasis on the role of chemocatalysis in using LA as a platform molecule for different chemicals. As the catalytic conversion of LA is not within the scope of this review we refer the reader to this work for catalytic details [10]. Here, we will discuss the position of frequently reported chemicals produced from LA in the modified *van Krevelen* diagram, as introduced in the first chapter of this volume [23], and, together with the atom economy, this allows us to select the most feasible chemicals from LA. Figure 2 shows that lactic acid transformation in nearly every direction on the modified *van Krevelen* plot is feasible.

To start with, a single dehydration of LA – typically a shift to the left on the plot – leads to acrylic acid as the major product. Such dehydration is interesting as it preserves the functionality per carbon in the molecule (2). The atom economy (AE) of this dehydration is as high as 80% with water as the only by-product. The reaction is usually catalyzed in the gas phase by phosphates, sulfates, clays, modified zeolites, or oxides with dual mild acid–base character, yielding acrylic acid at around 65% [30–34]. The current synthesis of acrylic acid is performed via the partial oxidation of petrochemical propene (1) in the presence of steam with equal atom economy [35].

$$C_3H_6 + \frac{3}{2}O_2 \rightarrow C_3H_4O_2 + H_2O \quad \text{(partial oxidation)} \quad (AE = 80\%) \qquad (1)$$

$$C_3H_6O_3 \rightarrow C_3H_4O_2 + H_2O \quad \text{(dehydration)} \quad (AE = 80\%) \qquad (2)$$

Even though the functional nature of LA is perfectly suited to synthesize acrylic acid in one step, the main driver for replacing the acrylic petrochemical route is an economic one and this demands a lower production price of lactic acid. At the moment, prices of both acids are similar, in the range of 1.2–2.0 USD kg^{-1}.

Other chemicals, like 2,3-pentanedione [36] and acetaldehyde [37], are synthesized from LA via loss of water combined with decarbox(n)ylation. A side product of these reactions is CO_2 or CO, which presents a loss in one carbon atom and thus a drop in atom economy. 2,3-Pentanedione is a valuable fine chemical, used as an aroma and flavor, but other properties, like its condensation with two phenols, might lead to interesting plasticizers (as an alternative to bisphenol A) and polycarbonate building blocks [168]. Although its functionality index of 0.8 is seriously lowered with respect to LA, as seen in Fig. 2, its structure contains very useful vicinal keto groups. In our analysis found in Chap. 1 of this volume [23], 2,3-pentanedione was found in the second circle of the F:C plot ranked per carbon number, indicating that the synthesis routes from carbohydrate and petrochemistry likely compete with each other. Its current small scale production is via extraction from dairy waste. The carbohydrate-based route via LA follows a reaction mechanism entailing a sequence of Claisen-condensation, decarboxylation, and dehydration. The reaction runs in the gas phase over supported phosphates [36] or alkali hydroxides [38], typically yielding 50% of 2,3-pentanedione. This reaction has a low atom economy of 55% calculated from LA (or from glucose), as seen in (3). A petrochemical route has been patented, via reacting hydroxyacetone with acetaldehyde (4), which is more atom efficient [39]. In the light of choosing between the carbohydrate path via LA or the fossil path, a detailed cost analysis should shed light on the economics. Given the availability of cheap LA, the renewable option could be a feasible route, but only after carefully comparing feedstock (acetaldehyde + hydroxyacetone vs LA) and process costs. The lower AE of the LA route is a big disadvantage. Note, however, that hydroxyacetone and acetaldehyde can also be made from renewable glycerol [40].

$$2C_3H_6O_3 \rightarrow C_5H_8O_2 + CO_2 + 2H_2O \quad (AE = 55\%) \qquad (3)$$

$$C_3H_6O_2 + C_2H_4O \rightarrow C_5H_8O_2 + 2H_2O \quad (AE = 85\%) \qquad (4)$$

LA-based acetaldehyde is formed by a dehydration and decarbonylation sequence according to (5) and is shown in Fig. 2. The functionality index F:C of acetaldehyde is 1 and thus predicts a decent reaction, but the atom economy of 49% is very low. To keep the H-index constant for only a slightly lower functionality, the reaction indeed requires the elution of CO as side-product. Decomposition of LA to acetaldehyde, H_2, and CO_2 is also known. Katryniok et al. showed that the reaction can be catalyzed via decarbonylation in the gas phase with supported heteropolyacids at 275°C in excellent yield (90%) at high conversion (92%) [37]. Although acetaldehyde is fairly functional, it is only a 2-carbon unit, and petrochemical routes via selective oxidation (of natural gas derived ethene) will likely always dominate.

$$C_3H_6O_3 \rightarrow C_2H_4O + CO + H_2O \quad (AE = 49\%) \tag{5}$$

Another common reaction with α-hydroxy acids such as LA is the self-condensation under water-removal circumstances, forming lactoyl lactate, oligomers, or lactide – the cyclic di-ester of LA (6) [10, 41]. These condensation reactions are dehydrations and thus retain the full functionality in the molecule, while lowering the H index; see Fig. 2. Moreover, the atom economy of 80% is high with only water as side-product. Lactide is the true industrial precursor of PLA and the atom economy of 80% is equally valid for PLA [42].

$$2C_3H_6O_3 \rightarrow C_6H_8O_4 + 2H_2O \quad (AE = 80\%) \tag{6}$$

$$C_3H_6O_3 + C_2H_6O \rightarrow C_5H_{10}O_3 + H_2O \quad (AE = 87\%) \tag{7}$$

Besides self-esterification, alkyl lactates can easily be made under acid or base catalysis in excess of alcohol or via water removal according to (7) [43]. The atom economy is high and the conversion only slightly reduces the F:C index with respect to LA. This reduction is larger with increasing length of alkyl chain in the alcohol part, but in absolute functionality (F, definition see Chap. 1 [23]), no loss is at hand. These alkyl esters harness realistic potential as green solvents. Ethyl lactate in particular, with its high boiling point, low vapor pressure, and low surface tension, receives increasing attention for solvent application and is currently commercially available at a medium scale. The molecule is moreover composed of two bio-derived platform chemicals and fully biodegradable [43, 44].

Another intensively studied reaction is the conversion of LA into 1,2-propanediol (propylene glycol) with hydrogen according to (8) [45]. This reaction proceeds with a significant loss of functionality, viz. from 1.33 to 0.67, while the atom economy of 81% is medium high. It can be written as a hydrogenation and subsequent hydrogenolysis, as seen in Fig. 2. Propylene glycol is mainly used as an antifreeze, solvent, and polyester precursor. Its commercial production involves the epoxidation of propylene with peroxides in the presence of Lewis acid catalysts like Mo or Ti, followed by hydration of the oxirane function [46]. If H_2O_2 is used as oxidant, a 100% atom economy is achieved with the petrochemical oxidation route according to (9). It will be difficult for the carbohydrate-based route to propanediol via LA (or glucose) to compete with this established route, due to the higher feedstock cost and lower atom economy. Our analysis in Chap. 1 of this volume predicted this competition, as propylene glycol is found on the edge of the third circle in the F:C vs carbon number plot, found in Fig. 9 of [23].

$$C_3H_6O_3 + 2H_2 \rightarrow C_3H_8O_2 + H_2O \quad (AE = 81\%) \tag{8}$$

$$C_3H_6 + H_2O_2 \rightarrow C_3H_8O_2 \quad (AE = 100\%) \tag{9}$$

Finally, the oxidation of LA to pyruvic acid has been described with both heterogeneous and bio-catalysts. Pyruvic acid is the simplest α-keto acid and may be considered as an added-value fine chemical for the synthesis of drugs and agrochemicals, as well as being an antioxidant [29, 47]. The oxidation of the α-hydroxyl of LA corresponds to an upward shift in the modified *van Krevelen* plot along a +1/2 slope (Fig. 2), yielding a high F:C of 1.67 with a high atom economy (10). The commercial production at the moment is already carbohydrate-based, as predicted again by Chap. 1 of this volume [23], via the fermentation of glucose with free or immobilized enzyme catalysis [48]. Pyruvic acid is the end product of the energy providing glycolysis pathway in living cells.

$$C_3H_6O_3 + \frac{1}{2}O_2 \rightarrow C_3H_4O_3 + H_2O \quad (AE = 83\%) \tag{10}$$

To conclude, LA has tremendous potential as a high volume platform chemical to produce a range of different products. Our assessment along the guidelines of Chap. 1 of this volume [23] is summarized in Fig. 3, presenting the atom economy as a function of the functionality index. The dashed box indicates the zone with higher or equal F:C as LA and at least 80% of AE. This plot thus predicts the following order of importance/viability of lactic acid-derived chemicals: pyruvic acid > lactide = PLA = acrylic acid > methyl and ethyl lactate >> propylene glycol > 2,3-pentanedione > acetaldehyde. The drop in atom economy for the last two chemicals is too severe, and more AE efficient petrochemical precursors can be found. In the case of pyruvic acid, LA has to compete with a direct carbohydrate-based bio-catalytic process [48]. In fact, Fig. 3 would be identical if the reactions were considered to start from glucose or cellulose itself, because the atom economy for making LA from this feedstock is 100%.

The versatile nature of the multifunctional LA is thus ideal to offer the chemical side of the biorefinery the necessary adaptability to product demand and market prices, and it is encouraging to see the high selectivity of the conversions with heterogeneous catalysis [10]. Today, only the synthesis of esters and polymers (via lactide) among the described reactions is processed on a commercial scale. Lactide is, for instance, produced commercially by Purac (Corbion) and Galactic (Futerro, with Total), the largest producers of lactic acid in the world, as well as by Natureworks, the largest PLA producer. Commercialization of other chemicals from LA is currently hampered by the cost of its fermentative production from glucose, as explained in the next section. The interest in bio-derived chemicals and PLA bioplastics, and the rising demand for green solvent ethyl lactate, should be a stimulus for researchers to develop novel chemocatalytic pathways to LA or to

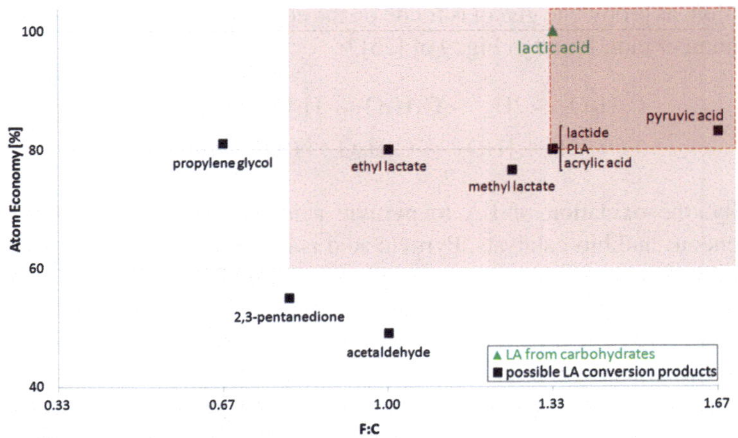

Fig. 3 Atom economy vs F:C for lactic acid conversion routes

improve current fermentation processes, preferably direct from cellulosic biomass. After a short discussion on the current fermentation to LA and its use as a monomer for PLA, a comprehensive summary of the newly-developed chemocatalytic routes will be given.

3 Lactic Acid as Key Monomer for Biodegradable Polyesters

Lactic acid is the key building block of the second largest volume commercial bio-plastic, namely PLA. This renewable polymer is a biodegradable and biocompatible (in vivo) thermoplastic polyester [28, 49–51]. These unique features render the PLA polymer suitable for many custom applications, for instance in medicine, such as prostheses and in drug delivery, while PLA is also a suitable replacement for certain forms of polystyrene, polypropylene, and polyethylene-terephthalate, e.g., in packaging, fibers, and textiles. On the downside, two major bottlenecks hamper the worldwide megaton-scale breakthrough of PLA: (1) its production cost and (2) some of its properties such as its brittleness, certain barrier properties, too pronounced hydrophobicity, and lack of reactive side groups [52]. Nonetheless, up to 187,000 metric tons of LA were produced in 2011, with a tentative estimated forecast of up to 600,000 tons in 2020 [10, 53, 54].

The synthesis of PLA and its cost-driving bottlenecks are presented in Fig. 4. The first step is the fermentative synthesis of (usually L-) LA with bacteria or yeast. Elegant reviews on this anaerobic fermentation are available elsewhere [55–58]. LA is further converted to its cyclic dimeric ester lactide via a two-stage process consisting of a pre-polymerization and a backbiting reaction (second frame in Fig. 4). This L,L-lactide is the actual monomer for high molecular weight PLA.

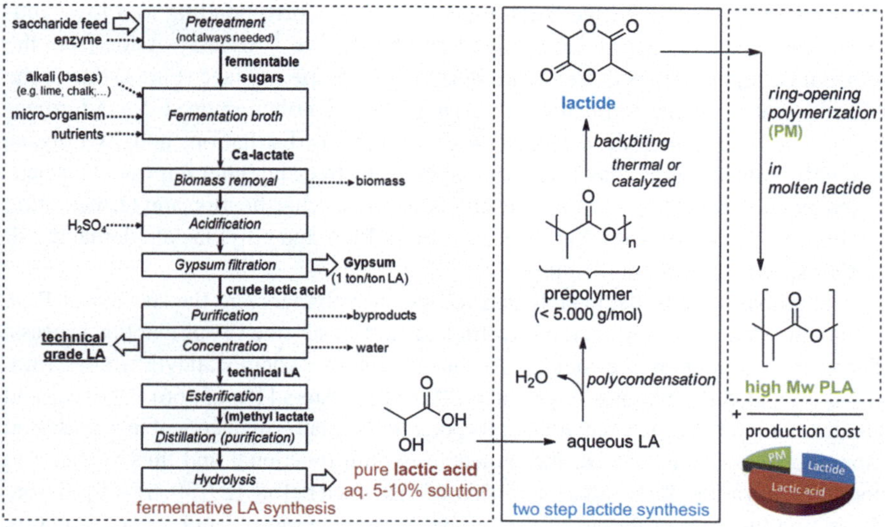

Fig. 4 Current industrial synthesis of (L-)PLA from carbohydrate feedstock: the fermentative synthesis of L-LA; the chemical two-step synthesis of the cyclic intermediate (L,L-) lactide and the polymerization to PLA. A pie chart shows a rough estimate of each of the frames' contribution to the total production cost of PLA, from sugar to pellet

PLA synthesis proceeds via ring-opening polymerization, usually carried out in molten lactide in the presence of minute amounts of metal catalyst (e.g., Sn, Al) [59]. The lactide route is industrially preferred over the direct synthesis of PLA via polycondensation, especially for the production of high molecular weight PLA. Such high molecular weight is important to ensure high performance in physical properties such as melt points and strengths. Polycondensation is a less controlled process and leads to shorter, more polydisperse chains and thus PLA of inferior quality [60]. In addition, such condensation is a tedious process because of the removal of water from a molten, highly viscous polycondensate [61]. The lower-right corner of Fig. 4 shows a pie chart with a rough estimate of the total PLA production cost, from the carbohydrate feedstock to the PLA plastic pellet. It is estimated that about 50% of the total cost is governed by the fermentative LA synthesis from glucose or sucrose. Another 30% is spent on the transformation of LA to the lactide, while the remaining cost is dedicated to the polymerization of lactide and the pelleting [169].

In order to facilitate the production of PLA on a megaton scale, as suited for a commodity plastic, it is imperative to focus on alternative production routes to LA, as this is clearly the largest cost factor for producing PLA. The main reason for this cost is displayed in Fig. 4 as well: the fermentation is a slow and laborious process, which requires continuous pH buffering with alkali, ending up with dilute Ca-lactate salt in water. To release LA from this salt, acid work-up with H_2SO_4, viz. the acidification step in Fig. 4, is needed but leads to the production of $CaSO_4$ in quantities up to 1 ton per ton of LA. Moreover, the batch process is slow, corresponding to a low volumetric productivity ranging between 0.3 and 5 g L^{-1} h^{-1} [62, 63]. Aside from the gypsum issue, numerous downstream purifying steps are in place (Fig. 4) due to the complex nature of the fermentation's living broth. To obtain pure LA via fermentation, final steps include a laborious esterification, distillation, and hydrolysis. Recently, some progress has been made in gypsum-free and other improved fermentation technologies [57], as well as in simultaneous saccharification and fermentation of cellulose feedstock [64], but the steep rise in PLA and ethyl lactate demand will further stress the fermentative process.

The global drive to degradable and recyclable polymers and the success of PLA herein thus demands a cheaper, more efficient, and waste-free LA production process. Therefore, researchers are looking for novel selective chemocatalytic routes from cellulose and carbohydrates to produce LA and its esters [10, 65, 66]. Cheaper and greater availability of LA will unlock its potential as platform chemical in a dedicated biorefinery approach. Indeed, the molecule is multifunctional and thus versatile to convert into various high value chemicals, as discussed before [29, 30, 67–69]. Before describing the state-of-the-art of the catalytic results in more detail (Sects. 5 and 6), the next section will be devoted to clarifying the mechanism and the catalytic requirements to convert cellulose into lactates.

4 Reaction Pathways and Catalytic Requirements

The price of refined sugar syrups of, e.g., glucose and sucrose, viz. around 250–500 USD ton^{-1}, is much higher than that of raw cellulose. Direct conversion of low-value cellulose into highly priced LA or lactates, the precursors of a booming bioplastics market, is thus desired but challenging since as many as six reactions need to be controlled in a cascade reaction fashion, as seen in Fig. 5.

Step 1 is the acid- (or base-) catalyzed hydrolysis of cellulose into D-glucose. This reaction is hampered by the recalcitrant nature of the cellulose structure due to the presence of semi-crystalline and apolar domains [8]. The cellulose structure comprises chains organized into planes as a result of many inter- and intra-chain hydrogen-bonds. These planes stack on top of each other, rendering cellulose quite crystalline and hydrophobic [8, 70]. Besides several one-pot approaches combining, for instance, hydrolysis of cellulose with other reactions, such as the hydrogenation of cellulose to sorbitol [5, 71–77] or the cascade to LA in Fig. 5, the production of glucose from cellulose itself is a field of great interest on its own, as it provides an entry of cheap glucose in the biorefinery. Several groups have obtained intriguing results for this reaction in different ways [9, 78–85]. With respect to cascade transformations directly from cellulose in water, Brønsted acid-catalyzed hydrolysis is most frequently used; it is performed by taking advantage of the increased proton concentration in hot water [86] or by mineral acids such as HCl [71], heteropoly acids [87], or solid acid catalysts [71, 73, 84, 88]. The glucose yields range from 50% to 80%, depending on the catalyst nature, cellulose type and

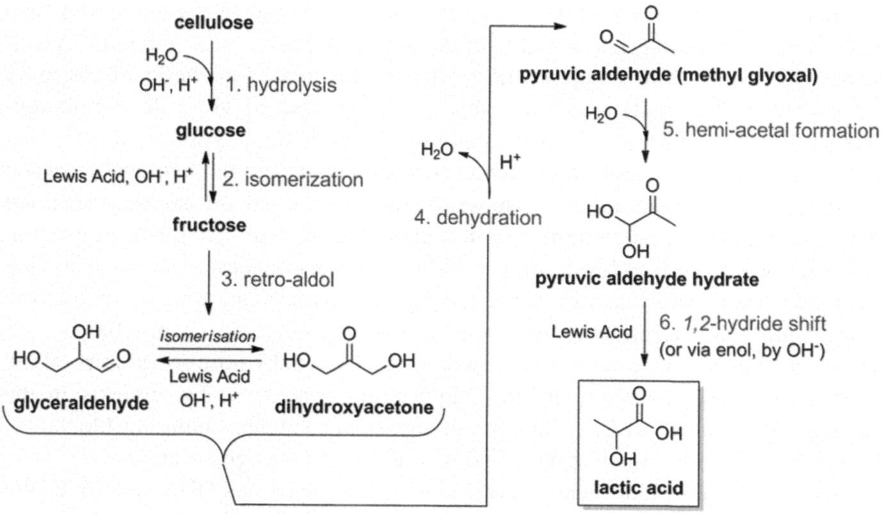

Fig. 5 Cascade reaction from cellulose to LA in water. Required catalysts are indicated. Bottlenecks are usually in step 1 (hydrolysis of solid cellulose) or step 3 (retro-aldol)

loading (usually as low as 1–5 wt% in water) and the reaction temperature (in the range of 120–180°C). The main difficulty in keeping the glucose yield high is the reactive nature of glucose, ultimately forming furans by acid catalysis and humins at higher temperatures and concentrations. Therefore, others have foreseen direct cellulose conversion to more stable molecules such as sorbitol, isosorbide [89–91], and levulinic acid [92, 93]. While (hydrolysis) reactions with cellulose in hot water have been investigated deeply, reactions in alcoholic solvents are underexplored. In addition to some heteropoly acids [94], organic acids can be used such as the combined $In(OTf)_3$ and *p*-toluenesulfonic acid system, reported in a cascade towards methyl levulinate with 74% yield at 180°C [95]. While alcoholysis of cellulose to more stable products like α-methyl-glycoside is obvious [94], it is usually the major side reaction towards methyl levulinate.

Step 2 is the isomerization of glucose to fructose. This reaction involves the conversion of the aldohexose into the 2-ketohexose. Retro-aldol reaction of the aldohexose leads to a C_4 and C_2 sugar, whereas the ketohexose leads to the two trioses, dihydroxyacetone (DHA) and glyceraldehyde (GLY). As the pathway to LA involves the trioses, selective glucose isomerization is essential, its conversion being limited by equilibrium in the operational temperature window. The isomerization of aldo- to ketoses can proceed via an acid-catalyzed hydride shift, a base-catalyzed mechanism with a proton shift (and intermediate enol), or via a concerted 1,2-hydride shift in neutral media [96, 97]. The latter isomerization mechanism occurs at mild temperatures (100°C) in the presence of Lewis acid catalysts, first reported by Davis and co-workers for hexoses with the use of Sn-containing Beta zeolites (BEA topology). The 1,2-hydride shift was proven with labeling experiments [96, 98, 99]. Our group recently noticed an increased isomerization activity per tin site for a Sn-Beta zeolite, which was synthesized via a dealumination-grafting sequence instead of via the common hydrothermal synthesis in the presence of HF [100]. A recent experimental comparison of reported homogeneous and heterogeneous catalysts for the glucose-to-fructose conversion has been reported [101]. Triose aldo-keto isomerization has also been assessed [97], but, in general, this reaction proceeds at lower temperatures than hexose isomerization, likely due the solely linear nature and higher reactivity of trioses. Since glucose-to-fructose isomerization is essential in the route to LA, the reaction should be fast, as otherwise glucose undergoes side reactions such as dehydrations or retro-aldol, leading to 2- and 4-carbon sugars. While being considered by-products in the pathway of cellulose to LA, these small sugars are valuable building blocks for the synthesis of other interesting α-hydroxy acids, as highlighted in Sect. 7.

Step 3 is the retro-aldol reaction of fructose into the trioses DHA and GLY. We have already described its atom economy and impact on functionality in Chap. 1 of this volume [23]. Retro-aldol is the reverse reaction of the better known aldol addition. The reaction is rare, but most famous in metabolic pathways like the third stage of glycolysis: the splitting of fructose-1,6-bisphosphate into DHA- and GLY-3-phosphate, catalyzed by aldolase A enzymes [102]. The reaction is catalyzed by a reactive lysine in the active pocket of the retro-aldolases [103, 104]. It also depends on equilibrium thermodynamics (with the inverse aldol) and reactions pulling trioses away from this. Currently, there is no chemocatalyst known that is able to perform the reaction at temperatures below 140°C. W, Pb, and Sn-based catalysts show retro-aldol activity above 150°C in water. Thermally uncatalyzed retro-aldol is demonstrated in (near) supercritical water [105, 106]. However, low selectivity is observed due to a plethora of other reactions with lower energy barriers, occurring in the high temperature conditions. One should keep in mind that C–C splitting in retro-aldol is indeed more demanding than, for instance, dehydration or hydrogenation of sugars. Development of active catalysts that selectively promote the retro-aldol reaction channel is therefore imperative.

A milestone publication in this respect came in 2010 from Holm et al., with a paper on the selective conversion of sucrose, fructose, and glucose to lactates with Sn-Beta zeolites [107]. Apparently, isolated Sn^{4+}, coordinated in the silica framework, is able to catalyze retro-aldol chemistry at temperatures around 160°C, besides being an efficient aldo-keto isomerization catalyst, as described above. The mechanistic details are, however, not fully elucidated. A tentative mechanism was given by Taarning et al., shown in Fig. 6 [108].

Step 4 in the route to LA from cellulose deals with the dehydration of the triose sugars, according to a retro-Michael mechanism, forming the unstable intermediate pyruvic aldehyde (or methyl glyoxal). This reaction is typically performed by Brønsted acid catalysis, although Lewis acids are also capable of this water

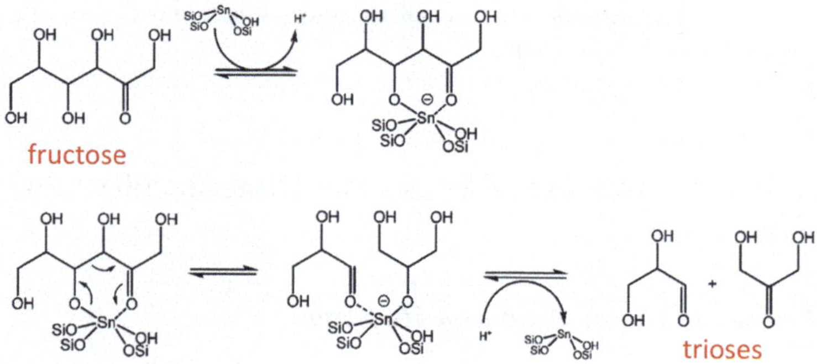

Fig. 6 Fructose undergoing retro-aldol reaction to glyceraldehyde and dihydroxyacetone, catalyzed by an isolated Sn^{4+} site in Sn-beta zeolite. Adapted from [108]

elimination [109]. More mechanistic details on retro-Michael mechanisms are found in the first chapter of this volume [23].

Step 5 converts the multifunctional and thus reactive pyruvic aldehyde into its hemiacetal or hydrate, when performed in water or alcohol. Brønsted acidity typically catalyzes these reactions. Strong acidity even leads to di-acetalization, e.g., pyruvic aldehyde diethyl acetal in ethanol [65, 110].

Step 6 is the final step in the cellulose-to-lactic acid cascade, involving the isomerization of the 2-keto-hemi-acetal (here: pyruvic aldehyde hydrate) into a 2-hydroxy-carboxylic acid. This reaction is known to proceed in basic media following a Cannizzaro reaction with 1,2-hydride shift [111]. Under mild conditions, Lewis acids are able to catalyze this vital step, which can also be seen as an Meerwein-Ponndorf-Verley reduction reaction mechanism. The 1,2-hydride shift has been demonstrated with deuterium labeled solvents [110, 112]. Attack of the solvent molecule (water or alcohol) on pyruvic aldehyde (step 5) and the hydride shift (step 6) might occur in a concerted mechanism, but the presence of the hemiacetal in ethanol has been demonstrated for pyruvic aldehyde with chromatography by Li et al. [113] and for 4-methoxy ethylglyoxal with in situ ^{13}C NMR by Dusselier et al. (see Sect. 7) [114].

To conclude, the one-pot conversion of cellulose-to-lactic acid (or lactate ester in alcoholic media) thus follows a complex cascade reaction network involving at least six reactions. These reactions have different catalytic needs, but, in general, the presence of both Lewis and Brønsted acidity are paramount for catalytic success. Brønsted acidity is key to the hydrolysis of cellulose (step 1) at mild temperatures (<200°C), and to some extent to the dehydration of triose (step 4), whereas Lewis acid sites play a vital role in the isomerization reaction of glucose-to-fructose (step 2), the retro-aldol (step 3), and the 1,2-hydride shift (step 6). Steps 4 and 5 are relatively less demanding; they are catalyzed by both acid types.

Rather than performing the whole cascade in one pot, many reports are available which describe the catalytic needs of some individual steps. For instance, the conversion of trioses to LA (and its esters) has been studied over various catalysts by several authors. Since the number of steps is reduced from six to three (only steps 4–6 in Fig. 5), an understanding of the mechanism and description of the catalytic criteria are more evident.

5 Catalytic Conversion of Mono- and Disaccharides into Lactates

5.1 Trioses to Lactic Acid and Its Esters

As triose sugars are not abundant in nature, they are not the preferred feedstock for LA production. Oxidation of glycerol might result in a mixture of trioses, but the selectivity at full conversion is currently not high enough [40, 115]. Yet reactions with trioses are crucial for mechanistic purposes.

Conversion of trioses to lactates has been achieved in near quantitative yields in both water and alcohols in the presence of a suitable catalyst. In water, the reaction is more tedious and slower, and catalyst deactivation owing to the produced LA is more pronounced. In alcohol, the formation of the di-acetal side-product is an issue. Figure 7 summarizes the triose-to-lactate pathways. Step 4, the initial dehydration of the triose, is thought to be rate-determining in this cascade when conducted in the presence of a Lewis acid [66], but generally, it strongly depends on the catalyst and temperature.

A first eye-opening report on the conversion of trioses to alkyl lactates in alcoholic media was produced by Hayashi and Sasaki in 2005 [112], after earlier attempts by Eriksen [116] and Kelly [117]. With a simple catalytic procedure using Lewis acidic $SnCl_2$ and $SnCl_4$ salts (10 mol% on 0.625 M of triose) in different alcohols they achieved yields above 80% of methyl, ethyl and butyl lactate, in a 1 h reaction at 90°C. Later, Rasrendra et al. explored this reaction in water, and, there, different catalytic trends were found [118]. Whereas Sn^{2+} and Sn^{4+} were clearly superior to Cr^{3+} and Al^{3+} chlorides in Hayashi's work in alcohol, in water Cr^{3+} and Al^{3+} were better catalysts for LA synthesis. They showed a 90% yield in 1.5 h at 140°C using 5 mol% of Al^{3+} on 0.1 M of triose. In our experience, working with Sn salts in water entails the formation of polymeric Sn-(hydr)oxide precipitate. With $AlCl_3$, the authors have assessed the activation energies for step 4, viz. the triose dehydration to pyruvic aldehyde, and the concerted steps 5 and 6, viz. LA from pyruvic aldehyde. Values of 93 and 58 kJ mol^{-1} were respectively calculated, proving dehydration (step 4) to be the rate-limiting step under mild conditions. In the presence of Al and Cr halides, the pH of the reaction medium was lowered to about 3, and it was shown that the released protons assisted in the rate-determining dehydration step [118]. Finally, alkaline earth metal hydroxides and especially Ca (OH_2) at 2:1 OH^-:DHA ratio in water, were found to be very active for the conversion of DHA to LA via the keto-enol route, with yields up to 59% at only 25°C for 1 h. However, such alkali approaches are corrosive and end up with lactate salts instead of the free acid, and the advantage of Lewis acid catalysis over fermentation in these basic conditions is thus lost [119]. Although these homogeneous studies are very informative, the search for heterogeneous catalysts has

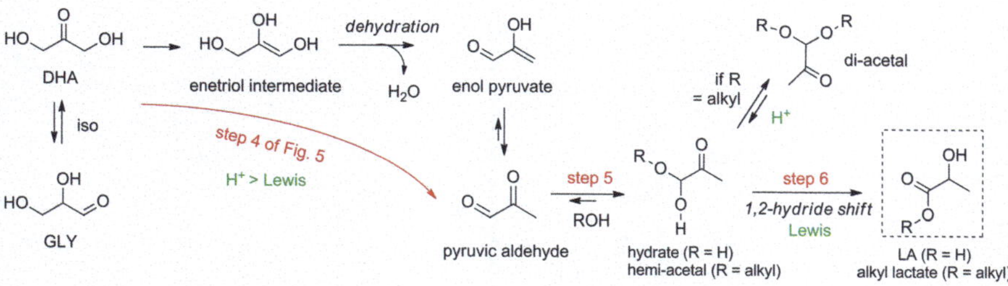

Fig. 7 Cascade reaction from triose to LA/lactates in water/alcohol respectively

received increasing attention since Sels and co-workers reported the first heterogeneous catalyzed triose conversion based on H-USY zeolites [120].

The use of heterogeneous catalysts has many advantages, such as their easy separation and reuse. Sels et al. have further unraveled the active sites of the USY zeolite and they have noticed a clear connection between the ethyl lactate selectivity and the amount of extra-framework aluminum [110]. While most framework Al in common zeolites delivers Brønsted acidity when counterbalanced with H^+, steam and acid treated zeolites like USYs also contain several types of extra-framework Al such as AlO(OH) and $Al(OH)^{2+}$ and they offer the required Lewis acidity. Thus, Brønsted acidity is required to dehydrate the trioses, while Lewis acidity favors the 1,2-hydride shift. Too strong Brønsted acidity should be avoided since it favors the formation of the di-alkyl acetals instead of lactates (see Fig. 7). High selectivity is thus obtained in the presence of a medium density of (weak) Brønsted sites, together with a high density of Lewis acidic sites. Such a balance was found by investigating the commercial CBV USY-zeolite series. Figure 8a shows that the optimal Lewis (extra-framework) vs Brønsted (framework) balance was present in CBV 600 – with 27% of framework Al – leading to the highest lactate selectivity of 77% at 90°C. Independently, West et al. have confirmed the use of USY for lactate synthesis from trioses, and also studied a continuous flow setup. Furthermore, they unraveled the reasons for catalyst deactivation [121]. In alcoholic media, the deactivation is nearly zero, but in water, the presence of LA (p*K*a 3.72) destroys the zeolite structure. Another reason for deactivation is the deposition of carbonaceous cokes, which is said to derive from side reactions with methylglyoxal or LA.

Alongside the classic Si- and Al-containing zeolites, the design and catalytic use of zeolites and other microporous materials with Lewis acidic heteroatoms has been reported, with a focus on substituted heteroatoms in zeolite frameworks. In 2009, Taarning et al. used Sn-Beta zeolites and reported LA and lactate yields of 90% and 99%, respectively, for the complete conversion of DHA in water and methanol, respectively, at either 100°C or 80°C with an Si:Sn ratio of 125 [65]. The reaction was near to completion after 6 h. The initial turnover frequency was calculated to be 45 mol mol_{Sn} h^{-1}. In comparison, Hayashi's soluble $Sn^{IV}Cl_4.5H_2O$ salt only reached about 4.2 mol mol_{Sn} h^{-1}. Since then, numerous reports have surfaced in the literature studying (among other catalysts) the use of Sn-MCM-41 [113], Sn-SBA-15 [122], Sn-MFI [123], Sn-montmorillonite [124], Sn-MWW [125],

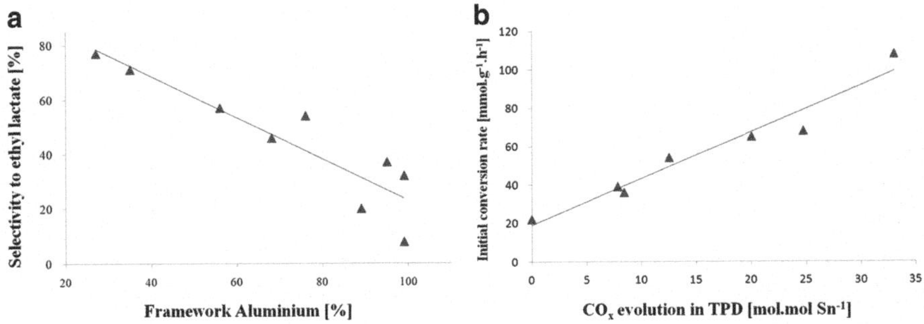

Fig. 8 Catalytic trends for the conversion of dihydroxyacetone at 90°C in ethanol. (**a**) Selectivities of different USY zeolites with varying amount of framework aluminum. (**b**) Sn-containing carbon-silica catalysts: correlation between *n*° of weak Brønsted acid sites (measured by CO_x release in TPD-experiments) and initial conversion rate. Based on data from [110] and [66]

Table 1 Comparison of various reported heteroatom-substituted zeolite catalysts for the conversion of DHA to LA or lactates

E	Catalyst	Solvent	*T* (°C)	Time (h)	Yield % di-acetal[a]	Yield % lactate[b]	References
1	Al-Beta	MeOH	115	24	72	2	[65]
2	Sn-Beta[c]	MeOH	80	6	0	97	[65]
3	Ti-Beta	MeOH	115	24	0	40	[65]
4	Zr-Beta	MeOH	115	24	0	39	[65]
5	Sn-Beta[c]	H_2O	125	24	–	90	[65]
6	Sn-Beta[d]	MeOH	90	24	0	99	[129]
7	Sn-MFI[e]	H_2O	90	3.5	–	65[e]	[123]
8	Sn-MWW	MeOH	120	24	0	99	[125]
9	desil-MFI[f]	H_2O	140	6	–	84	[126]
10	Ga-FAU[g]	EtOH	85	24	10	80	[127]
11	Al-FAU[g,h]	EtOH	90	6	18	59	[110]
12	Al-FAU[g,h]	EtOH	120	3	9	91	[110]

E = entry, Take note that the concentration of DHA varies between entries from different authors, as well as the sugar:catalyst or sugar:active metal ratios
[a]Only in alcohol can di-alkyl acetal side product be formed
[b]Yields in mol% of LA when solvent is H_2O and of alkyl lactate when solvent is alcohol
[c]Prepared via classic hydrothermal method using HF
[d]Prepared via dealumination of Al-Beta followed by solid state Sn ion-exchange
[e]When classic Sn-Beta was used under identical conditions, 54% of LA was analyzed
[f]Desilicated MFI zeolite with 0.6 M NaOH
[g]Ultrastable Y zeolites of FAU topology
[h]Framework Al: 27% = the most left point in Fig. 8a (at 90°C)

gallium oxides [113], desilicated MFI zeolites [126], GaUSY as well as other post-synthetic galliated-zeolites [127, 128], and novel Sn-Beta zeolites made via dealumination and post-synthetic Sn incorporation [129]. A comparison of some of the reported (metal-substituted) zeolites for the conversion of DHA in water or alcoholic media is given in Table 1. Entries 1–4, reported by Taarning et al., compare different Lewis acid-containing Beta zeolites (BEA) and a classic

Al-Beta for their activity in the conversion of DHA to methyl lactate in methanol. Clearly, Sn is the most active Lewis acid for this conversion in comparison to Zr and Ti. The Brønsted acidic Al-Beta mostly catalyzed the formation of the di-acetal side product in line with the high framework Al in the USY series of Sels et al. (see Fig. 8a) [110]. Not only is the conversion with Sn-Beta complete and the selectivity near 99%, the zeolite with Sn is also able to perform this reaction at a lower reaction temperature of 80°C, whereas the other metals required 115°C. A higher reaction temperature is needed in water to compensate for the slower reaction, but the yield remained high (entry 5, Table 1). Tsapatsis and co-workers compared hydrothermal Sn-MFI (small pore zeolite) and Sn-Beta (large pore zeolite) for the title conversion and noticed a higher yield for the MFI topology in their conditions (entry 7).

As hydrothermally synthesized Sn-BEA and Sn-MFI zeolites are difficult to synthesize, other synthesis methods following post-synthetic metal incorporation or modification were attempted. For a start, Hermans and co-workers have devised a new synthesis route to Sn-Beta via complete dealumination in strong acid, followed by solid state Sn exchange using Sn^{II}acetate. They noticed a high productivity of their catalyst due to the high Sn content (10 wt%, entry 6). Secondly, inspired by Hermans et al., Hensen and co-workers have reported the full deboronation of B-MWW and subsequent grafting with $SnCl_4.5H_2O$ in the presence of hexamethyleneimine [125]. The choice for MWW was for reasons of accessibility in anticipation of converting more bulky substrates such as glucose and fructose (see later) along with trioses. The MWW topology comprises a 10-membered ring interlayer pore opening connected to a 12-membered ring supercage and an independent intralayer sinusoidal 10-membered ring channel [130]. Sn-MWW exhibited comparable activities to Sn-Beta (entry 8). In light of avoiding Sn and inspired by the extra-framework Al Lewis acidic results of USY, Dapsens et al. have investigated the creation of Lewis acid sites from Al containing ZSM-5 (MFI) zeolites via alkaline desilication [126]. They have demonstrated a highly selective Lewis acidic site for making LA from trioses in water, but they also noted that the catalyst was in need of a high reaction temperature (entry 9, Table 1) and that the catalyst stability was limited. Their research showed leaching of Al species from the zeolite caused by the low pH of the LA product solution. Dapsens et al. continued their effort by working in ethanol and by creating gallium Lewis acid sites by metalation during desilication in alkaline conditions in USY zeolites [127, 128]. These zeolites are the most selective non-tin materials reported at temperatures below 100°C in alcoholic media; one catalytic experiment is illustrated in entry 10. Entries 11 and 12 display two entries of the CBV series of Sels and coworkers and the effect of temperature.

The importance of a balanced catalyst with appropriate number and strength of Lewis and Brønsted acid sites has been demonstrated by de Clippel et al. They have designed a porous carbon–silica composite with independently tunable mild Brønsted acidic sites and Lewis acid Sn [66]. The catalyst was composed of an Si-MCM-41, grafted with Lewis Sn sites (e.g., 0.5 wt%), and filled with intraporous carbon by furfural impregnation and pyrolysis under helium. A schematic representation of the catalyst structure is displayed in Fig. 9. The intraporous carbon compound has weak surface acidic groups such as –COOH and phenols, which could easily be tuned by changing the temperature in the pyrolysis or by a

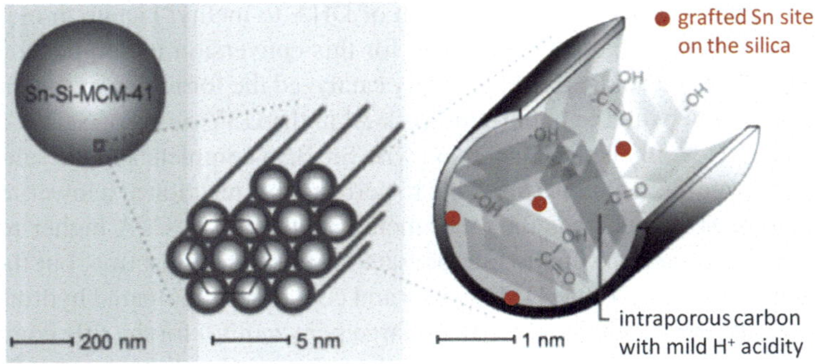

Fig. 9 Carbon silica composite catalyst design, based on Sn-grafted Si-MCM-41 and intraporous carbon with mild Brønsted acid sites. Adapted from [66]

post-synthetic oxidization treatment in air, while the Sn content is easily changed in the grafting step. In this way, for a given content of Sn, it was unambiguously shown that the rate could be enhanced sevenfold simply by increasing the mild Brønsted sites per Sn, in line with the acceleration of the rate-limiting triose dehydration step (4 in Fig. 7). The figure of merit is shown in Fig. 8b. The rise in initial reaction rate in function of the amount of Brønsted acidity, measured as the release of CO_x species in separate TPD analysis, is apparent. Comparing the initial turnover per Sn for the parent Sn-MCM, viz. 41 mol mol_{Sn} h^{-1}, in close agreement with that of Sn-Beta, with the record of 289 mol mol_{Sn} h^{-1} for the optimal hybrid catalyst, validates the mechanistic proposal [66]. More information on the design and synthesis of nanohybrid and nanocomposite catalysts for use in multi-functional cascades – as often needed in selective catalytic approaches for biomass conversion – is found in our recent review [131].

5.2 *Hexose-Based Sugars: Sucrose, Glucose, Fructose*

The conversion of hexoses to LA is one of the most atom efficient cascade transformations out there, fully preserving (and even enhancing) chemical functionality (F:C of 1.17 for hexoses, vs 1.33 for LA, see [23]). However, the reshuffle of O atoms in the molecule and the selective C–C breakage requires multiple steps, and a selective catalyst. There is currently good insight into the reaction network with the details presented in Figs. 5 and 11.

5.2.1 Reactions in Alcoholic Solvents

Realization of high alkyl lactate yields from trioses in various alcohols is thus possible. For hexoses this implies that steps 2 and 3 of the reaction cascade in Fig. 5 require extra attention, before the known triose chemistry can ensue. The

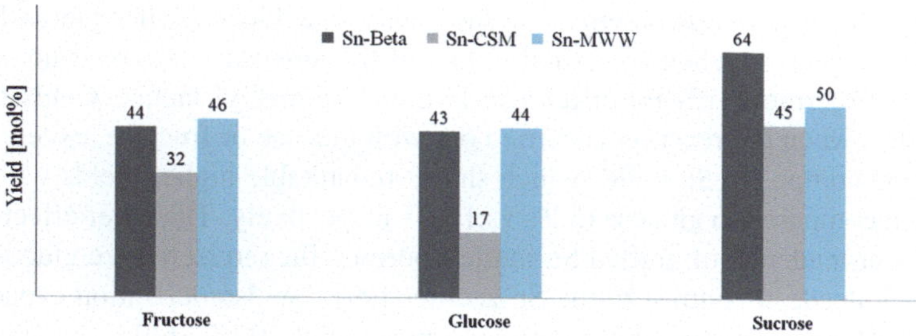

Fig. 10 Comparison of catalytic yields of methyl lactate for the conversion of hexose-based sugars in methanol with different heterogeneous catalysts. CSM = carbon silica composite. Conditions: 160°C, 20 h, 22.5 mg_{sugar} mL^{-1}, 16 $mg_{catalyst}$ mL^{-1} (slightly different for CSM: respectively 155°C, 30 and 21 mg mL^{-1}). All conversions >95%. Data from [66, 107, 125]

isomerization in step 2 is not difficult, but one encounters the equilibrium situation between glucose and fructose. As a consequence, the substrate selectivity in step 3 is probably most challenging since fructose, and not glucose, should be converted via retro-aldol. Besides, other side reactions like sugar dehydration and sugar alkylation (when performed in alcohol) are competitive as well. Lobo has pointed out the resemblances of the glucose-to-lactate route with the natural glycolysis pathway in living cells [132]. The conversion of trioses into LA via pyruvic aldehyde in turn presents a biomimetic of the enzymatic glyoxalase system, as suggested by Sels et al. [120].

The major breakthrough in lactate synthesis from hexoses was made in 2010 by Holm et al., who applied Sn-Beta to convert sucrose at 160°C in methanol, yielding 64% methyl lactate. Figure 10 summarizes the key catalytic results in black. As mechanistically proposed in Fig. 6, Sn^{4+} is capable of coordinating both a hydroxy and the keto group of fructose, thereby facilitating retro-aldol cleavage in the β position (between C_3 and C_4) to the carbonyl. This reaction is attributed to the known tendency of Lewis acid Sn^{IV} to coordinate –OH bonds rather than to form Sn–O–C bonds [133, 134]. Besides being active for this particular C–C scission, equally important, Sn-Beta Lewis acid sites do not easily dehydrate hexoses or form stable methyl glucosides, two competitive side reactions [135].

de Clippel et al. also tested their bifunctional carbon–silica composite material (Sn-CSM) for this reaction with slightly lower methyl lactate yield (up to 45% from sucrose), as seen in Fig. 10 in grey. They figured out that the Brønsted acidity, which was beneficial to accelerate the triose reactions, caused multiple side reactions such as methyl glucoside formation and dehydration towards levulinate esters. The use of composites with less Brønsted acidity was necessary to obtain a high methyl lactate yield. The presence of low amounts of these sites was, however, still beneficial, as parent siliceous Sn-MCM only yielded 18% [66].

Hensen and co-workers have used the Sn-MWW zeolite (see above) to convert hexoses to methyl lactate and they achieved identical results for glucose and fructose as with Sn-Beta, but somewhat lower yields (50%) were analyzed for reactions with sucrose (blue bars in Fig. 10) [125].

In Fig. 10, two trends emerge: (1) the catalysts achieve higher yields for the disaccharide sucrose when compared to that of the monosaccharides, fructose and glucose; furthermore, (2) no difference is found in methyl lactate yields for the Sn-zeolites, when the reaction is carried out with glucose or fructose, except in the case of the composite Sn-CSM, which shows remarkable higher yields with fructose when compared to glucose (32% vs 17%), respectively. The latter effect is due to the lower tendency of grafted Sn in mesoporous silica to isomerize glucose into fructose. Indeed, in contrast to the Sn zeolites [99], low isomerization capacity of the Sn-CSM was demonstrated for glucose in water at 100°C [66].

The notable difference in methyl lactate yield from sucrose or hexoses is likely attributed to the higher thermostability and hydrolytic resistance of the disaccharide. Sucrose is a non-reducing sugar with both carbonyl groups protected in the α,β-(1 → 2) glycosidic bond [136]. Consequently, the slow release of hexoses from sucrose in solution prevents unwanted side reactions with the more labile glucose and fructose. Such effects have also been witnessed in the conversion of cellulose to sorbitol and ethylene glycol, for which the highest selectivity is obtained when cellulose hydrolysis is rate-determining [72, 137]. An elegant solution to this phenomenon is the use of a fed-batch reactor system with glucose feedstock, mimicking the slow release of glucose from cellulose hydrolysis and thus minimizing side reactions [138, 139]. Such endeavors remain unexplored for lactate synthesis, and one could envision the use of more recalcitrant inulin polymers (containing anhydrofructose units) or a fed-batch reactor feeding hexoses.

Aside from the superior catalysts mentioned above, Osmundsen et al. have reported the use of hydrothermally synthesized Sn-SBA-15, Sn-MCM-41, and Sn-MFI for sucrose to methyl lactate in methanol, but their yields ranged between 20% and 30% [122]. A similar 18% yield was reported for the parent material of the Sn-CSM catalysts. This mesoporous stannosilicate was made via Sn grafting procedure [66]. Murillo et al. reported a yield of 43% of methyl lactate from glucose, with a different hydrothermal Sn-MCM-41 [140]. Apart from tin-catalysts, Liu et al. reported on the use of basic MgO in MeOH, but only 30% of methyl lactate was formed for glucose and other hexoses at a high 200°C [141].

To conclude, 4 years of research since the original report still demonstrate the superior activity of the original Sn-Beta to convert sucrose directly to methyl lactate in methanol [107]. The catalyst is recyclable at least up to six times by intermittent calcination, and the methyl lactate yield remains 60% at full conversion after 20 h, even when more concentrated sucrose (10 wt%) was used, corresponding to a volumetric productivity of 3.3 $g_{lactate}\ L_{reactor}^{-1}\ h^{-1}$. Since both this heterogeneous and the fermentative approach mainly run the process on sucrose or hexoses, a comparison is made in Table 2 for a set of parameters. It may be clear that the heterogeneous route is clearly a less complex process with many advantages, but mainly the lower selectivity and volume productivity need to be improved in order to overtake the fermentation. Moreover, racemic lactates are produced which can be seen as a disadvantage for direct L-PLA synthesis, but interestingly also provide an equal source of D-isomer, given a successful enantioseparation, as explained in Sect. 8.

Table 2 Comparison of heterogeneous catalytic routes vs classic (bacterial) fermentation

Criterion	Heterogeneous [107]	Fermentation [55, 63]
Media	Alcohol	Aqueous broth
Feedstock	Glucose, sucrose	Glucose, sucrose
Concentration feed	Up to 10 wt%	Up to 15 wt%
Catalyst	Sn-Beta zeolite	Bacteria (yeast)
Co-reagents	None	Alkali + nutrients
Major product	Alkyl lactates	Ca-lactate salt
Stereopurity product	Racemic	L
Selectivity (mol%)	Up to 65%	Up to 95%
Productivity	Up to 3.3 g $L_{reactor}{}^{-1}$ h^{-1}	0.3–5 g $L_{reactor}{}^{-1}$ h^{-1}
Gypsum co-product	No	1 kg per kg of LA
Work-up	Easy: filtration, distillation. alcohol reuse	Complex: acidification, two filtrations, purification, esterification, distillation
Catalyst reuse	Yes	No (in a complex broth)

Despite the unique ability of Sn-zeolites, their performance in terms of activity and selectivity is the main issue to improve before competing with the fermentation route. In our belief, the race to search for better heterogeneous catalysts has just started. A profound study of the formation of the side-products is the first step towards the design of outstanding catalysts. All identified side products seen when using Sn-CSM and Sn-Beta in methanol are incorporated in Fig. 11, with their respective origins indicated in specific colors: red (methylation of sugars), blue (retro-aldol derivatives of glucose), and black (dehydration products, usually via fructose).

To start with, in methanol – which always contains some water unless extreme solvent drying is pursued – sucrose either undergoes hydrolysis or solvolysis. The latter is indicated in red in Fig. 11 and usually yields methyl-D-glucopyranoside and fructose from sucrose instead of methyl-fructoside and glucose. These methylated sugars are non-reducing and chemically more stable; they are considered unreactive to retro-aldol chemistry, since their ether bond should be hydrolyzed prior to be of use in the lactate synthesis.

Second, the retro-aldol reaction applied to glucose (indicated in blue) is the major side reaction, yielding C_2 and C_4 sugars, which in turn undergo multiple chemical transformations. The most common products are 4-carbon backbone α-hydroxy acid esters: methyl-4-methoxy-2-hydroxybutanoate (MMHB), methyl vinyl glycolate (MVG), and methyl-2-hydroxybutanoate (MHB). The formation of these intriguing and useful esters from tetroses was recently unraveled by Dusselier et al., and the reaction networks are discussed in more detail in Sect. 7. Their formation bears a close resemblance to that of the lactate system, but important differences exist. Besides the four carbon products, the diose glycolaldehyde is converted to methyl glycolate or its di-acetal GADMA. For instance, in the conversion of sucrose with the optimal Sn-containing CSM composite, in total 10% of the input sugar was converted according to the blue pathway [66].

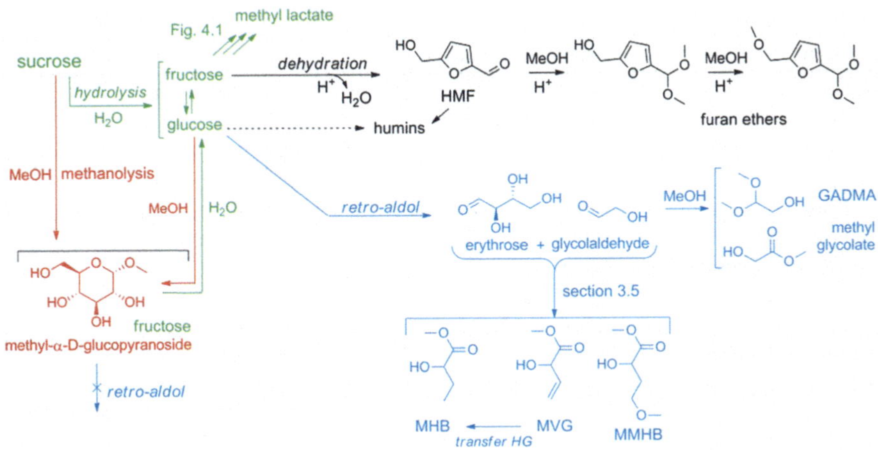

Fig. 11 Side products in the sucrose-to-lactate (*green*) conversion in MeOH and their origin: *blue*: retro-aldol of glucose (see Sect. 7); *red*: methanolysis of sucrose or acetalization of glucose; *black*: dehydration of sugars to HMF, furan ethers and humins. MMHB: methyl-4-methoxy-2-hydroxybutanoate, MVG: methyl vinyl glycolate, MHB: methyl-2-hydroxybutanoate, GADMA: glycolaldehyde dimethyl acetal. HG = hydrogenation

Lastly, dehydration of hexoses produces side products (indicated in black). Both di- and tri-methoxylated acetals of HMF, as well as methyl levulinate, were analyzed and their formation is usually enhanced by the presence of Brønsted acidity. de Clippel et al., for instance, noticed an increase in these products from <1% with the optimal Sn-CSM composite to about 18% with an oxidized CSM, bearing a high density of Brønsted acid sites. Excess of these sites was less problematic in the case of triose conversion (see Fig. 8b), due to the lower process temperature and the less complicated cascade. For Sn-Beta, alongside 8% of MVG, no major side product was identified. The deficiency in carbon is probably due to some humin formation, which can proceed from glucose as well as from HMF, as recently demonstrated [142, 143]. Their presence indicates that the retro-aldol reaction is not fast enough, thereby allowing the free aldehyde groups of the sugars to undergo unwanted side reactions.

5.2.2 Reactions in Water

The conversion of sucrose to LA in water was assessed with Sn-Beta zeolite, but only low yields of LA of less than 30% were encountered [107]. Besides LA, HMF and levulinic acid were analyzed in the product mixture. One reason of the significant change in product spectrum was related to the auto-catalytic effect of LA. Its acidity lowers the pH of the medium, thereby enhancing the dehydration rate with an increase of HMF and the likes in the product spectrum as a result [107]. Pronounced carbonaceous deposits on the catalysts confirmed its deactivation by LA, similar to the triose reaction in water.

Besides the above-mentioned heterogeneous catalytic approaches in alcoholic or aqueous media, the use of soluble Lewis acidic salts in water has recently been investigated under catalytic conditions for glucose and fructose. Heeres and co-workers explored the use of Al^{3+} and Cr^{3+} salts at 140°C in 2010, but they have found only about 20% of LA with $AlCl_3$ [144]. Lanthanide triflates such as $Er(OTf)_3$ have been reported to yield around 50% of LA from the common hexose sugars. This reaction was, however, carried out at a high 240°C [145]. A breakthrough for this reaction in water was made very recently by Wang et al., who discovered the remarkable catalytic nature of Pb^{2+} for the reaction of hexoses (and even cellulose; see below) to LA. In the presence of 7 mM of $PbNO_3$, for instance, a 25.5 mM solution of glucose or fructose was transformed, respectively, into 65% and 74% of LA at 170°C, with unparalleled selectivity [146]. Fructose was noticed as the intermediate when reacting glucose, and the reaction mechanism was explained in agreement with steps 2–6 in Fig. 5. Computational results indicated that the presence of the lead ions significantly lowered the activation barrier for the retro-aldol reaction of fructose. This research is considered an important result, as the group of very active (at mild temperature) retro-aldol elements now contains an additional member, albeit a very toxic one. Besides these mild hydrothermal catalytic approaches, near critical water has been investigated as reactive medium for LA production as well, both from sugars and cellulose, with several homogeneous and heterogeneous catalysts. These often alkaline approaches, are not the scope of this review, but noteworthy studies in high temperature flow setups have been reported by Bicker et al., using $ZnSO_4$, [170] and Esposito et al [171]. The latter reported LA yields up to 57 % on a 0.025 M glucose solution with 0.1 M of $Ba(OH)_2$ at 250°C for 3 minutes.

5.3 *Other Carbohydrates, Pseudo-hemicellulose, and Glycolaldehyde*

Holm et al. further exploited Sn-Beta for the conversion of hemicellulosic sugars such as pentoses in near identical conditions as in Fig. 10. For one, they obtained a yield of 42% of methyl lactate for xylose [147]. Retro-aldol here evidently delivers glycolaldehyde along with a triose and, therefore, 5% of GADMA (structure in Fig. 11) was found, alongside 7% of MVG. Furthermore, a pseudo-hemicellulosic model mixture was assessed, composed of 1/6 of each of following sugars: xylose, arabinose, ribose, glucose, mannose, and galactose. Their conversion with Sn-Beta yielded 43% of methyl lactate in methanol, in line with the linear combinations of the yields attained for each sugar individually. Interestingly, the conversion of glycolaldehyde resulted in 16% methyl lactate, 27% MVG, 7% GADMA, and 6% MMHB. This is an intriguing result, as (1) the formation of 4-carbon backbone esters MMHB and MVG from the C_2 sugar implies Sn-Beta catalyzes aldol reactions and (2) the formation of lactate implies that retro-aldol reaction occurs simultaneously under these conditions (160°C, methanol). Both observations indicate that hexoses were formed from subsequent aldol reactions with glycolaldehyde. A detailed reaction pathway to convert glycolaldehyde to methyl

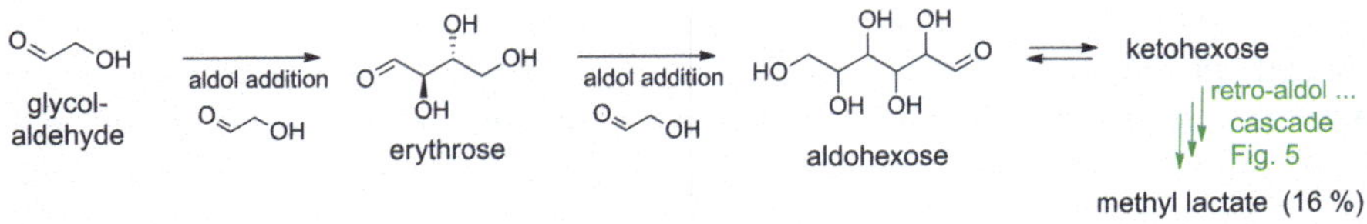

Fig. 12 Sn-Beta catalyzed conversion of glycolaldehyde to methyl lactate in MeOH

lactate is presented in Fig. 12. The presence of the intermediately formed hexose sugars was supported by the presence of trace amounts of HMF. Later, the selective conversion of glycolaldehyde to the useful 4-carbon backbone α-hydroxy esters MVG and MMHB instead of lactates was targeted by Dusselier et al. as discussed in Sect. 7 [148].

6 Direct Catalytic Conversion of Cellulose into Lactic Acid

Despite promising results with sugars, the direct use of cheaper cellulose is of course more appealing. A first attempt towards converting cellulose was made by feeding cellobiose in presence of Sn-Beta [147]. Cellobiose is a disaccharide of glucose with a β-(1-4) glycosidic linkage similar to that of cellulose. The authors have reported that only 13% of methyl lactate yield was obtained after 44 h of reaction at 160°C. Furthermore, in contrast to the near complete conversion for sucrose after only 20 h, 39% of unconverted sugars were recovered for cellobiose, resulting in a lactate selectivity of only 21%. This points to a low hydrolysis activity of the Sn Lewis acid sites (step 1 in the cascade in Fig. 5) or a limited accessibility to the glycosidic bond. The direct use of cellulose in catalytic conditions has been reported as well, but yields of LA (or esters) never exceeded 30% [149]. A first heterogeneous example of this was given by Chambon et al., using AlW (tungstated alumina) with a yield up to 28% of LA with 60% selectivity at 190°C for 24 h [150].

In 2013, remarkable progress was made when two research groups independently published on cellulose conversion, either in the presence of lanthanide triflates [145] or Pb^{2+} ions [146], showing LA yields of 89% and 68%, respectively. Soluble lanthanide triflates are famous Lewis acid catalysts because they retain their pure Lewis acid property in H_2O, in contrast to soluble Al and Sn salts (which provide Brønsted acidity as well or decompose). Moreover, for homogeneous catalyst standards, they are easy to recover. Wang et al. have screened a range of triflates for the conversion of untreated cellulose at 240°C under an inert 2 MPa N_2 atmosphere [145]. Despite the severe temperature, all triflates produced LA yields above 50%. Lanthanides with the smallest ionic radius such as erbium, ytterbium, and lutetium provide the best results (see Fig. 13). The authors stated that the smaller cation radius of a metal explains its stronger ability to coordinate wih the hydroxyl groups and that this lead to the higher catalytic activity.

The hydrolysis of cellulose was ascribed in this report to be promoted by hydroxonium ions from water auto-protolysis. Since Wang et al. worked at

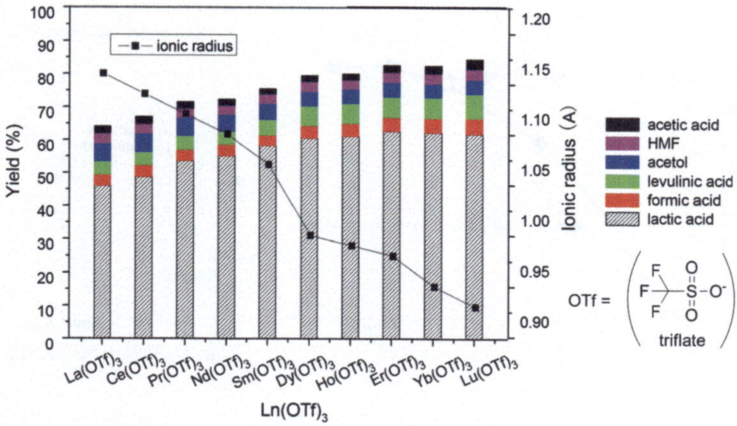

Fig. 13 Different lanthanide triflates for the conversion of cellulose to LA. 240°C, 2 MPa N_2, 30 min, 0.3 g cellulose, 0.05 g triflate, 30 mL H_2O. Adapted from [145]

240°C, this is in agreement with other reports showing Brønsted acid catalysis in water above 200°C [86]. Moreover, the yield and conversion dropped to only 14% and 25%, respectively, at 200°C, confirming the need for Brønsted acidity from the hot water and thus the lack of hydrolysis capacity of the La triflates. In a search for industrial relevance, the authors tested higher cellulose loadings. For 0.5 g cellulose in 30 mL water, under the conditions of Fig. 13, they achieved a LA yield of 50%. A quick calculation reveals a corresponding volumetric productivity of 17.7 g_{LA} $L_{reactor}{}^{-1}$ h^{-1}. At 0.1 g cellulose, their yield mounted to 89% of LA, but at a lower productivity of 6.3 g_{LA} $L_{reactor}{}^{-1}$ h^{-1}. These volumetric productivities are considerably high when compared to those of the fermentation and heterogeneous approach on hexoses in Table 2. Reactions with soluble catalysts are often more laborious though: the lanthanide catalysts were recovered via vacuum distillation, diethyl ether dissolution, and solvent evaporation, but could be reused several times. The ether dissolution was needed to separate the catalyst from the humins formed. This report is very interesting because it demonstrates high yields at short reaction times, with the need for hot water to promote cellulose hydrolysis [145].

Direct conversion of cellulose to LA was also achieved under milder conditions in the presence of Pb(II) ions [146]. As hydrolysis was too slow at 190°C for crystalline cellulose, ball-milling was used to enhance its reactivity. Such mechanical treatment is an effective way to decrease the biopolymer's crystallinity and to enhance the accessibility for chemicals (and catalysts) by increasing the surface area [8, 9, 78, 151–153]. Here, ball-milling reduced the crystallinity from 85% to 33%, as ascertained by XRD analysis. Among a series of other ions (7 mM) assessed at this temperature, Pb was highly selective to LA, yielding 68% after 4 h for 0.5 wt% cellulose solutions in water. Second best was Al^{3+}, in line with the catalytic results of Heeres et al. [144], showing 35% LA yield. Untreated cellulose required 15 h at 190°C to reach complete conversion with a LA yield of 62%. This reaction rate is quite fast, indicating that Pb^{2+} salt itself, in contrast to the previous

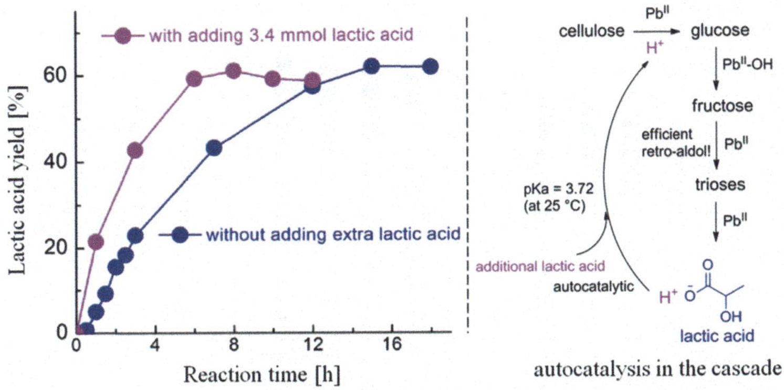

Fig. 14 Enhancement of the rate of LA formation from microcrystalline cellulose in the presence of Pb^{2+} with addition of LA. 190°C, 3 MPa N_2, 30 min, 0.5 g cellulose, 7 mM $Pb^{II}NO_3$, 100 mL H_2O. Adapted from [146]

lanthanides, also assists the hydrolysis step (step 1 in Scheme 5). A more concentrated 2.5 wt% cellulose slurry of milled cellulose was converted in 10 h in the presence of 6 mM of Pb^{2+} at 190°C, yielding 55% LA, corresponding to a volumetric productivity of 1.5 $g_{LA}\ L_{reactor}{}^{-1}\ h^{-1}$. Besides purified cellulose, several raw biomass feedstocks were converted and LA yields were always around 40%. Interestingly, the authors demonstrated an autocatalytic assistance of LA to fasten the hydrolysis of cellulose, as shown in Fig. 14. The intentional addition of LA significantly shortened the reaction time (by a factor of 3) to achieve 60% yield of LA in the conversion of microcrystalline cellulose.

Besides these experimental achievements, the authors have discussed the reaction pathway thoroughly; their reaction network is in line with the reaction scheme in Fig. 5. Thermodynamics were calculated and DFT calculations suggested that the coordination of Pb^{II} with three oxygen atoms of fructose increases the positive charge of the C4-OH, facilitating its proton transfer to C2 = O, and thus the retro-aldol of fructose. Apart from this, Pb(II)-OH was postulated to be the active species in solution. Although Pb^{2+} could be completely recovered from the reaction, it is a very toxic cation. Nevertheless, this research demonstrates the potential of chemocatalysis to convert cellulose to LA and should therefore motivate the search for more environmentally friendly catalysts [146].

7 Catalytic Synthesis of Other Biomass Derived α-Hydroxy Acids

7.1 C2: Glycolic Acid

Besides LA, other α-hydroxy acids (AHA) are very intriguing molecules as well, because of their multifunctionality and their relevance for polyesters. For instance, the biodegradable polymer polyglycolic acid is of commercial interest. Its monomer, glycolic acid (or 2-hydroxy acetic acid, GA) is, however, not derived from renewable resources at the moment, but via the carbonylation of formaldehyde with H_2SO_4 [154]. A route via the hydrolysis of molten monochloroacetic acid with NaOH is also known [155]. It is used in dyeing and printing and it acts as a Ca^{2+} ion chelator and antibacterial agent. In 2011, about 40 ktons were produced worldwide [154]. The synthesis of polyglycolic acid from glycolic acid is analogous to PLA from LA (Fig. 4) and thus proceeds via ring-opening polymerization of glycolide, its cyclic dimer [59].

As the current synthesis of GA is non-renewable, the development of biomass derived and more efficient routes is an interesting topic. Two catalytic carbohydrate-based approaches deserve some attention. In light of the valorization of cellulose to LA discussed above, the approach by Zhang et al. is of interest as they have reported the conversion of α-cellulose into GA in one pot over a phosphomolybdic acid catalyst ($H_3PMo_{12}O_{40}$). This reaction proceeded in water at 180°C under light oxygen pressure (0.6 MPa) and is shown in Fig. 15 [156]. At best, 0.2 g of cellulose was converted in 20 mL of water containing 15 mM of heteropoly acid over the course of 1 h, yielding 49% of glycolic acid and 10% of formic acid [156]. This is impressive, considering the conditions and the fact that the cascade seen in Fig. 15 is as complicated as that in the formation of LA; it involves not only hydrolysis of cellulose and multiple retro-aldols of many sugars, but also the oxidation of glycolaldehyde and formaldehyde. The F:C index of glycolic acid is 2, and this route from carbohydrates has a theoretical atom economy of 100%, indicating a realistic alternative to the petrochemical route for GA [23].

Beside the direct approach from cellulose presented above, Dapsens et al. recently proposed a route to convert glyoxal to GA under mild conditions with heterogeneous catalysis in both batch and flow conditions [154]. Glyoxal is currently obtained through oxidation of acetaldehyde or ethylene glycol [157], but it is also present in pyrolysis bio-oil, albeit in low quantities ranging from 1 to 3 wt%. Some catalysts known for their LA synthesis from trioses were tested, such as extra-framework aluminum containing USY zeolites and Sn-MFI. The latter, for instance, produced about 91% yield of GA at 90°C after 6 h, using 0.2 g of glyoxal in 3.8 mL H_2O. The reaction mechanism was studied via deuterium labeling and was found to proceed exclusively via a 1,2-hydride shift mechanism of the monohydrate form of glyoxal as seen in Fig. 16. This is in complete analogy with the

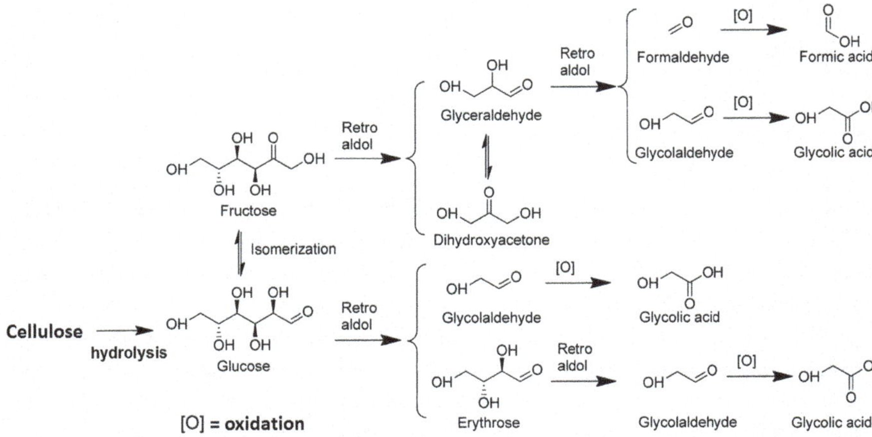

Fig. 15 Proposed pathway for converting cellulose to glycolic acid. Adapted from [156]

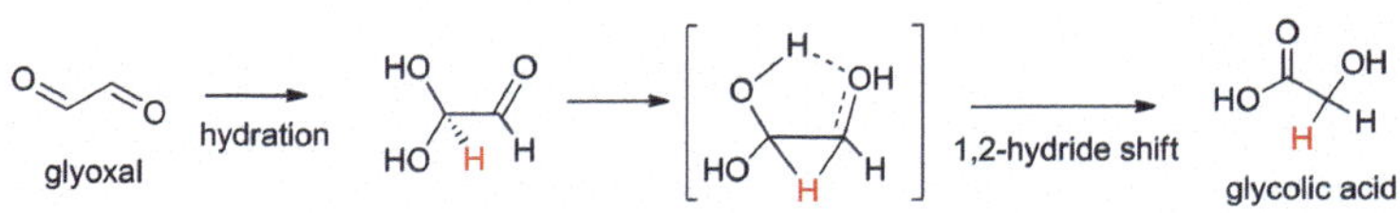

Fig. 16 Path for the conversion of glyoxal into glycolic acid with Lewis acidic zeolites

conversion of trioses to lactates [110]. The conversion of GA into alkyl glycolates was demonstrated in alcoholic media as well [154].

7.2 *C4: 4-Alkoxy-2-Hydroxybutanoates and Vinyl Glycolic Acid*

Inspired by the easy conversion of trioses to lactates, and determined to unravel the faith of the retro-aldol products of glucose, viz. erythrose and glycolaldehyde (the blue pathway in Fig. 11), Dusselier et al. set out to convert tetrose sugars with a focus on 4-carbon AHA esters in order to present a mechanistic proposal [114]. They have discovered the unique catalytic property of Sn halides to form methyl vinylglycolate (MVG) and methyl-4-methoxy-2-hydroxybutanoate (MMHB) from tetroses. For instance, in 1 h at 80°C, with 10 mol% of $SnCl_4.5H_2O$, 83% of MMHB and 2% of MVG were analyzed in a reaction using 0.63 M erythrulose in methanol. In situ NMR spectroscopy, deuterium labeling, and control experiments with intermediates revealed the individual pathways to both 4-carbon AHA esters. The cascade, shown in Fig. 17, starts with two consecutive retro-Michael dehydrations, of which the mechanistic details are visualized in the first chapter of this volume [23]. They lead to the highly reactive proposed intermediate

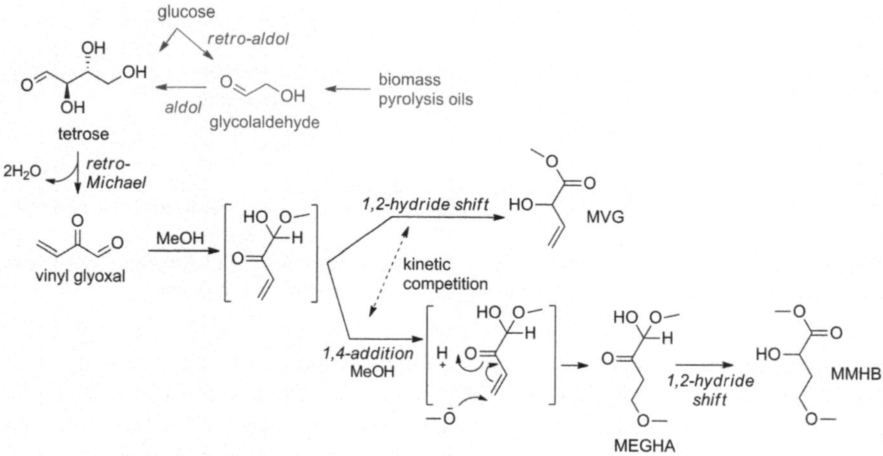

Fig. 17 Path for converting tetroses and glycolaldehyde (glucose retro-aldol products) into MMHB and MVG with $SnCl_x$ in MeOH. Glycolaldehyde is found in pyrolysis bio-oils

vinyl glyoxal. Its reactive aldehyde group quickly transforms into a hemi-acetal, as seen in Fig. 17, identical to the case of pyruvic aldehyde in lactate synthesis. The reactivity of this molecule is decisive for the reaction selectivity: either a 1,4-addition of MeOH or a Lewis acid catalyzed 1,2-hydride shift ensues. The former leads to the in situ observed 4-methoxy-ethylglyoxal-hemi-acetal (MEGHA in Fig. 17) intermediate, which then undergoes a 1,2-hydride shift into MMHB. However, occurrence of the 1,2-hydride shift of the hemi-acetal of vinyl glyoxal prohibits the 1,4-addition and MVG is formed instead. The kinetic competition between the hydride shift and the 1,4-addition is thus key to the product outcome. In methanol, the addition is fast and thus MMHB is found as the major product. In isopropanol, the 1,4-addition is more hindered, whereas the intramolecular hydride shift is not affected by the alkyl chain of the hemi-acetal and thus isopropyl vinyl glycolate was found as the major product [114]. Hydrolysis of these ester mixtures renders access to different 4-alkoxy-2-hydroxy-butanoic acids and/or vinyl glycolic acid (VGA).

The fate of glycolaldehyde under similar conditions was the subject of another study [148]. This 2-carbon sugar feedstock is of more relevance due to its pronounced presence, up to 10 wt%, in cellulose pyrolysis oils [158–160]. The authors proved that, under the same catalytic conditions, tin salts catalyzed the aldol reaction of two glycolaldehyde units into tetrose sugars, as seen in grey in Fig. 17, as well as the ensuing reactions. The rate-determining step of the entire glycolaldehyde to MMHB and MVG cascade was, however, not found in the aldol addition. Due to the alcoholic solvent, glycolaldehyde dimethylacetal (GADMA, seen in Fig. 11) was formed immediately after dissolving the 2-carbon sugar. In order to allow aldol, hydrolysis of GADMA to unprotect the active carbonyl group is necessary and rate-determining. By adding small amounts of water to methanol (1:7), the reaction rate was enhanced almost fivefold. Finally, Dusselier et al. also

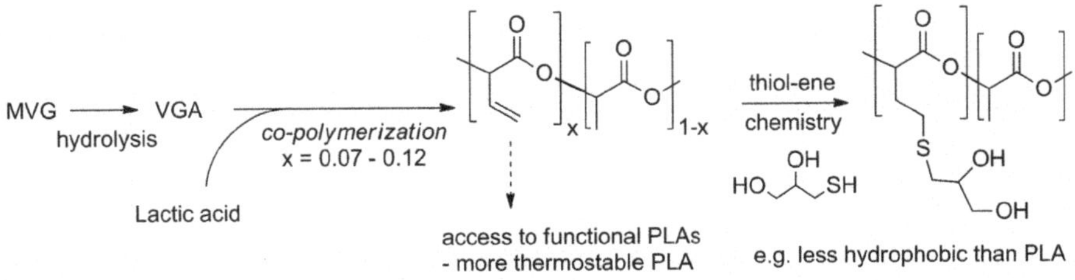

Fig. 18 Successful use of vinyl glycolic acid (VGA) as a monomer for PLA based co-polymers (via hydrolysis of MVG). The vinyl containing polymer was active for thiol-ene

presented the first proof of the usefulness of these monomers by incorporating vinyl glycolic acid (VGA, via hydrolysis of MVG) together with LA into a co-polymer by polycondensation, as seen in Fig. 18. The resulting polymer had an enhanced thermal stability, a function of the amount of VGA incorporated (denoted as x). The vinyl side group was preserved during the polymerization. This PLA polymer with vinyl side groups has tremendous potential, as it makes it easy for one to modify or create more complex PLA-based polyesters. The vinyl group proved, for instance, accessible to thiol-ene functionalization chemistry, which rendered a PLA with enhanced hydrophilic properties with respect to identically made pure L-PLA [148]. Moreover, VGA, with its F:C value of 1.25 and a theoretical atom economy of 85% (via glucose, considering retro-aldol and C2-aldol coupling [23]) is an excellent and viable biomass target chemical, which is difficult to prepare from fossil resources.

7.3 C6: Furyl Glycolic Acid

The group of Dumesic recently reported on a novel and inventive route for the renewable production of furyl glycolic acid (FA) [161]. This is a pseudo-aromatic AHA, likely suitable for co-polymerization with LA. An inspiring combination of both enzymatic and heterogeneous catalysis proved key in the new route (Fig. 19). In short, glucose is converted into cortalcerone via glucosone using recombinant *Escherichia coli* strains expressing pyranose-2-oxidase and aldos-2-ulose dehydratase in whole cell catalysis. Cortalcerone is then dehydrated towards furyl glyoxal hydrate over a Brønsted acidic Al-Beta zeolite. The final step in the hybrid cascade is the 1,2-hydride shift turning the hydrate into FA with the Lewis acidic Sn-Beta zeolite. The last two steps of this cascade were also attempted in one pot using a methanol/water mixture in presence of an Al-containing Sn-Beta zeolite. This material, possessing both Brønsted and Lewis acid sites, achieved 42% selectivity to FA at 53% cortalcerone conversion at 85°C in 0.5 h [161].

This example nicely illustrates the integration of bio- and chemocatalysis in creating novel added-value chemicals such as FA from biomass [162, 163]. FA has a very high functionality index of 1.33 and the reaction runs at moderate atom economy of 67% (see Chap. 1 of this volume [23]). Moreover, petrochemical-based

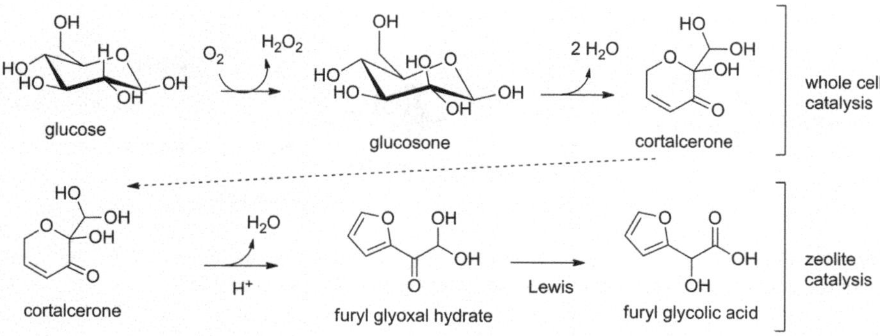

Fig. 19 Pathway for the conversion of glucose to furyl glycolic acid (FA), via enzymatic cortalcerone synthesis, Brønsted acid, and Lewis acid catalysis. Based on [161]

routes to this compound seem highly unlikely. The co-polymerization of FA with LA or GA should be looked into, as interesting polymer properties are bound to surface due to the unique position of the furan ring as side group in the PLA based co-polyester. Next to the aforementioned vinyl glycolic acid – with proven polymer potential – this monomer could present new opportunities for functional PLA or bio-derived polyesters.

8 Note on the Stereochemistry of Chemically Produced Lactates

The above-mentioned chemical catalytic routes lead to racemic AHA mixtures. For the direct use of LA (or its esters) as a solvent or platform molecule for achiral molecules like acrylic acid and pyruvic acid, stereochemistry does not matter. The properties of the polyester PLA, the major application of LA, however, suffer tremendously if D and L isomers are built in irregularly [28]. This is exemplified by atactic PLA, made from racemic LA, which is an amorphous polymer with low performance and limited application. However, when L- and D-lactic acid are processed separately into their respective isotactic L- and D-PLA, as discovered by Tsuji et al., a stereocomplex is formed upon blending these polymers. This polymer exhibits enhanced mechanical and thermal properties [28, 164]. A productive route to D-lactic acid is, however, missing today. If the chemocatalytic routes to LA are to become viable, enantiomer resolution of the racemate needs to be performed. Given separation success, a cheap source of D-lactic acid will be unlocked immediately, providing an additional advantage over the fermentation route (cfr. Table 2).

The separation has proved to be difficult, although methods based on supported liquid membranes with chiral selectors [165] and selective enzymatic oxidation to pyruvic acid have been reported [166]. Either the enantiomer selectivity is too low or the loss of one of the lactate isomers is limiting. To overcome this, our group

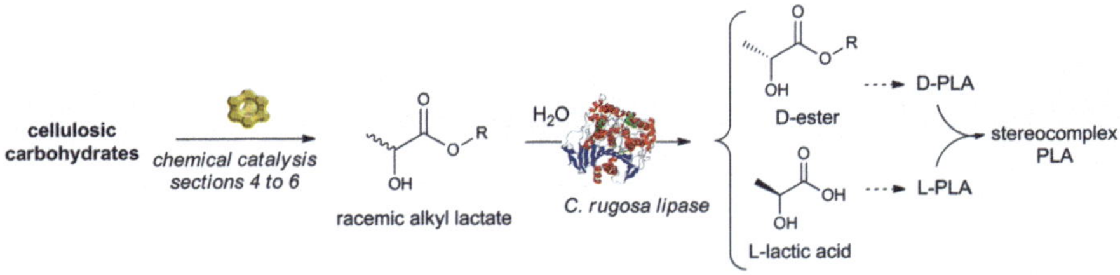

Fig. 20 Integration of catalytic production of racemic lactates (Sects. 4–6) with an enantioselective enzymatic hydrolysis as proposed by Van Wouwe et al. [167]

recently reported a very productive approach for enantioseparation, based on the hydrolysis of racemic lactates with *Candida rugosa* lipase, as seen in Fig. 20. Van Wouwe et al. reported volumetric productivities of about 18 g L^{-1} h^{-1} for ethyl lactate hydrolysis at 45°C after only 2 h, with the molar enantiomeric purity of the product at 95% (L-lactic acid divided by total LA). The separation of the D-ester from L-lactic acid in water is straightforward. The hydrolysis of a series of alkyl lactates and other AHA esters (based on GA or 4-carbon AHAs) was demonstrated [167].

9 Summary, Conclusions, Outlook

α-Hydroxy acids (AHA) are very useful molecules with a wide applicability, especially in the production of renewable and degradable polyesters, but also as a platform chemical in the biorefinery approach and in the portfolio of green solvents (e.g., ethyl lactate) and plasticizers (e.g., butyl glycolate). The F:C analysis in Sect. 2 showed that these molecules retain the high functionality present in the carbohydrate feedstock, and their synthesis proceeds according to a very high atom economy (100 % for lactic acid). The great potential of AHAs is exemplified by lactic acid, the most studied molecule of its kind. The rising demand for lactic acid is driven by the need for the above-mentioned green applications and, in particular, the ever-growing popularity of the commercial bioplastic PLA. This success, however, stresses the current fermentative production route for its synthesis, which requires long reaction times and costly purification steps, besides producing gypsum waste. Novel routes for the chemical synthesis of lactic acid or alkyl lactates starting from trioses, hexose sugars, and even cellulose are therefore progressively under development, and this progress is usually based on the use, design, and fine-tuning of Lewis and Brønsted acidic heterogeneous and homogeneous catalytic sites. The ultimate goal, the conversion of cellulose to lactic acid, requires many individual reaction steps in a one-pot cascade fashion and these steps face different catalytic needs. In this chapter we have reviewed and highlighted some of the most promising chemocatalytic approaches and pointed out the greatest bottleneck, namely combining a selective Lewis acid catalyzed retro-aldol of hexose sugars with the Brønsted acidic hydrolysis of cellulose. Care has to be taken with the latter acidic sites, as they usually also promote the dehydration of hexoses to HMF and humin formation. Besides LA, other AHA of interest include glycolic acid, vinyl glycolic acid, methoxy-2-hydroxy-butanoic acid, and furyl glycolic acid. These mostly novel molecules with high functionality are only just beginning to receive attention. Their bio-based production with inventive catalytic approaches, perhaps incorporating enzymatic catalysis, for instance, for enantioseparation, harnesses great potential. Besides, their synthesis from carbohydrates usually entails a high atom economy. The co-polymerization of vinyl glycolic acid with lactic acid showed that the search for new AHA monomers is worthwhile, as they are easily incorporated in PLA-based polymers, hereby leading to novel functional polyesters. The great advantage of having an additional functional group in the side chain of a PLA-type polyester was immediately proven, by its easy post-synthetic functionalization, which allowed subtle tuning of the hydrophilic properties of the polyester. A multitude of variations are yet about to be discovered.

Acknowledgements M.D. acknowledges FWO Vlaanderen (Research Foundation - Flanders) for a post-doctoral fellowship. B.F.S thanks the Research Council of the KU Leuven (IDO-3E090504) for financial support, as well as the Belgian government for its funding through IAP (Belspo).

References

1. Ragauskas AJ, Williams CK, Davison BH, Britovsek G, Cairney J, Eckert CA, Frederick WJ, Hallett JP, Leak DJ, Liotta CL, Mielenz JR, Murphy R, Templer R, Tschaplinski T (2006) The path forward for biofuels and biomaterials. Science 311:484–489
2. Vennestrøm PNR, Osmundsen CM, Christensen CH, Taarning E (2011) Beyond petrochemicals: the renewable chemicals industry. Angew Chem Int Ed 50(45):10502–10509
3. Climent MJ, Corma A, Iborra S (2014) Conversion of biomass platform molecules into fuel additives and liquid hydrocarbon fuels. Green Chem 16:516–547
4. Corma A, Iborra S, Velty A (2007) Chemical routes for the transformation of biomass into chemicals. Chem Rev 107:2411–2502
5. Ruppert AM, Weinberg K, Palkovits R (2012) Hydrogenolysis goes bio: from carbohydrates and sugar alcohols to platform chemicals. Angew Chem Int Ed 51(11):2564–2601
6. Kobayashi H, Fukuoka A (2013) Synthesis and utilisation of sugar compounds derived from lignocellulosic biomass. Green Chem 15(7):1740–1763
7. Serrano-Ruiz JC, Dumesic JA (2011) Catalytic routes for the conversion of biomass into liquid hydrocarbon transportation fuels. Energy Environ Sci 4(1):83–99
8. Van de Vyver S, Geboers J, Jacobs PA, Sels BF (2011) Recent advances in the catalytic conversion of cellulose. ChemCatChem 3(1):82–94
9. Geboers JA, Van de Vyver S, Ooms R, Op de Beeck B, Jacobs PA, Sels BF (2011) Chemocatalytic conversion of cellulose: opportunities, advances and pitfalls. Catal Sci Technol 1(5):714–726
10. Dusselier M, Van Wouwe P, Dewaele A, Makshina E, Sels BF (2013) Lactic acid as a platform chemical in the biobased economy: the role of chemocatalysis. Energy Environ Sci 6(5):1415–1442
11. Alonso DM, Wettstein SG, Dumesic JA (2013) Gamma-valerolactone, a sustainable platform molecule derived from lignocellulosic biomass. Green Chem 15(3):584–595

12. Alonso DM, Wettstein SG, Dumesic JA (2012) Bimetallic catalysts for upgrading of biomass to fuels and chemicals. Chem Soc Rev 41(24):8075–8098
13. Sheldon RA (2014) Green and sustainable manufacture of chemicals from biomass: state of the art. Green Chem 16:950–963
14. Besson M, Gallezot P, Pinel C (2013) Conversion of biomass into chemicals over metal catalysts. Chem Rev 114(3):1827–1870
15. Gallezot P (2012) Conversion of biomass to selected chemical products. Chem Soc Rev 41: 1538–1558
16. Lange J-P, van der Heide E, van Buijtenen J, Price R (2012) Furfural—a promising platform for lignocellulosic biofuels. ChemSusChem 5(1):150–166
17. Sheldon RA (2011) Utilisation of biomass for sustainable fuels and chemicals: molecules, methods and metrics. Catal Today 167(1):3–13
18. Kromus S, Kamm B, Kamm M, Fowler P, Narodoslawsky M (2008) Green biorefineries: the green biorefinery concept – fundamentals and potential. In: Kamm B, Kamm M, Gruber P (eds) Biorefineries-industrial processes and products. Wiley-VCH, Verlag GmbH, pp 253–294
19. Song J, Fan H, Ma J, Han B (2013) Conversion of glucose and cellulose into value-added products in water and ionic liquids. Green Chem 15(10):2619–2635
20. Kamm B, Kamm M, Gruber PR, Kromus S (2008) Biorefinery systems – an overview. In: Kamm B, Kamm M, Gruber P (eds) Biorefineries-industrial processes and products. Wiley-VCH, Verlag GmbH, pp 1–40
21. Kamm B (2007) Production of platform chemicals and synthesis gas from biomass. Angew Chem Int Ed 46(27):5056–5058
22. Centi G, van Santen RA (eds) (2007) Catalysis for renewables: from feedstock to energy production. Wiley-VCH, Weinheim
23. Dusselier M, Mascal M, Sels BF (2014) Top chemical opportunities from carbohydrate biomass – a chemist's view of the biorefinery. Top Curr Chem. doi:10.1007/128_2014_544
24. Bozell JJ, Petersen GR (2010) Technology development for the production of biobased products from biorefinery carbohydrates-the US Department of Energy's "Top 10" revisited. Green Chem 12:539–554
25. Chahal SP, Starr JN (2000) Lactic acid. In: Ullmann's encyclopedia of industrial chemistry. Wiley-VCH, Weinheim
26. Nattrass L, Higson A (2010). National Non-Food Crops Centre, Renewable chemicals factsheet: lactic acid. http://www.nnfcc.co.uk/publications/nnfcc-renewable-chemicals-factsheet-lactic-acid
27. Castillo Martinez FA, Balciunas EM, Salgado JM, Domínguez González JM, Converti A, Oliveira RPDS (2013) Lactic acid properties, applications and production: a review. Trends Food Sci Technol 30(1):70–83
28. Auras R, Lim LT, Selke SEM, Tsuji H (eds) (2010) Poly(lactic acid): synthesis, structures, properties, processing, and applications. John Wiley & Sons, Inc., Hoboken, New Jersey
29. Fan Y, Zhou C, Zhu X (2009) Selective catalysis of lactic acid to produce commodity chemicals. Catal Rev Sci Eng 51:293–324
30. Peng J, Li X, Tang C, Bai W (2014) Barium sulphate catalyzed dehydration of lactic acid to acrylic acid. Green Chem 16(1):108–111
31. Zhang J, Zhao Y, Pan M, Feng X, Ji W, Au C-T (2011) Efficient acrylic acid production through bio lactic acid dehydration over NaY zeolite modified by alkali phosphates. ACS Catal 1:32–41
32. Sun P, Yu D, Tang Z, Li H, Huang H (2010) NaY zeolites catalyze dehydration of lactic acid to acrylic acid: studies on the effects of anions in potassium salts. Ind Eng Chem Res 49:9082–9087
33. Sun P, Yu D, Fu K, Gu M, Wang Y, Huang H, Ying H (2009) Potassium modified NaY: a selective and durable catalyst for dehydration of lactic acid to acrylic acid. Catal Commun 10:1345–1349

34. Holmen RE (1958) Acrylates by catalytic dehydration of lactic acid and lactates. US Patent 2,859,240
35. Ohara T, Sato T, Shimizu N, Prescher G, Schwind H, Weiberg O, Marten K, Greim H (2000) Acrylic acid and derivatives. In: Ullmann's encyclopedia of industrial chemistry. Wiley-VCH, Weinheim
36. Gunter GC, Miller DJ, Jackson JE (1994) Formation of 2,3-pentanedione from lactic acid over supported phosphate catalysts. J Catal 148(1):252–260
37. Katryniok B, Paul S, Dumeignil F (2010) Highly efficient catalyst for the decarbonylation of lactic acid to acetaldehyde. Green Chem 12(11):1910–1913
38. Tam MS, Craciun R, Miller DJ, Jackson JE (1998) Reaction and kinetic studies of lactic acid conversion over alkali-metal salts. Ind Eng Chem Res 37(6):2360–2366
39. Lambrecht S, Franke O, Zahlmann K (2003) Preparation of 2,3-pentanedione by reacting hydroxyacetone with paraldehyde in the presence of a phase transfer catalyst and an acid. EP Patent 1,310,476A1
40. Sels B, D'Hondt E, Jacobs P (2007) Catalytic transformation of glycerol. In: Catalysis for renewables. Wiley-VCH, Weinheim, pp 223–255
41. Vu DT, Kolah AK, Asthana NS, Peereboom L, Lira CT, Miller DJ (2005) Oligomer distribution in concentrated lactic acid solutions. Fluid Phase Equilib 236:125–135
42. Inkinen S, Hakkarainen M, Albertsson A-C, Sodergard A (2011) From lactic acid to poly (lactic acid) (PLA): characterization and analysis of PLA and its precursors. Biomacromolecules 12:523–532
43. Pereira CSM, Silva VMTM, Rodrigues AE (2011) Ethyl lactate as a solvent: properties, applications and production processes – a review. Green Chem 13:2658–2671
44. Aparicio S, Alcalde R (2009) The green solvent ethyl lactate: an experimental and theoretical characterization. Green Chem 11(1):65–78
45. Cortright RD, Sanchez-Castillo M, Dumesic JA (2002) Conversion of biomass to 1,2-propanediol by selective catalytic hydrogenation of lactic acid over silica-supported copper. Appl Catal B 39:353–359
46. Sullivan CJ (2000) Propanediols. In: Ullmann's encyclopedia of industrial chemistry. Wiley-VCH, Weinheim
47. Gao C, Ma C, Xu P (2011) Biotechnological routes based on lactic acid production from biomass. Biotechnol Adv 29:930–939
48. Xu P, Qiu J, Gao C, Ma C (2008) Biotechnological routes to pyruvate production. J Biosci Bioeng 105(3):169–175
49. Gross RA, Kalra B (2002) Biodegradable polymers for the environment. Science 297(5582): 803–807
50. Drumright RE, Gruber PR, Henton DE (2000) Polylactic acid technology. Adv Mater 12: 1841–1846
51. Gruber P, Henton DE, Starr J (2008) Polylactic acid from renewable resources. In: Kamm B, Kamm M, Gruber P (eds) Biorefineries-industrial processes and products. Wiley-VCH, Verlag GmbH, pp 381–407
52. Rasal RM, Janorkar AV, Hirt DE (2010) Poly(lactic acid) modifications. Prog Polym Sci 35(3):338–356
53. Carus M (2012) Growth in PLA bioplastics: a production capacity of over 800,000 tonnes expected by 2020. Nova-Institute, Hürth
54. European Bioplastics (2012) Driving the evolution of plastics. http://en.european-bioplastics.org/multimedia/
55. Groot W, van Krieken J, Sliekersl O, de Vos S (2010) Production and purification of lactic acid and lactide. In: Poly(lactic acid): synthesis, structures properties, processing, and applications, John Wiley & Sons, Inc., Hoboken, New Jersey, pp 1–18
56. John RP, Anisha GS, Nampoothiri KM, Pandey A (2009) Direct lactic acid fermentation: focus on simultaneous saccharification and lactic acid production. Biotechnol Adv 27(2): 145–152

57. Datta R, Henry M (2006) Lactic acid: recent advances in products, processes and technologies – a review. J Chem Technol Biotechnol 81(7):1119–1129
58. Okano K, Tanaka T, Ogino C, Fukuda H, Kondo A (2010) Biotechnological production of enantiomeric pure lactic acid from renewable resources: recent achievements, perspectives, and limits. Appl Microbiol Biotechnol 85:413–423
59. Dechy-Cabaret O, Martin-Vaca B, Bourissou D (2004) Controlled ring-opening polymerization of lactide and glycolide. Chem Rev 104(12):6147–6176
60. Asthana NS, Kolah AK, Vu DT, Lira CT, Miller DJ (2006) A kinetic model for the esterification of lactic acid and its oligomers. Ind Eng Chem Res 45:5251–5257
61. Kim KW, Woo SI (2002) Synthesis of high-molecular-weight poly(L-lactic acid) by direct polycondensation. Macromol Chem Phys 203(15):2245–2250
62. Vijayakumar J, Aravindan R, Viruthagiri T (2008) Recent trends in the production, purification and application of lactic acid. Chem Biochem Eng Q 22:245–264
63. Wee Y-J, Kim J-N, Ryu H-W (2006) Biotechnological production of lactic acid and its recent applications. Food Technol Biotechnol 44:163–172
64. Adsul MG, Varma AJ, Gokhale DV (2007) Lactic acid production from waste sugarcane bagasse derived cellulose. Green Chem 9(1):58–62
65. Taarning E, Saravanamurugan S, Spangsberg HM, Xiong J, West RM, Christensen CH (2009) Zeolite-catalyzed isomerization of triose sugars. ChemSusChem 2:625–627
66. de Clippel F, Dusselier M, Van Rompaey R, Vanelderen P, Dijkmans J, Makshina E, Giebeler L, Oswald S, Baron GV, Denayer JFM, Pescarmona PP, Jacobs PA, Sels BF (2012) Fast and selective sugar conversion to alkyl lactate and lactic acid with bifunctional carbon–silica catalysts. J Am Chem Soc 134(24):10089–10101
67. Serrano-Ruiz JC, Dumesic JA (2009) Catalytic processing of lactic acid over Pt/Nb_2O_5. ChemSusChem 2(6):581–586
68. Simonov MN, Zaikin PA, Simakova IL (2012) Highly selective catalytic propylene glycol synthesis from alkyl lactate over copper on silica: performance and mechanism. Appl Catal B 119–120:340–347
69. Ai M (2002) Catalytic activity of iron phosphate doped with a small amount of molybdenum in the oxidative dehydrogenation of lactic acid to pyruvic acid. Appl Catal A 234(1–2): 235–243
70. Nishiyama Y, Sugiyama J, Chanzy H, Langan P (2003) Crystal structure and hydrogen bonding system in cellulose Iα from synchrotron X-ray and neutron fiber diffraction. J Am Chem Soc 125(47):14300–14306
71. Geboers J, Van de Vyver S, Carpentier K, Jacobs P, Sels B (2011) Efficient hydrolytic hydrogenation of cellulose in the presence of Ru-loaded zeolites and trace amounts of mineral acid. Chem Commun 47(19):5590–5592
72. Van de Vyver S, Geboers J, Dusselier M, Schepers H, Vosch T, Zhang L, Van Tendeloo G, Jacobs PA, Sels BF (2010) Selective bifunctional catalytic conversion of cellulose over reshaped Ni particles at the tip of carbon nanofibers. ChemSusChem 3(6):698–701
73. Van de Vyver S, Geboers J, Schutyser W, Dusselier M, Eloy P, Dornez E, Seo JW, Courtin CM, Gaigneaux EM, Jacobs PA, Sels BF (2012) Tuning the acid/metal balance of carbon nanofiber-supported nickel catalysts for hydrolytic hydrogenation of cellulose. ChemSusChem 5(8):1549–1558
74. Kobayashi H, Ito Y, Komanoya T, Hosaka Y, Dhepe PL, Kasai K, Hara K, Fukuoka A (2011) Synthesis of sugar alcohols by hydrolytic hydrogenation of cellulose over supported metal catalysts. Green Chem 13(2):326–333
75. Palkovits R, Tajvidi K, Ruppert AM, Procelewska J (2011) Heteropoly acids as efficient acid catalysts in the one-step conversion of cellulose to sugar alcohols. Chem Commun 47(1): 576–578
76. Pang J, Wang A, Zheng M, Zhang Y, Huang Y, Chen X, Zhang T (2012) Catalytic conversion of cellulose to hexitols with mesoporous carbon supported Ni-based bimetallic catalysts. Green Chem 14(3):614–617

77. Geboers J, Van de Vyver S, Carpentier K, Jacobs P, Sels B (2011) Hydrolytic hydrogenation of cellulose with hydrotreated caesium salts of heteropoly acids and Ru/C. Green Chem 13(8):2167–2174
78. Meine N, Rinaldi R, Schüth F (2012) Solvent-free catalytic depolymerization of cellulose to water-soluble oligosaccharides. ChemSusChem 5(8):1449–1454
79. Benoit M, Rodrigues A, Zhang Q, Fourré E, De Oliveira VK, Tatibouët J-M, Jérôme F (2011) Depolymerization of cellulose assisted by a nonthermal atmospheric plasma. Angew Chem Int Ed 50(38):8964–8967
80. Rinaldi R, Engel P, Büchs J, Spiess AC, Schüth F (2010) An integrated catalytic approach to fermentable sugars from cellulose. ChemSusChem 3(10):1151–1153
81. Rinaldi R, Schüth F (2009) Acid hydrolysis of cellulose as the entry point into biorefinery schemes. ChemSusChem 2(12):1096–1107
82. Onda A, Ochi T, Yanagisawa K (2008) Selective hydrolysis of cellulose into glucose over solid acid catalysts. Green Chem 10(10):1033–1037
83. Van de Vyver S, Peng L, Geboers J, Schepers H, de Clippel F, Gommes CJ, Goderis B, Jacobs PA, Sels BF (2010) Sulfonated silica/carbon nanocomposites as novel catalysts for hydrolysis of cellulose to glucose. Green Chem 12(9):1560–1563
84. Huang Y-B, Fu Y (2013) Hydrolysis of cellulose to glucose by solid acid catalysts. Green Chem 15(5):1095–1111
85. Shimizu K-I, Furukawa H, Kobayashi N, Itaya Y, Satsuma A (2009) Effects of Bronsted and Lewis acidities on activity and selectivity of heteropolyacid-based catalysts for hydrolysis of cellobiose and cellulose. Green Chem 11(10):1627–1632
86. Luo C, Wang S, Liu H (2007) Cellulose conversion into polyols catalyzed by reversibly formed acids and supported ruthenium clusters in hot water. Angew Chem Int Ed 46(40): 7636–7639
87. Geboers J, Van de Vyver S, Carpentier K, de Blochouse K, Jacobs P, Sels B (2010) Efficient catalytic conversion of concentrated cellulose feeds to hexitols with heteropoly acids and Ru on carbon. Chem Commun 46(20):3577–3579
88. Kobayashi H, Komanoya T, Hara K, Fukuoka A (2010) Water-tolerant mesoporous-carbon-supported ruthenium catalysts for the hydrolysis of cellulose to glucose. ChemSusChem 3(4):440–443
89. Op de Beeck B, Geboers J, Van de Vyver S, Van Lishout J, Snelders J, Huijgen WJJ, Courtin CM, Jacobs PA, Sels BF (2013) Conversion of (ligno)cellulose feeds to isosorbide with heteropoly acids and Ru on carbon. ChemSusChem 6(1):199–208
90. Sun P, Long X, He H, Xia C, Li F (2013) Conversion of cellulose into isosorbide over bifunctional ruthenium nanoparticles supported on niobium phosphate. ChemSusChem 6(11):2190–2197
91. Liang G, Wu C, He L, Ming J, Cheng H, Zhuo L, Zhao F (2011) Selective conversion of concentrated microcrystalline cellulose to isosorbide over Ru/C catalyst. Green Chem 13(4): 839–842
92. Weingarten R, Conner WC, Huber GW (2012) Production of levulinic acid from cellulose by hydrothermal decomposition combined with aqueous phase dehydration with a solid acid catalyst. Energy Environ Sci 5(6):7559–7574
93. Van de Vyver S, Thomas J, Geboers J, Keyzer S, Smet M, Dehaen W, Jacobs PA, Sels BF (2011) Catalytic production of levulinic acid from cellulose and other biomass-derived carbohydrates with sulfonated hyperbranched poly(arylene oxindole)s. Energy Environ Sci 4(9):3601–3610
94. Deng W, Liu M, Zhang Q, Tan X, Wang Y (2010) Acid-catalysed direct transformation of cellulose into methyl glucosides in methanol at moderate temperatures. Chem Commun 46(15):2668–2670
95. Tominaga K-I, Mori A, Fukushima Y, Shimada S, Sato K (2011) Mixed-acid systems for the catalytic synthesis of methyl levulinate from cellulose. Green Chem 13(4):810–812

96. Roman-Leshkov Y, Moliner M, Labinger JA, Davis ME (2010) Mechanism of glucose isomerization using a solid Lewis acid catalyst in water. Angew Chem Int Ed 49:8954–8957
97. Assary RS, Curtiss LA (2011) Theoretical study of 1,2-hydride shift associated with the isomerization of glyceraldehyde to dihydroxyacetone by Lewis acid active site models. J Phys Chem A 115:8754–8760
98. Bermejo-Deval R, Assary RS, Nikolla E, Moliner M, Román-Leshkov Y, Hwang S-J, Palsdottir A, Silverman D, Lobo RF, Curtiss LA, Davis ME (2012) Metalloenzyme-like catalyzed isomerizations of sugars by Lewis acid zeolites. Proc Natl Acad Sci U S A 109(25): 9727–9732
99. Moliner M, Roman-Leshkov Y, Davis ME (2010) Tin-containing zeolites are highly active catalysts for the isomerization of glucose in water. Proc Natl Acad Sci U S A 107:6164–6168
100. Dijkmans J, Gabriels D, Dusselier M, de Clippel F, Vanelderen P, Houthoofd K, Malfliet A, Pontikes Y, Sels BF (2013) Productive sugar isomerization with highly active Sn in dealuminated [small beta] zeolites. Green Chem 15(10):2777–2785
101. Choudhary V, Pinar AB, Lobo RF, Vlachos DG, Sandler SI (2013) Comparison of homogeneous and heterogeneous catalysts for glucose-to-fructose isomerization in aqueous media. ChemSusChem 6:2369–2376
102. Madigan M, Martinko J, Stahl D, Clark D (2010) Brock biology of microorganisms, 13th edn. Benjamin Cummings, San Francisco
103. Giger L, Caner S, Obexer R, Kast P, Baker D, Ban N, Hilvert D (2013) Evolution of a designed retro-aldolase leads to complete active site remodeling. Nat Chem Biol 9(8): 494–498
104. Jiang L, Althoff EA, Clemente FR, Doyle L, Röthlisberger D, Zanghellini A, Gallaher JL, Betker JL, Tanaka F, Barbas CF, Hilvert D, Houk KN, Stoddard BL, Baker D (2008) De novo computational design of retro-aldol enzymes. Science 319(5868):1387–1391
105. Aida TM, Tajima K, Watanabe M, Saito Y, Kuroda K, Nonaka T, Hattori H, Smith RL Jr, Arai K (2007) Reactions of d-fructose in water at temperatures up to 400°C and pressures up to 100 MPa. J Supercrit Fluids 42(1):110–119
106. Jin F, Enomoto H (2011) Rapid and highly selective conversion of biomass into value-added products in hydrothermal conditions: chemistry of acid/base-catalysed and oxidation reactions. Energy Environ Sci 4(2):382–397
107. Holm MS, Saravanamurugan S, Taarning E (2010) Conversion of sugars to lactic acid derivatives using heterogeneous zeotype catalysts. Science 328:602–605
108. Taarning E, Osmundsen CM, Yang X, Voss B, Andersen SI, Christensen CH (2011) Zeolite-catalyzed biomass conversion to fuels and chemicals. Energy Environ Sci 4(3):793–804
109. De SK, Gibbs RA (2004) Ruthenium(III) chloride-catalyzed chemoselective synthesis of acetals from aldehydes. Tetrahedron Lett 45(44):8141–8144
110. Pescarmona PP, Janssen KPF, Delaet C, Stroobants C, Houthoofd K, Philippaerts A, De Jonghe C, Paul JS, Jacobs PA, Sels BF (2010) Zeolite-catalysed conversion of C3 sugars to alkyl lactates. Green Chem 12:1083–1089
111. Smith MB, March J (2007) March's advanced organic chemistry: reactions, mechanisms, and structure, 7 edn. John Wiley & Sons, Inc., Hoboken, New Jersey
112. Hayashi Y, Sasaki Y (2005) Tin-catalyzed conversion of trioses to alkyl lactates in alcohol solution. Chem Commun 21:2716–2718
113. Li L, Stroobants C, Lin K, Jacobs PA, Sels BF, Pescarmona PP (2011) Selective conversion of trioses to lactates over Lewis acid heterogeneous catalysts. Green Chem 13:1175–1181
114. Dusselier M, Van Wouwe P, de Clippel F, Dijkmans J, Gammon DW, Sels BF (2013) Mechanistic insight into the conversion of tetrose sugars to novel α-hydroxy acid platform molecules. ChemCatChem 5(2):569–575
115. Painter RM, Pearson DM, Waymouth RM (2010) Selective catalytic oxidation of glycerol to dihydroxyacetone. Angew Chem Int Ed 49(49):9456–9459
116. Eriksen J, Monsted O, Monsted L (1998) Mechanism of lactic acid formation catalyzed by tetraamine rhodium(III) complexes. Transition Met Chem 23:783–787

117. Kelly RL (1991) Production of hydroxy carboxylic compounds. EU Patent 0460,831A2
118. Rasrendra CB, Fachri BA, Makertihartha IGBN, Adisasmito S, Heeres HJ (2011) Catalytic conversion of dihydroxyacetone to lactic acid using metal salts in water. ChemSusChem 4:768–777
119. Lux S, Siebenhofer M (2013) Synthesis of lactic acid from dihydroxyacetone: use of alkaline-earth metal hydroxides. Catal Sci Technol 3(5):1380–1385
120. Janssen KPF, Paul JS, Sels BF, Jacobs PA (2007) Glyoxylase biomimics: zeolite catalyzed conversion of trioses. Stud Surf Sci Catal 170B:1222–1227
121. West RM, Holm MS, Saravanamurugan S, Xiong J, Beversdorf Z, Taarning E, Christensen CH (2010) Zeolite H-USY for the production of lactic acid and methyl lactate from C3-sugars. J Catal 269:122–130
122. Osmundsen CM, Holm MS, Dahl S, Taarning E (2012) Tin-containing silicates: structure–activity relations. Proc R Soc A Math Phys Eng Sci 468(2143):2000–2016
123. Lew CM, Rajabbeigi N, Tsapatsis M (2012) Tin-containing zeolite for the isomerization of cellulosic sugars. Microporous Mesoporous Mater 153:55–58
124. Wang J, Masui Y, Onaka M (2011) Conversion of triose sugars with alcohols to alkyl lactates catalyzed by Bronsted acid tin ion-exchanged montmorillonite. Appl Catal B 107:135–139
125. Guo Q, Fan F, Pidko EA, van der Graaff WNP, Feng Z, Li C, Hensen EJM (2013) Highly active and recyclable Sn-MWW zeolite catalyst for sugar conversion to methyl lactate and lactic acid. ChemSusChem 6(8):1352–1356
126. Dapsens PY, Mondelli C, Pérez-Ramírez J (2013) Highly selective Lewis acid sites in desilicated MFI zeolites for dihydroxyacetone isomerization to lactic acid. ChemSusChem 6(5):831–839
127. Dapsens PY, Menart MJ, Mondelli C, Perez-Ramirez J (2014) Production of bio-derived ethyl lactate on GaUSY zeolites prepared by post-synthetic galliation. Green Chem 16: 589–593
128. Dapsens PY, Kusema BT, Mondelli C, Pérez-Ramírez J (2013) Gallium-modified zeolites for the selective conversion of bio-based dihydroxyacetone into C1–C4 alkyl lactates. J Mol Catal A Chem. http://dx.doi.org/10.1016/j.molcata.2013.09.032
129. Hammond C, Conrad S, Hermans I (2012) Simple and scalable preparation of highly active Lewis acidic Sn-β. Angew Chem Int Ed 51(47):11736–11739
130. IZA-Structure-Commission. database of zeolite structures. http://izasc.fos.su.se/fmi/xsl/IZA-SC/ft.xsl. Accessed 23 Jan 2014
131. de Clippel F, Dusselier M, Van de Vyver S, Peng L, Jacobs PA, Sels BF (2013) Tailoring nanohybrids and nanocomposites for catalytic applications. Green Chem 15(6):1398–1430
132. Lobo RF (2010) Synthetic glycolysis. ChemSusChem 3(11):1237–1240
133. Blunden SJ, Cusack PA, Smith PJ (1987) The use of tin compounds in carbohydrate and nucleoside chemistry. J Organomet Chem 325:141–152
134. Roman-Leshkov Y, Davis ME (2011) Activation of carbonyl-containing molecules with solid Lewis acids in aqueous media. ACS Catal 1:1566–1580
135. Nikolla E, Román-Leshkov Y, Moliner M, Davis ME (2011) “One-pot” synthesis of 5-(hydroxymethyl)furfural from carbohydrates using tin-beta zeolite. ACS Catal 1(4): 408–410
136. Hurtta M, Pitkänen I, Knuutinen J (2004) Melting behaviour of D-sucrose, D-glucose and D-fructose. Carbohydr Res 339(13):2267–2273
137. Ji N, Zhang T, Zheng M, Wang A, Wang H, Wang X, Chen JG (2008) Direct catalytic conversion of cellulose into ethylene glycol using nickel-promoted tungsten carbide catalysts. Angew Chem 120(44):8638–8641
138. Zhao G, Zheng M, Zhang J, Wang A, Zhang T (2013) Catalytic conversion of concentrated glucose to ethylene glycol with semicontinuous reaction system. Ind Eng Chem Res 52(28): 9566–9572
139. Ooms R, Dusselier M, Geboers JA, Op de Beeck B, Verhaeven R, Gobechiya E, Martens J, Redl A, Sels BF (2014) Conversion of sugars to ethylene glycol with nickel tungsten carbide

in a fed-batch reactor: high productivity and reaction network elucidation. Green Chem 16:695–707
140. Murillo B, Sánchez A, Sebastián V, Casado-Coterillo C, de la Iglesia O, López-Ram- de-Viu MP, Téllez C, Coronas J (2013) Conversion of glucose to lactic acid derivatives with mesoporous Sn-MCM-41 and microporous titanosilicates. J Chem Technol Biot. doi:10.1002/jctb.4210
141. Liu Z, Li W, Pan C, Chen P, Lou H, Zheng X (2011) Conversion of biomass-derived carbohydrates to methyl lactate using solid base catalysts. Catal Commun 15:82–87
142. van Zandvoort I, Wang Y, Rasrendra CB, van Eck ERH, Bruijnincx PCA, Heeres HJ, Weckhuysen BM (2013) Formation, molecular structure, and morphology of humins in biomass conversion: influence of feedstock and processing conditions. ChemSusChem 6(9):1745–1758
143. Patil SKR, Heltzel J, Lund CRF (2012) Comparison of structural features of humins formed catalytically from glucose, fructose, and 5-hydroxymethylfurfuraldehyde. Energy Fuel 26(8): 5281–5293
144. Rasrendra CB, Makertihartha IGBN, Adisasmito S, Heeres HJ (2010) Green chemicals from D-glucose: systematic studies on catalytic effects of inorganic salts on the chemo-selectivity and yield in aqueous solutions. Top Catal 53:1241–1247
145. Wang F-F, Liu C-L, Dong W-S (2013) Highly efficient production of lactic acid from cellulose using lanthanide triflate catalysts. Green Chem 15(8):2091–2095
146. Wang Y, Deng W, Wang B, Zhang Q, Wan X, Tang Z, Wang Y, Zhu C, Cao Z, Wang G, Wan H (2013) Chemical synthesis of lactic acid from cellulose catalysed by lead(II) ions in water. Nat Commun 4:2141
147. Holm MS, Pagan-Torres YJ, Saravanamurugan S, Riisager A, Dumesic JA, Taarning E (2012) Sn-Beta catalysed conversion of hemicellulosic sugars. Green Chem 14(3):702–706
148. Dusselier M, Van Wouwe P, De Smet S, De Clercq R, Verbelen L, Van Puyvelde P, Du Prez FE, Sels BF (2013) Toward functional polyester building blocks from renewable glycolaldehyde with Sn cascade catalysis. ACS Catal 3:1786–1800
149. dos Santos JB, da Silva FL, Altino FMRS, da Silva Moreira T, Meneghetti MR, Meneghetti SMP (2013) Cellulose conversion in the presence of catalysts based on Sn(IV). Catal Sci Technol 3(3):673–678
150. Chambon F, Rataboul F, Pinel C, Cabiac A, Guillon E, Essayem N (2011) Cellulose hydrothermal conversion promoted by heterogeneous Brønsted and Lewis acids: remarkable efficiency of solid Lewis acids to produce lactic acid. Appl Catal B 105(1–2):171–181
151. Carrasquillo-Flores R, Käldström M, Schüth F, Dumesic JA, Rinaldi R (2013) Mechanocatalytic depolymerization of dry (ligno)cellulose as an entry process for high-yield production of furfurals. ACS Catal 3(5):993–997
152. Hilgert J, Meine N, Rinaldi R, Schuth F (2013) Mechanocatalytic depolymerization of cellulose combined with hydrogenolysis as a highly efficient pathway to sugar alcohols. Energy Environ Sci 6(1):92–96
153. Zhang Q, Jérôme F (2013) Mechanocatalytic deconstruction of cellulose: an emerging entry into biorefinery. ChemSusChem 6(11):2042–2044
154. Dapsens PY, Mondelli C, Kusema B, Verel R, Perez-Ramirez J (2013) Continuous process for glyoxal valorisation using tailored Lewis-acid zeolite catalysts. Green Chem
155. Miltenberger K (2000) Hydroxycarboxylic acids, aliphatic. In: Ullmann's encyclopedia of industrial chemistry. Wiley-VCH, Weinheim
156. Zhang J, Liu X, Sun M, Ma X, Han Y (2012) Direct conversion of cellulose to glycolic acid with a phosphomolybdic acid catalyst in a water medium. ACS Catal 2(8):1698–1702
157. Mattioda G, Blanc A (2000) Glyoxal. In: Ullmann's encyclopedia of industrial chemistry. Wiley-VCH, Weinheim
158. Vinu R, Broadbelt LJ (2012) A mechanistic model of fast pyrolysis of glucose-based carbohydrates to predict bio-oil composition. Energy Environ Sci 5(12):9808–9826

159. Richards GN (1987) Glycolaldehyde from pyrolysis of cellulose. J Anal Appl Pyrolysis 10: 251–255
160. Vitasari CR, Meindersma GW, de Haan AB (2012) Laboratory scale conceptual process development for the isolation of renewable glycolaldehyde from pyrolysis oil to produce fermentation feedstock. Green Chem 14:321–325
161. Schwartz TJ, Goodman SM, Osmundsen CM, Taarning E, Mozuch MD, Gaskell J, Cullen D, Kersten PJ, Dumesic JA (2013) Integration of chemical and biological catalysis: production of furylglycolic acid from glucose via cortalcerone. ACS Catal 3(12):2689–2693
162. Anbarasan P, Baer ZC, Sreekumar S, Gross E, Binder JB, Blanch HW, Clark DS, Toste FD (2012) Integration of chemical catalysis with extractive fermentation to produce fuels. Nature 491(7423):235–239
163. Vennestrøm PNR, Taarning E, Christensen CH, Pedersen S, Grunwaldt J-D, Woodley JM (2010) Chemoenzymatic combination of glucose oxidase with titanium silicalite-1. ChemCatChem 2(8):943–945
164. Tsuji H (2005) Poly(lactide) stereocomplexes: formation, structure, properties, degradation, and applications. Macromol Biosci 5:569–597
165. Yang Q, Chung T-S (2007) Modification of the commercial carrier in supported liquid membrane system to enhance lactic acid flux and to separate L, D-lactic acid enantiomers. J Membr Sci 294:127–131
166. Gao C, Qiu J, Li J, Ma C, Tang H, Xu P (2009) Enantioselective oxidation of racemic lactic acid to D-lactic acid and pyruvic acid by Pseudomonas stutzeri SDM. Bioresour Technol 100(5):1878–1880
167. Van Wouwe P, Dusselier M, Basic A, Sels BF (2013) Bridging racemic lactate esters with stereoselective polylactic acid using commercial lipase catalysis. Green Chem 15(10): 2817–2824
168. Schutyser W et al (2014) Regioselective synthesis of renewable bisphenols from 2,3-pentanedione and their application as plasticizers. Green Chem 16(4):1999–2007
169. Shen L, Worrell E, Patel M (2010) Present and future development in plastics from biomass. Biofuels Bioprod Bioref 4(1):25–40
170. Bicker M, Endres S, Ott L, Vogel H (2005) Catalytical conversion of carbohydrates in subcritical water: a new chemical process for lactic acid production. J Mol Catal A Chem 239:151–157
171. Esposito D, Antonietti M (2013) Chemical conversion of sugars to lactic acid by alkaline hydrothermal processes. ChemSusChem 6:989–992

Top Curr Chem (2014) 353: 127–162
DOI: 10.1007/128_2014_538

Published online: 4 April 2014

Selective Hydrogenolysis of C–O Bonds Using the Interaction of the Catalyst Surface and OH Groups

Keiichi Tomishige, Yoshinao Nakagawa, and Masazumi Tamura

Abstract Hydrogenolysis of C–O bonds is becoming more and more important for the production of biomass-derived chemicals. Since substrates originated from biomass usually have high oxygen content and various kinds of C–O bonds, selective hydrogenolysis is required. Rhenium or molybdenum oxide modified rhodium and iridium metal catalysts (Rh-ReO_x, Rh-MoO_x, and Ir-ReO_x) have been reported to be effective for selective hydrogenolysis. This review introduces the catalytic performance and reaction kinetics of Rh-ReO_x, Rh-MoO_x, and Ir-ReO_x in the hydrogenolysis of various substrates, where selectivity is especially characteristic. Based the model structure of the catalysts and the reaction mechanism, the role of the oxide components is to make the interaction between the OH groups in the substrates and the catalyst surface, and the role of metal components is to dissociate hydrogen molecule heterolytically to give hydride and proton.

Keywords Alkane · Biomass · Conformation · Cyclic ether · Diol · Hydride · Hydrogenolysis · Ir · Mo · Polyol · Re · Rh · S_N2

Contents

K. Tomishige (✉), Y. Nakagawa, and M. Tamura
Department of Applied Chemistry, School of Engineering, Tohoku University, Sendai, Japan
e-mail: tomi@erec.che.tohoku.ac.jp

1 Introduction

1.1 Production of Biomass-Derived Chemicals

A lot of chemicals have been produced from petroleum-based raw materials such as ethylene, propylene, benzene, *p*-xylene, and so on. The chemical composition of the feedstock from fossil and renewable resources is plotted in Fig. 1. In addition, the composition of some diols is also plotted as an example of the useful oxygen-containing chemicals "oxygenates." Since petroleum-based raw materials have low oxygen content, oxygenates are produced by oxidation with air and hydration. On the other hand, biomass-derived cellulose, hemi-cellulose, glucose, and glycerol have very high oxygen content as shown in Fig. 1. Therefore, in the case of diols being produced from biomass-related raw materials, it is essential to decrease the oxygen content using reductive reactions [1]. The composition of the building blocks for biomass refineries proposed by the US Department of Energy [2] is also plotted in Fig. 1. The typical building blocks for the biomass refinery are 5-hydroxymethyl-2-furaldehyde (HMF) and furfural, which can be derived from cellulose and hemi-cellulose by dehydration. The dehydration reaction decreases both H/C and O/C at the same time in the substrate. When HMF and furfural are converted to 1,6-hexanediol and 1,5-pentanediol, hydrogenation and hydrogenolysis are needed. The biomass with high oxygen content can be completely converted to useful chemicals by dehydration and subsequent hydrogenation/hydrogenolysis via intermediates. Another example is the conversion of glycerol, which is derived from the biodiesel production by the transesterification of vegetable oils with methanol [3]. A value-added target of glycerol conversion is propanediols by the hydrogenolysis of C–O bonds. Hydrogenolysis is one of the important reactions for the conversion of carbohydrates and sugar alcohols to platform chemicals [4].

This review focuses on the catalytic hydrogenolysis of C–O bonds in biomass-derived substrates, in particular sugar alcohols and cyclic ethers including glycerol, erythritol, xylitol, sorbitol, tetrahydrofurfuryl alcohol, and so on. The catalytic hydrogenolysis of C–O bonds has been studied, recently, new catalyst systems consisting of noble metals (Rh and Ir) and metal oxides (Re, Mo, and W) have been reported to be effective in hydrogenolysis utilizing the interaction between the catalyst surface and the substrate OH groups, and it is also suggested that the C–O

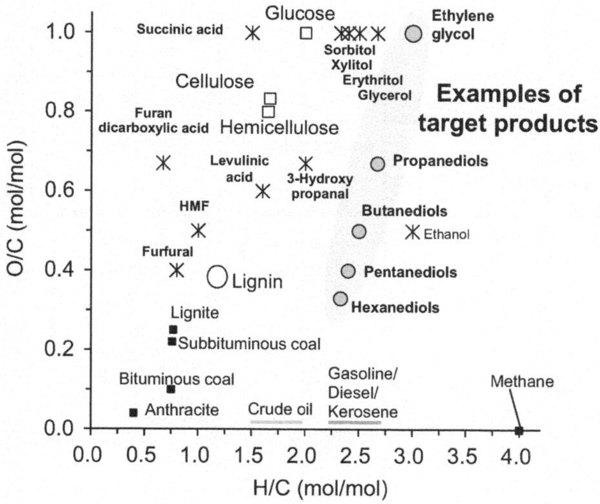

Fig. 1 Composition of feedstocks, biomass refinery building blocks, and target chemicals

hydrogenolysis reaction mechanism over the new catalyst systems is different from the conventional mechanism.

1.2 Conventional Hydrogenolysis of Glycerol and Tetrahydrofurfuryl Alcohol (THFA)

Hydrogenolysis of glycerol to 1,2- and 1,3-propanediols (PrD) seems to be simple judging from the reaction formula as below (1, 2) because the hydrogenolysis of C–O bonds apparently means the dissociation of C–O bonds and insertion of hydrogen atoms.

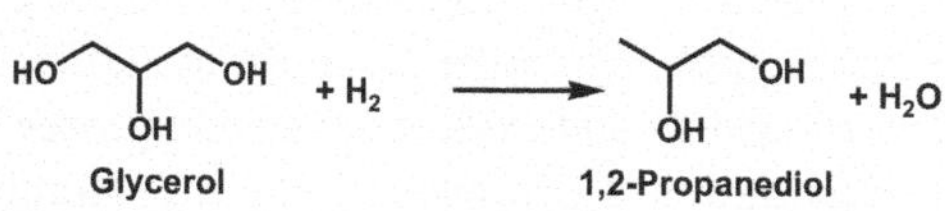

(1)

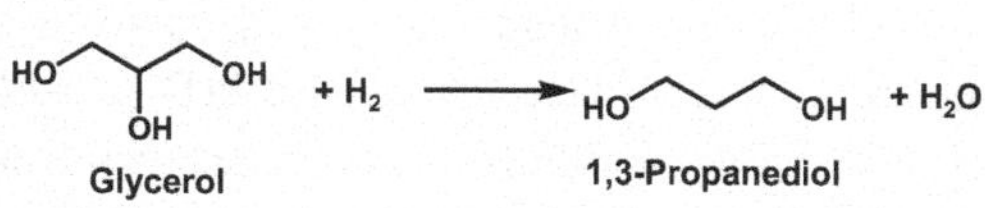

(2)

However, it is known that the glycerol hydrogenolysis actually consists of a few reaction steps such as dehydration, dehydrogenation, hydrogenation, and so on. Two reaction routes have been proposed [5]. One is the dehydration + hydrogenation route and the other is the dehydrogenation + dehydration + hydrogenation route, as shown in Fig. 2. Regarding the dehydration + hydrogenation route, glycerol dehydration gives acetol and 3-hydroxypropanal, and the dehydration proceeds preferentially under more acidic conditions. The consecutive

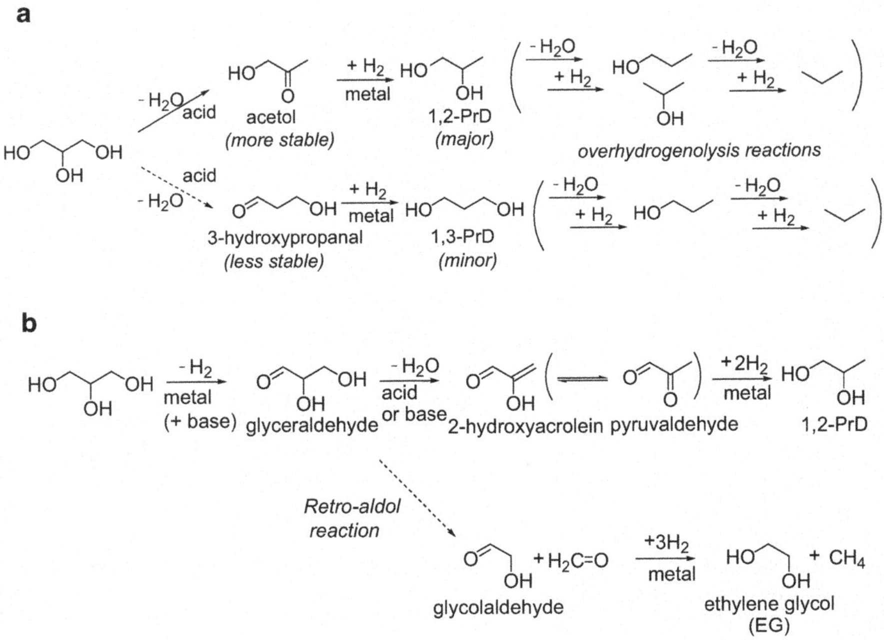

Fig. 2 Reaction mechanism of the glycerol hydrogenolysis [5]. (**a**) Dehydration + hydrogenation. (**b**) Dehydrogenation + dehydration + hydrogenation. Reprinted from RSC [5]

hydrogenation of acetol and 3-hydroxypropanal gives 1,2- and 1,3-PrDs, respectively. Here, the formation of acetol is more thermodynamically favorable than the formation of 3-hydroxypropanal, and therefore the main product tends to be 1,2-PrD. It has been reported that the combination of Ru/C (a heterogeneous hydrogenation catalyst) with an ion exchange resin (a solid catalyst) promoted the formation of 1,2-PrD strongly in the glycerol reaction [6–9]. The dehydration + hydrogenation route can be applied to other alcohols, including 1,2-PrD. Therefore over-hydrogenolysis to propanols and even propane can proceed over the catalysts that can work by the dehydration + hydrogenation route. Homogeneous Ru and Ir complexes combined with external acid are typical catalysts for production of propanols or propane from glycerol [10, 11] and 1,2-PrD [12–14].

In contrast, under more basic conditions over heterogeneous metal catalysts, the dehydrogenation + dehydration + hydrogenation route is preferred to the dehydration + hydrogenation route. Glycerol dehydrogenation can give two products: glyceraldehyde and dihydroxyacetone. Dehydration of glyceraldehyde can proceed because the acidity of C–H neighboring C=O is enhanced. On the other hand, dihydroxyacetone is difficult to dehydrate. Therefore the reaction route via glyceraldehyde is possible and 1,2-propanediol is given by the consecutive hydrogenation. It should be noted that retro-aldol reaction of glyceraldehyde can give the C–C cracking products as unfavorable by-products. An important point is that both the dehydration + hydrogenation and the dehydrogenation + dehydration + hydrogenation routes give 1,2-propanediol as a main product. In addition, glycerol is one of the most important building blocks and various other conversion routes of glycerol to useful chemicals have also been attempted over heterogeneous metal catalysts [5, 15–17].

It should be noted that some homogeneous complex catalysts without acids enable the double-dissociation of C–O bonds in polyols such as 1,2-hexanediol to *n*-hexane with Ru complex/H_2 [18, 19] and glycerol to allyl alcohol over CH_3ReO_3/alcohol [20]. Different mechanisms from those shown in Fig. 2 have been proposed for these systems, which are discussed in another chapter of this volume by Boucher-Jacobs and Nicholas.

THFA can be synthesized by the total hydrogenation of furfural [21, 22], which has been produced from hemicellulose, and, therefore, THFA is regarded as a biomass-derived intermediate [23]. It has been reported that THFA hydrogenolysis was not selective and both 1,2-pentanediol (PeD) and 1,5-PeD were formed (3) [24]:

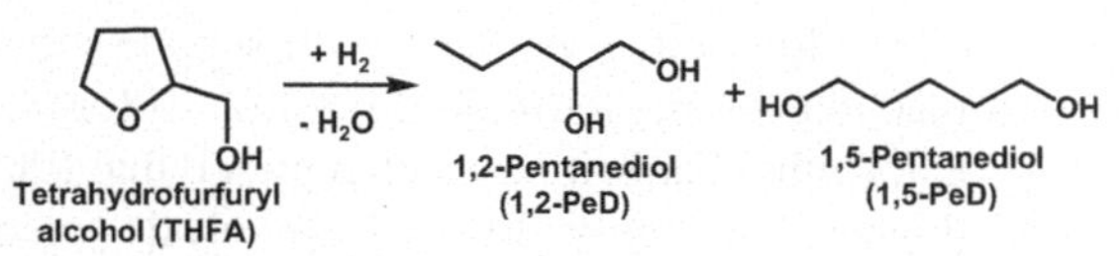

(3)

Therefore, in order to obtain 1,5-PeD selectively, the multi-step method has been reported (4) [25]. This system is composed of three separated steps including dehydration of THFA to dihydropyran, hydration of dihydropyran to δ-hydroxyvaleraldehyde, and hydrogenation of δ-hydroxyvaleraldehyde to 1,5-PeD. The system requires the isolation and purification of the intermediates and the overall yield was 70%. Hydrogenolysis of THFA to 1,5-PeD also seems to be simple judging from the reaction formula (3), just like the case of the glycerol hydrogenolysis; however, the selective synthesis of 1,5-PeD from THFA had not been realized.

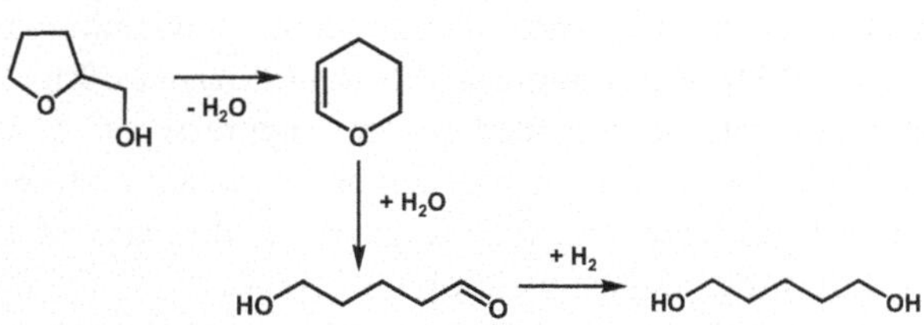

(4)

As mentioned above, conventional hydrogenolysis reactions of glycerol and THFA typically proceed by multi-steps, and this is why it is not easy to apply to substrates with more complex structure such as carbohydrates. Carbohydrates, including cellulose, are very important feedstocks, and reductive conversions of carbohydrates have been heavily investigated in spite of difficulties [4, 26–28]. Typical products are sugar alcohols such as sorbitol, glycerol, and ethylene glycol. 1,2-PrD, which can be produced from glycerol, is also frequently observed as a product. Production of sugar alcohols from carbohydrates does not involve C–O hydrogenolysis (hydrogenation and C–C hydrogenolysis, as well as hydration in the case of polysaccharides, are involved instead), and therefore we excluded this reaction from the scope of this review. On the other hand, direct C–O hydrogenolysis, where the hydrogenolysis reaction proceeds as expressed in the reaction equation, would be applied to a wide range of substrates, including those with complex structure. This review introduces the development of catalysts for this direct hydrogenolysis of C–O bonds, and explains the catalyst structure and proposed reaction mechanism.

2 Hydrogenolysis of Glycerol and Tetrahydrofurfuryl Alcohol (THFA) Using Rh-MO_x Catalysts

It has been reported that supported Ru catalysts were effective for the glycerol hydrogenolysis to 1,2-PrD. However, it is not easy to suppress the C–C cracking reaction (degradation reaction) by the retro-aldol reaction to form ethylene glycol and C1 products [7]. On the other hand, it has been reported that Rh/SiO_2 exhibited higher activity and higher selectivity to hydrogenolysis products such as propanediols and propanols than Ru/C catalysts at low temperature (393 K) by the suppression of the degradation reaction [29]. Based on these results, the modification of Rh/SiO_2 catalysts was attempted. Accidentally, at that time, our group studied the catalyst systems with strong metal-support interaction and strong metal oxide interaction, for example, Rh-VO_x and Rh-MoO_x [30–32], and these catalysts were first applied to the hydrogenolysis reactions. It is found that the addition of Re, Mo, and W to Rh/SiO_2 enhanced the catalytic activity of the glycerol hydrogenolysis (Fig. 3) [33–36]. It has also been reported that the combination of Re with Pd gave high hydrogenation activity of higher saturated carboxylic acids to the corresponding alcohols [37].

The modification with Re gave the highest conversion and yield of 1,3-propanediol (1,3-PrD). The optimized Rh-ReO_x/SiO_2 (Re/Rh = 0.5) catalyst maintained high selectivity to propanediols and suppressed C–C bond breaking even under low H_2 pressure and high reaction temperature, where Rh/SiO_2 is relatively active in C–C bond breaking (Fig. 4). In the case of Rh-ReO_x/SiO_2 (Re/Rh = 0.5) catalyst, the highest yield of 1,3-PrD was 11% (79.0% conversion, 13.8% selectivity to 1,3-PrD), not so high. In order to enhance the yield of 1,3-PrD, the initial selectivity to 1,3-PrD should be increased and the consecutive hydrogenolysis of 1,3-PrD to 1-propanol should be suppressed more.

An interesting point is that the catalysts effective in glycerol hydrogenolysis were applied to the hydrogenolysis of THFA to 1,5-PeD. Silica- or carbon-supported rhodium catalysts modified with Re, Mo, or W show high activity and selectivity [38–44], while commercial hydrogenation catalysts such as Ru/C, copper chromite, and Raney Ni show much lower activity and selectivity (Table 1). In

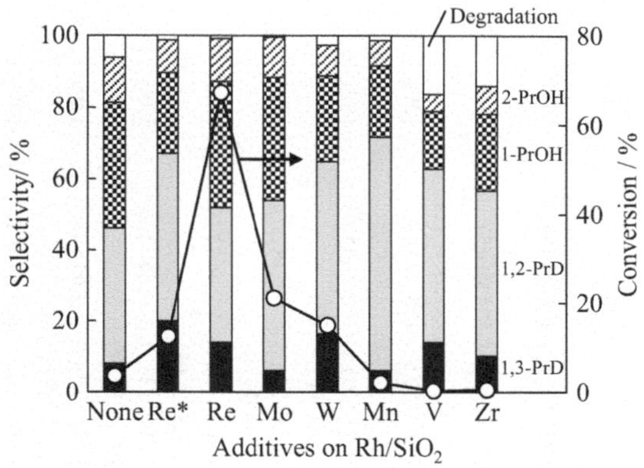

Fig. 3 Comparison of the catalytic the performances in the glycerol hydrogenolysis over modified Rh/SiO_2 [33]. Reaction conditions: glycerol 4 g, water 16 g, Rh-MO_x/SiO_2 (M/Rh = 0.25) 0.15 g, $P(H_2) = 8$ MPa, $T = 393$ K, $t = 5$ h. *Re/Rh = 0.5, $t = 1$ h. *PrD* propanediol, *PrOH* propanol. Degradation: ethylene glycol + ethanol + methanol + methane. Reprinted with permission from Elsevier [33]

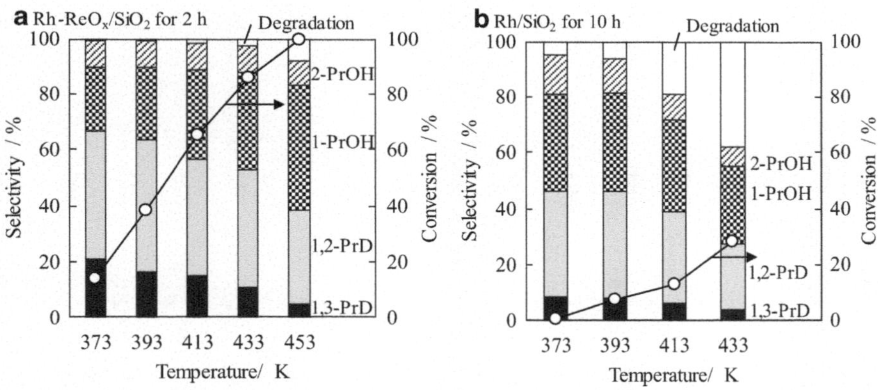

Fig. 4 Reaction temperature dependence of the glycerol hydrogenolysis over Rh-ReO_x/SiO_2 (Re/Rh = 0.5) [33] (**a**) and Rh/SiO_2 (**b**). Reaction conditions: Glycerol 4 g, water 16 g, $W_{cat} = 0.15$ g, $P(H_2) = 8$ MPa, $t = 2$ h (Rh-ReO_x/SiO_2) or 10 h (Rh/SiO_2). *PrD* propanediol, *PrOH* propanol. Degradation: ethylene glycol + ethanol + methanol + methane. Reprinted with permission from Elsevier [33]

the hydrogenolysis of THFA, Rh-ReO_x/SiO_2 (Re/Rh = 0.5) gave 86% yield of 1,5-PeD (Table 1, entry 11), and this yield is clearly higher than that of the reported multi-step method. In terms of the yield of 1,5-PeD, Rh-ReO_x/C (Re/Rh = 0.5) gave 94% yield (Table 1, entry 13). High initial selectivity to 1,5-PeD at low THFA conversion (>95%) and low activity of consecutive hydrogenolysis of 1,5-PeD to 1-PeOH enabled a high yield of the target product compared to the case of glycerol hydrogenolysis.

Table 1 Selected examples of hydrogenolysis of THFA [44]

Entry	Catalyst	H_2 (MPa)	THFA/water/catalyst (g)	Temperature (K)	Time (h)	Conv. (%)	Products (selectivity (%))	References
1	Rh/SiO_2	8	1/19/0.05	393	4	5.7	1,5-PeD (18), 1,2-PeD (62), 1-PeOH (6)	[38, 39]
2	$Rh\text{-}ReO_x/SiO_2$ (Re/Rh = 0.5)	8	1/19/0.05	393	4	57	1,5-PeD (94), 1-PeOH (4)	[38, 39]
3	$Rh\text{-}ReO_x/SiO_2$ (Re/Rh = 0.13)	8	1/19/0.05	393	4	26	1,5-PeD (84), 1-PeOH (11)	[39]
4	$Rh\text{-}MoO_x/SiO_2$ (Mo/Rh = 0.13)	8	1/19/0.05	393	4	50	1,5-PeD (96), 1-PeOH (4)	[39]
5	$Rh\text{-}MoO_x/SiO_2$ (Mo/Rh = 0.6)	8	1/19/0.05	393	4	40	1,5-PeD (93), 1-PeOH (5)	[39]
6	$Rh\text{-}WO_x/SiO_2$ (W/Rh = 0.13)	8	1/19/0.05	393	4	30	1,5-PeD (85), 1-PeOH (6)	[39]
7	ReO_x/SiO_2	8	1/19/0.05	393	4	<0.1	1,5-PeD (<0.1), 1,2-PeD (31), 1-PeOH (5)	[38]
8	Ru/C	8	1/19/0.05	393	4	5	1,5-PeD (15), 1,2-PeD (26), 1-PeOH (8)	[38]
9	Copper chromite	8	1/19/0.5	453	4	0.6	1,5-PeD (<0.1), 1,2-PeD (9), 1-PeOH (3)	[38]
10	Raney Ni	8	1/19/0.5	453	4	0.2	1,5-PeD (13), 1,2-PeD (13), 1-PeOH (9)	[38]
11	$Rh\text{-}ReO_x/SiO_2$ (Re/Rh = 0.5)	8	1/19/0.1	373	36	96	1,5-PeD (90), 1-PeOH (8)	[39]
12	$Rh\text{-}MoO_x/SiO_2$ (Mo/Rh = 0.13)	8	1/19/0.1	373	24	94	1,5-PeD (90), 1-PeOH (9)	[39]
13	$Rh\text{-}ReO_x/C$ (Re/Rh = 0.25)	8	1/19/0.1	373	24	99	1,5-PeD (95), 1-PeOH (4)	[40]
14	$Rh\text{-}ReO_x/C$ (Re/Rh = 0.5)	3.4	9/171/1	393	4	47	1,5-PeD (97), 1-PeOH (3)	[45]
15	$Rh\text{-}MoO_x/C$ (Mo/Rh = 0.1)	3.4	7/133/2	393	4	52	1,5-PeD (91), 1-PeOH (3)	[45]

Rh-MO_x/SiO_2 (M = Re, Mo, W) showed much higher performance than monometallic Rh/SiO_2 and MO_x/SiO_2 in both the glycerol and THFA hydrogenolysis. The additive effect of the MO_x on the activity was remarkable. Another interesting point is the additive effect on selectivity. In the THFA hydrogenolysis on Rh/SiO_2, the main product is 1,2-PeD. In contrast, in the case of Rh-MO_x/SiO_2, 1,5-PeD was selectively formed. These results suggest that the modification changes the product selectivity. The reaction order with respect to the THFA concentration over Rh-ReO_x/SiO_2 and Rh/SiO_2 was estimated to be zero and 0.4, respectively [39]. In particular, the reaction order indicated that Rh-ReO_x/SiO_2 adsorbs THFA more strongly than does Rh/SiO_2, and this suggests that ReO_x addition is responsible for the strong adsorption of THFA on Rh-ReO_x/SiO_2 [39]. In addition, the reaction orders with respect to H_2 on Rh/SiO_2 and Rh-ReO_x/SiO_2 were determined to be 0.5 and 1, respectively, and it is suggested that the modification can also change the mechanism of H_2 activation, as discussed in detail later. The indirect process of the conversion of THFA to 1,5-PeD via dihydropyran and δ-hydroxyvaleraldehyde is shown in (4). The reaction of dihydropyran was tested on Rh-ReO_x/SiO_2 (Re/Rh = 0.5) in order to evaluate the contribution of this reaction route [39]. The result shows that the formation of 1,5-PeD dehydration accompanies the tetrahydropyran formation (12% selectivity). However, the formation of tetrahydropyran is actually below the detection limit in the THFA hydrogenolysis over Rh-ReO_x/SiO_2 (Re/Rh = 0.5). Therefore, the reaction route over Rh-ReO_x/SiO_2 (Re/Rh = 0.5) is different from the indirect route via dihydropyran.

According to the previous report, the reaction time dependence of the THFA conversion over Rh-ReO_x/SiO_2 (Re/Rh = 0.5) is estimated to be proportional to $C_{\text{THFA}}/(C_{\text{THFA}} + 2C_{1,5\text{-PeD}})$ [39]. This equation means that the reaction rate at high THFA conversion is suppressed by the presence of 1,5-PeD product and the suppressing factor is second, which is interpreted by two primary OH groups in 1,5-PeD. Strong interaction between the primary OH group and the catalyst surface can be connected to zero reaction order with respect to THFA at low conversion and the suppressing factor due to 1,5-PeD at high conversion.

3 Structure of Rh-ReO_x/SiO_2 and Rh-MoO_x/SiO_2 Catalysts

The dependence of the activity on the additive amount of Re and Mo to Rh/SiO_2 is interesting [42]. Figure 5 shows the effect of the additive amount of Re and Mo to Rh/SiO_2 in the THFA hydrogenolysis, indicating that the optimum amounts of Re and Mo were Re/Rh = 0.5 and Mo/Rh = 0.13, respectively.

According to the catalyst characterization by TEM and XRD, both Rh-ReO_x/SiO_2 and Rh-MoO_x/SiO_2 with various additive amounts had similar Rh particle sizes (Fig. 6a) [42]. On the other hand, according to the result of the measurement of CO adsorption amounts by a volumetric method, the amount of CO adsorption decreased steadily with increase in the additive amounts of Re and Mo. It should be noted that the added Re and Mo exist in a low valence state like +2 to +4 judging

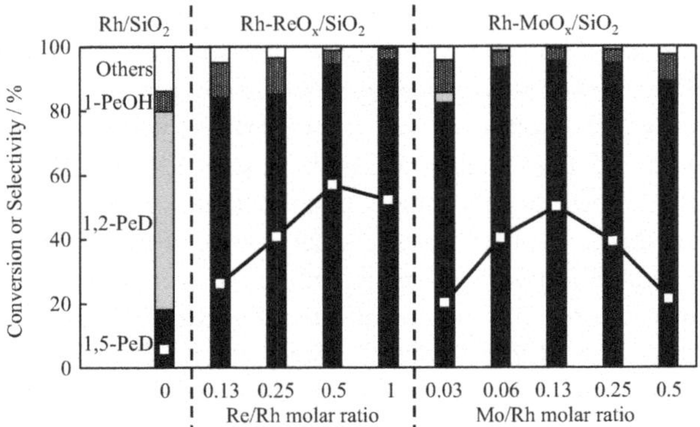

Fig. 5 Effect of additive amount of Re and Mo to Rh/SiO_2 in THFA hydrogenolysis [42]. Reaction conditions: THFA 1 g water 19 g, $W_{cat} = 0.05$ g, $P(H_2) = 8$ MPa, $t = 4$ h. *Squares*: conversion, *bars*: selectivity. Reprinted with permission from Elsevier [42]

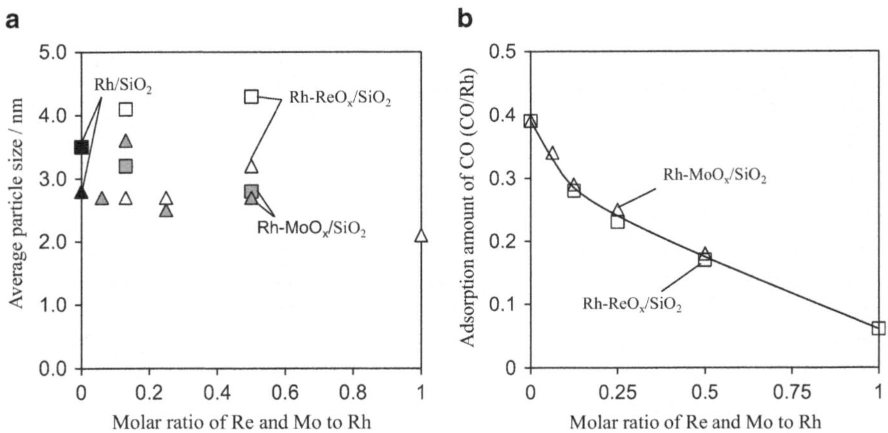

Fig. 6 Catalyst properties as a function of additive amount of Re and Mo. (**a**) Rh metal particle size from XRD (*squares*) and TEM (*triangles*) after reaction. (**b**) CO adsorption amount (the molar ratio of adsorbed CO to Rh) after reduction

from the results of temperature-programmed reduction with H_2. The Re and Mo species in the oxidized state cannot adsorb CO at room temperature. Therefore it is interpreted that the small amounts of CO adsorption on Rh-ReO_x/SiO_2 and Rh-MoO_x/SiO_2 for the size of Rh metal particle can be due to the partial coverage of the Rh metal surface with ReO_x and MoO_x species. In addition, the amounts of CO adsorption on Rh-ReO_x/SiO_2 and Rh-MoO_x/SiO_2 were almost the same, indicating that ReO_x and MoO_x species cover Rh metal surface in a similar way and one Re or Mo atom covers one CO adsorption site on the Rh surface, particularly in the range of M/Rh $\leqq 0.13$ (Fig. 6b) [42].

An important point is that the optimum amounts of added Re and Mo were different in terms of the catalytic activity, although the CO adsorption amount on

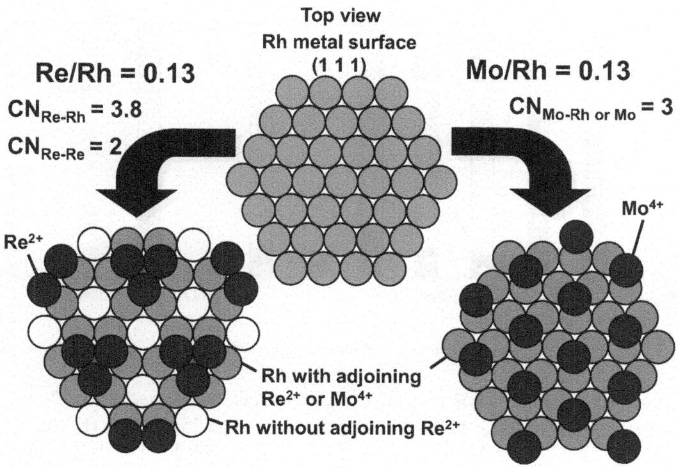

Fig. 7 Model structure of Rh-MO_x/SiO_2 (M/Rh = 0.13, M = Mo and Re) [42]. Reprinted with permission from Elsevier [42]

the same amount of added Re and Mo was almost the same. This tendency suggests that the structure of ReO_x and MoO_x on Rh metal surface is different. Extended X-ray absorption fine structure (EXAFS) is a useful tool for the determination of local structure around the absorbing atoms. Based on Re L_3- and Mo K-edge EXAFS analysis, the proposed model structure of Rh-MO_x/SiO_2 (M/Rh = 0.13, M = Re and Mo) is shown in Fig. 7 [42]. In the case of Rh-MoO_x/SiO_2 (Mo/Rh = 0.13), all the surface Rh atoms are modified with isolated MoO_x, and, as a result, the optimum Mo/Rh is 0.13. In contrast, the surface Rh atoms without the modification with ReO_x clusters are present on Rh-ReO_x/SiO_2 (Re/Rh = 0.13). Therefore, more added Re is needed, and it is thought that all the surface atoms are modified ReO_x when the optimum Re/Rh (Re/Rh = 0.5).

4 Glycerol Hydrogenolysis to 1,3-Propanediol Over Ir-ReO_x/SiO_2

In the THFA hydrogenolysis, both Rh-ReO_x/SiO_2 and Rh-MoO_x/SiO_2 were effective; however, it is clear that Rh-ReO_x/SiO_2 was more effective than Rh-MoO_x/SiO_2 in terms of the glycerol hydrogenolysis to 1,3-PrD [33]. Therefore we tested various metals-ReO_x catalysts supported on SiO_2 in the glycerol hydrogenolysis (Fig. 8) [46].

The Rh-ReO_x catalyst gave higher activity than other catalysts, although its selectivity to 1,3-PrD was far from the satisfactory. It is found that Ir-ReO_x catalyst showed much higher selectivity to 1,3-PrD and sufficient activity [46–50]. Another interesting point regarding the high performance of Ir-ReO_x/SiO_2 is that monometallic Ir/SiO_2 and ReO_x/SiO_2 had almost no activity in the glycerol hydrogenolysis.

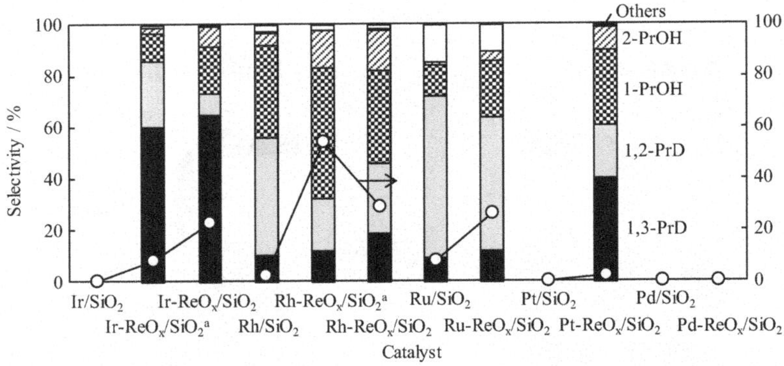

Fig. 8 Hydrogenolysis of glycerol on various Metal–ReO_x supported on SiO_2 [46]. Reaction conditions: glycerol 4 g, water 2 g, $W_{cat} = 0.15$ g, M/Re = 0.25, H_2SO_4 (H^+/Ir = 1), $P(H_2) =$ 8MPa, $T = 393$ K, $t = 12$ h. Reduction conditions: $T = 473$ K, $t = 1$ h, $P(H_2) = 8$ MPa. [a]The reaction temperature is 393 K. Reprinted with permission from Elsevier [46]

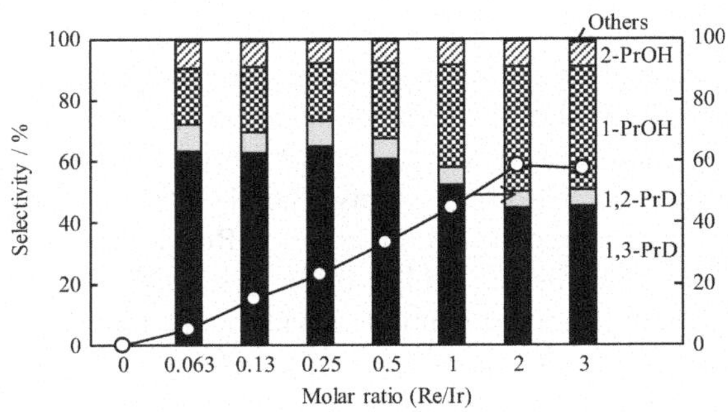

Fig. 9 Effect of Re amount over Ir-ReO_x/SiO_2 in glycerol hydrogenolysis [46]. Reaction conditions: glycerol 4 g, water 2 g, Ir-ReO_x/SiO_2 0.15 g, H_2SO_4 (H^+/Ir = 1), $P(H_2) = 8$ MPa, $T = 393$ K, $t = 12$ h. Reprinted with permission from Elsevier [46]

It is not easy to find excellent bimetallic catalysts consisting of two components with almost no activity. This point is different from the case of Rh-ReO_x where Rh is a good component for the hydrogenolysis reaction. As a result, it is concluded that the synergy between Ir and ReO_x generates the catalysis of the glycerol hydrogenolysis to 1,3-PrD.

Figure 9 shows the effect of the Re amount over Ir-ReO_x/SiO_2 in glycerol hydrogenolysis. Glycerol conversion increased steadily with increasing Re amount until the amount of Re was comparable to that of Ir [46]. Even small additions of ReO_x to Ir/SiO_2 led to high selectivity to 1,3-PrD. No activity of unmodified Ir is associated with high selectivity. If the unmodified part has some activity, the selectivity can be decreased. High selectivity is related to the much higher activity of Ir-ReO_x than Ir and ReO_x. It is characteristic of the higher molar ratio of MO_x to metal on Ir-ReO_x/SiO_2 than that on Rh-MoO_x and Rh-ReO_x. Characterization results indicate that Ir metal particles and low-valence Re species are formed on

Table 2 Comparison of metal particle size and dispersion on Ir-ReO_x/SiO_2 [46]

Re/Ir	Ir particle size (XRD) (nm)	Ir metal dispersion (%) XRD	CO adsorption
3	1.9	58	19
2	1.9	58	15
1	2.1	52	16
0.5	2.7	41	16
0.25	3.0	37	20
0.13	3.4	32	–
0.063	3.4	32	–
0	3.6	31	23

The catalyst was reduced at 473 K (Ir-ReO_x/SiO_2) or 573 K (Ir/SiO_2) for 1 h. Reprinted with permission from Elsevier [46]

Ir-ReO_x/SiO_2 [46, 49]. Table 2 lists the dependence of Ir metal dispersion from XRD and CO adsorption on the additive amount of ReO_x. The Ir metal particle size determined from XRD analysis on Ir-ReO_x/SiO_2 decreased gradually with increasing added Re. This tendency is different from that on Rh-ReO_x/SiO_2 and Rh-MoO_x/SiO_2. It is thought that this difference could be due to the formation mechanism during the reduction pretreatment. From the results of quick-scanning XAFS of Rh-ReO_x/SiO_2 during the reduction treatment, it is found that the reduction of Re species followed the reduction of Rh species, although a single peak was observed in the TPR profile [43]. The presence of reduced Rh species promoted the reduction of Re species, and this phenomenon is manifested by the hydrogen spillover effect [51–54]. This behavior can explain the similar Rh metal particle size of Rh-ReO_x/SiO_2. In contrast, from the results of quick-scanning XAFS of Ir-ReO_x/SiO_2 during the reduction treatment, it is found that the Ir and Re species were reduced simultaneously (Ir: +4 to 0, Re: +7 to +2), and the particle size of Ir metal can be influenced by the suppression of the aggregation by the reduced Re species [46, 49].

Another important point is that the calculated dispersion from XRD was much larger than that from CO adsorption on Ir-ReO_x/SiO_2. This suggests that Ir metal particles can be partially covered with ReO_x species. On the other hand, according to the Re L_3-edge EXAFS analysis of Ir-ReO_x/SiO_2 (Re/Ir = 1), the Re–O and Re–metal (Ir or Re) bonds were assigned, and the coordination numbers were determined to be 1.6 and 6.5, respectively [49]. However, the particle size of Ir metal is about 2 nm, the ratio of surface Ir atoms to the total is about 50%, and the ratio of the CO-accessible surface Ir atoms to the total is 16%. Based on the molar ratio of the added Re to Ir (Re/Ir = 1), this can be explained by the model structure of Ir-ReO_x with Ir metal particles covered with three-dimensional clusters of low-valence Re species. Figure 10 shows the model structure of Ir-ReO_x together with Rh-ReO_x and Rh-MoO_x. Here, the MoO_x species on the Rh metal surface is isolated monomer, the ReO_x species on the Rh metal surface is a two-dimensional cluster, and the ReO_x species on Ir metal surface is a three-dimensional cluster. The structure of ReO_x species on Pt metal particles has also been reported [55, 56]. Later, the relation between the structure of oxides on metal surfaces and the selectivity in the hydrogenolysis reactions is discussed.

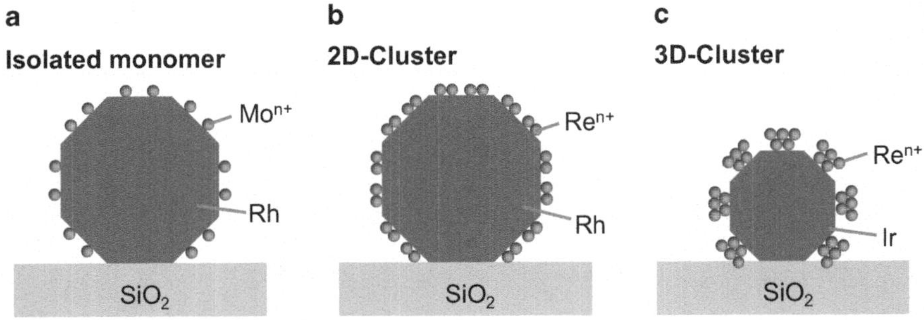

Fig. 10 Model structure of Rh-MoO$_x$/SiO$_2$ (**a**), Rh-ReO$_x$/SiO$_2$ (**b**), and Ir-ReO$_x$/SiO$_2$ (**c**)

5 Substrate Scope of Ir-ReO$_x$/SiO$_2$ and the Reaction Kinetics in the Hydrogenolysis Reaction

Table 3 lists the results of the hydrogenolysis reaction of various substrates including cyclic ethers over Ir-ReO$_x$/SiO$_2$ catalyst [57]. Ir-ReO$_x$/SiO$_2$ catalyzed the hydrogenolysis of THFA and tetrahydropyran-2-methanol to 1,5-PeD and 1,6-hexanediol, respectively (Table 3, entries 1and 2). In addition, it catalyzed the hydrogenolysis of 3-hydroxytetrahydrofuran and 3-hydroxytetrahydropyran to 1,3-butanediol and 1,4-pentanediol, respectively (Table 3, entries 3 and 4). These four substrates are reactive and have a common structure of HO–C–C–O–, where the underlined C–O bond is dissociated by the hydrogenolysis reaction. In particular, the substrates with different OH positions and no OH groups are much less reactive and this tendency supports the role of the common structure (Table 3, entries 5 and 6). Furthermore, Ir-ReO$_x$/SiO$_2$ also catalyzed the hydrogenolysis of erythritol to 1,4- and 1,3-butanediols (Table 3, entry 7), and the details regarding erythritol are introduced in another section. The hydrogenolysis of *cis*-1,2-cyclohexanediol to cyclohexanol was also catalyzed (Table 3, entry 8); however, the behavior of the reaction of *cis*-1,2-cyclohexanediol was rather different from other reactive substrates.

Figure 11 shows the kinetics of the hydrogenolysis of various substrates regarding the concentration of the substrates over Ir-ReO$_x$/SiO$_2$ [46–50, 57, 58]. It is characteristic that the substrates with primary OH groups like glycerol, THFA, and erythritol gave about zero reaction orders with respect to the substrate concentration. In contrast, the substrates with secondary OH group like 3-hydroxytetrahydrofuran and *cis*-1,2-cyclohexanediol gave around 0.5 reaction orders. One possible interpretation is that the interaction of primary OH groups with the catalyst surface is stronger than that of secondary OH group, so that the coverage of the substrates with primary OH groups can be high and be connected to zero reaction orders.

Figure 12 shows kinetics of the hydrogenolysis of various substrates regarding H_2 pressure over Ir-ReO$_x$/SiO$_2$ [46–50, 57, 58]. The hydrogenolysis of the

Table 3 Results of the hydrogenolysis of various substrates over Ir-ReO_x/SiO_2 (Re/Ir = 1) [57]

Entry	Substrate	Product	*t* (h)	Conv. (%)	Selectivity (%)	TOF (h^{-1})
1	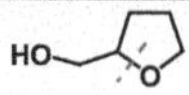		2	58.2	95.8	609
			8	100	82.0	–
2	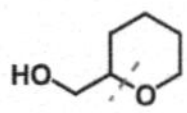	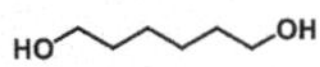	4	40.4	87.7	186
3	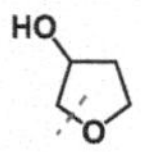	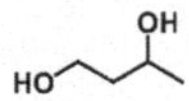	6	23.9	81.1	96
4	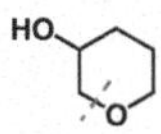	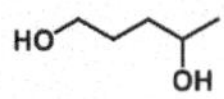	6	35.0	76.4	122
5	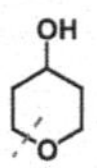	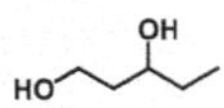	6	0.0	0.0	0
6		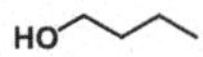	4	2.7	90.5	18
7[a]	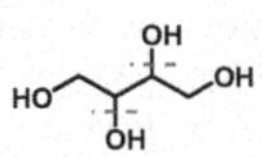	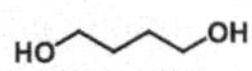	24	74.0	33.0	–
		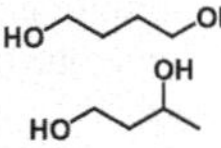			12.0	
8	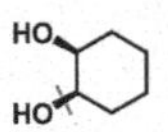	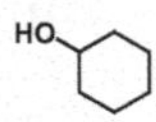	1	9.3	64.3	170

Reaction conditions: substrate 1 g, water 4 g, Ir-ReO_x/SiO_2 0.15 g, $P(H_2)$ = 8 MPa, T = 373 K. *Dotted lines* mean the position of C–O hydrogenolysis. [a]The amount of Ir-ReO_x/SiO_2 is 0.3 g, and H_2SO_4 (H^+/Ir=1) is added. Reprinted with permission from Elsevier [57]

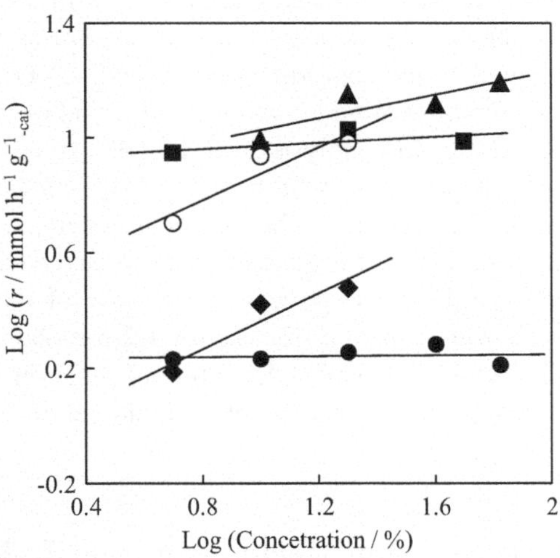

Fig. 11 Kinetics of the hydrogenolysis of various substrates regarding the concentration of the substates over Ir-ReO_x/SiO_2. *Open circles*: *cis*-1,2-cyclohexanediol; substrate 1 g, water 4–19 g, W_{cat} = 0.15 g, $P(H_2)$ = 2 MPa, T = 373 K, t = 1 h. *Filled circles*: erythritol; substrate 1 g, water 0.67–19 g, W_{cat} = 0.3 g, $P(H_2)$ = 8 MPa, T = 373 K, t = 4 h. *Filled triangles*: glycerol, substrate 1–4 g, water 2–19 g, W_{cat} = 0.15 g, $P(H_2)$ = 8 MPa, T = 393 K, t = 4 h. *Filled squares*: THFA; substrate 1 g, water 1–19 g, W_{cat} = 0.15 g, $P(H_2)$ = 8 MPa, T = 373 K, t = 1 h. *Filled diamonds*: 3-HTHF; substrate 1 g, water 4–19 g, W_{cat} = 0.15 g, $P(H_2)$ = 8 MPa, T = 373 K, t = 6 h

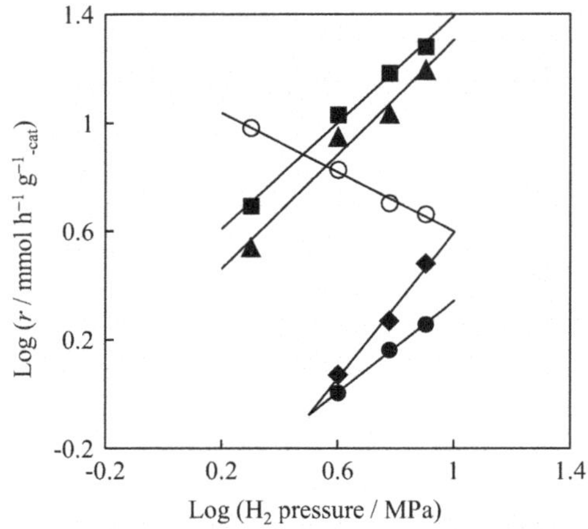

Fig. 12 Kinetics of the hydrogenolysis of various substrates regarding H_2 pressure over Ir-ReO_x/SiO_2. Reactant 1 g, water 4 g. *Open circles*: *cis*-1,2-cyclohexanediol; $W_{cat} = 0.15$ g, $P(H_2) = 2$ MPa, $T = 373$ K, $t = 1$ h. *Filled circles*: erythritol; $W_{cat} = 0.3$ g, $P(H_2) = 8$ MPa, $T = 373$ K, $t = 4$ h. *Filled triangles*: glycerol, $W_{cat} = 0.15$ g, $P(H_2) = 8$ MPa, $T = 393$ K, $t = 4$ h. *Filled squares*: THFA; $W_{cat} = 0.15$ g, $P(H_2) = 8$ MPa, $T = 373$ K, $t = 2$ h. *Filled diamonds*: 3-HTHF; $W_{cat} = 0.15$ g, $P(H_2) = 8$ MPa, $T = 373$ K, $t = 6$ h

substrates like glycerol, THFA, erythritol, and 3-hydroxytetrahydrofuran gave first reaction order with respect to H_2 pressure. An important point is that *cis*-1,2-cyclohexanediol showed negative reaction order (−0.5) with respect to H_2 pressure. At present, it is thought that the trend of the reaction orders with respect to H_2 pressure can be explained by the conformation of HO–C–C–O– structure [57].

Figure 13 shows the conformation around the C–C bond of HO–C–C–O– in the reactive substrates in the hydrogenolysis over Ir-ReO_x/SiO_2. In the case of the substrates giving first reaction order with respect to H_2, the cleavage of the C–O bond in the hydrogenolysis reaction can be located at the anti-conformation to the OH group, which can contribute to the interaction between the OH group and the Re species on the catalyst surface. In contrast, in the case of the substrates giving negative reaction order with respect to H_2 like *cis*-1,2-cyclohexanediol, the cleavage of the C–O bond cannot be located at the anti-conformation. This interpretation can be connected to the reaction mechanism of the hydrogenolysis of the C–O bond.

6 Hydrogenolysis Reaction Mechanism Over Ir-ReO_x/SiO_2

Figure 14 shows the reaction mechanism of the hydrogenolysis of the above substrates giving first reaction orders with respect to H_2. On the basis that the strong synergy between Ir metal and ReO_x clusters generates the hydrogenolysis activity, the catalytically active site can be the interface between Ir metal surface and three-dimensional ReO_x clusters as described in Fig. 14(I). Judging from the

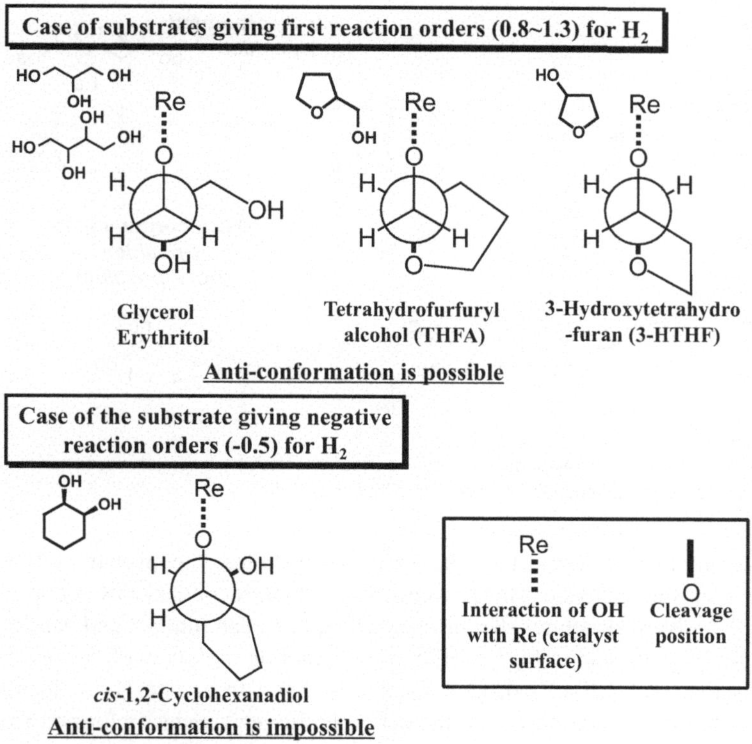

Fig. 13 Conformation of HO–C–C–O– in the reactive substrates in the hydrogenolysis over Ir-ReO$_x$/SiO$_2$

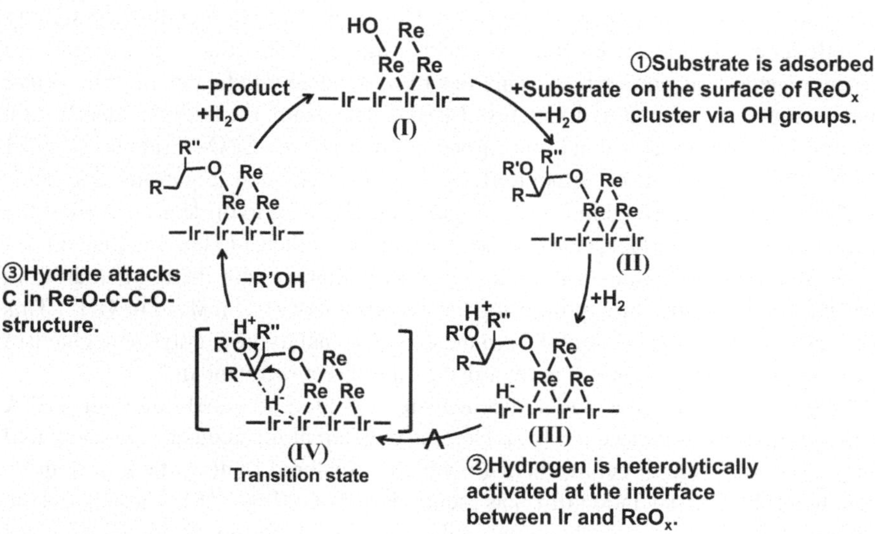

Fig. 14 Proposed reaction mechanism of the hydrogenolysis reaction over Ir-ReO$_x$/SiO$_2$

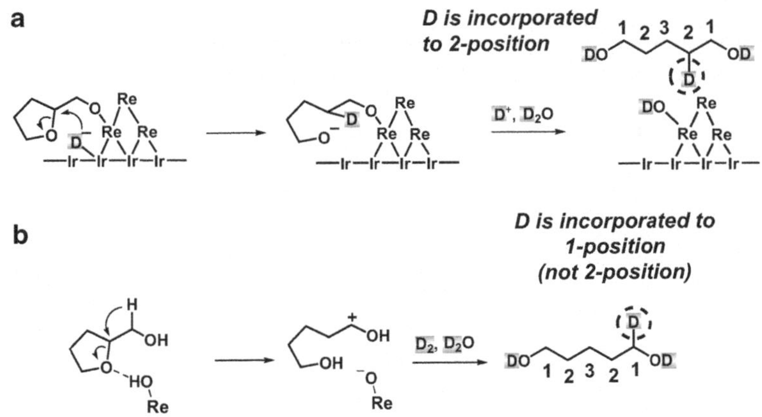

Fig. 15 Deuterium insertion position in the reaction of THFA hydrogenolysis. (**a**) Direct mechanism. (**b**) Concerted mechanism proposed in Chia et al. [45]

reaction orders with respect to the various substrates, the primary OH groups interact with the catalyst surface more strongly than secondary OH groups. In the case of 1,2-glycol structures, the primary OH groups can interact and can form the corresponding alkoxide (Fig. 14(II)). First reaction orders with respect to H_2 suggests that one active hydrogen species is given from one H_2 molecule, and one possible interpretation is that the active hydrogen species is formed from the heterolytic dissociation of H_2 to proton and hydride. If the active species is hydrogen radical, the reaction orders with respect to H_2 may be 1/2; however, this is not true for the present hydrogenolysis. In the case that the proton is the active hydrogen species, the proton can attack the oxygen atom of the secondary OH group to give a secondary carbocation. However, the hydrogenolysis reactivity of ethylene glycol to ethanol is much higher than that of glycerol to 1,3-propanediol, as shown later. This reactivity tendency indicates that the active species is not proton, but hydride (Fig. 14(III)). Therefore, the hydride adsorbed on the interface Ir atom can attack the carbon atom at HO–C–<u>C</u>–O– structure (Fig. 14 (IV)). This state can be supported by the interpretation regarding the anti-conformation as described in Fig. 13 because the S_N2-like hydride attacks from the side of the catalyst surface, causing the dissociation of the C–O bond in the HO–C–<u>C–O</u>– structure. It is thought that the proton can interact with the oxygen atom in the H<u>O</u>–C–C–O– structure, promoting the desorption of H_2O instead of OH^-. This proposed mechanism is different from the conventional hydrogenolysis mechanism mentioned in Sect. 1.2, and it is regarded as the direct mechanism.

In order to verify the direct mechanism, a deuterium label study in THFA hydrogenolysis was carried out [57]. Here, two different mechanisms are compared (Fig. 15). One is the direct mechanism, which is featured by the attack of hydride from heterolytic dissociation of H_2 as mentioned above (Fig. 15a). The other is the concerted mechanism, which is featured by attack of proton on Re species and

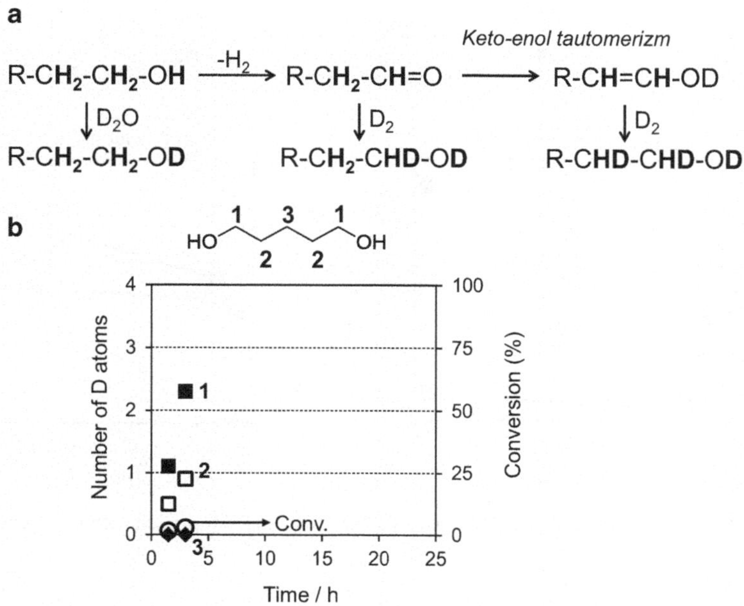

Fig. 16 (**a**) Mechanism of H–D Exchange (scrambling) between alcohols (substrates and products) and D_2/D_2O. (**b**) Measurement of the scrambling rate of 1,5-PeD. Reaction conditions: D_2 2 MPa, D_2O (2 g), Ir-ReO_x/SiO_2 (0.05 g), 373 K

subsequent intramolecular hydride shifts (Fig. 15b) proposed in Chia et al. [45]. The different reaction mechanism can give the different deuterium position in the product of 1,5-PeD. Before showing the details of the THFA reaction, we checked the scrambling rate of 1,5-PeD in the deuterium experiments (Fig. 16) [57]. Figure 16a shows the scheme of the scrambling reactions. The rate in 1,5-PeD is rather fast, and the rate at the 1-position was about twice as high as that at the 2-position.

Figure 17 shows the results of THFA + D_2 reaction in D_2O. Regarding the D atom in unreacted THFA, about 0.5 of an atom of D was incorporated only to the 1-position of THFA at the initial stage. Based on this result, the THFA molecule can be described as $C_4H_7O–CH_{1.5}D_{0.5}–OH$. On the other hand, regarding the D atom in the produced 1,5-PeD, about 1.5 atoms of D were incorporated in both the 1- and 2-positions of 1,5-PeD. Figure 18 shows the incorporation of D atoms in the reaction of THFA + D_2 in D_2O by two reaction mechanism. In the case of the direct mechanism, D^- attacks to C at the 2-position and, as a result, 0.5 atom D is incorporated to the 1-position and 1 atom D is incorporated to the 2-position. Considering the scrambling rate of the 1- and 2-positions in THFA, the experimental results of 1.5 atom D at 1-position and 1.5 atom D at 2-position can be explained by the direct mechanism. On the other hand, in the case of the concerted mechanism, 1.25 atom D is incorporated to the 1-position and 0.25 atom D is incorporated to the 2-position, which cannot explain the observed distribution of D atoms. Overall, the deuterium label study supports the direct reaction mechanism.

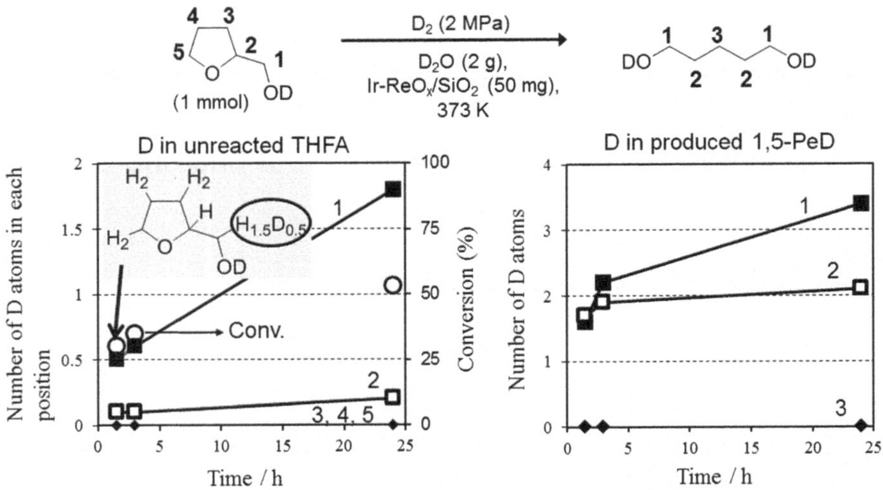

Fig. 17 Results of THFA + D_2 in D_2O over Ir-ReO_x/ SiO_2 [57]. Reprinted with permission from Elsevier [57]

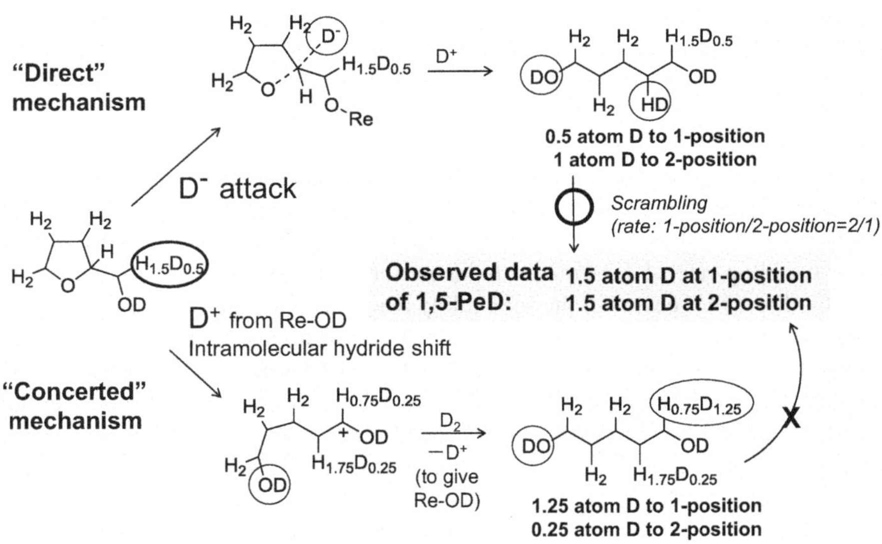

Fig. 18 Incorporation of D atoms in the reaction of THFA + D_2 in D_2O in two reaction mechanisms

7 Hydrogenolysis of Erythritol Using Ir-ReO_x/SiO_2

A large number of studies on the fermentation of glucose and glycerol to erythritol have been reported. The highest yield of erythritol from glucose is 61% [59]. Moreover, it has been already scaled up from laboratory scale to plant scale [60]. In

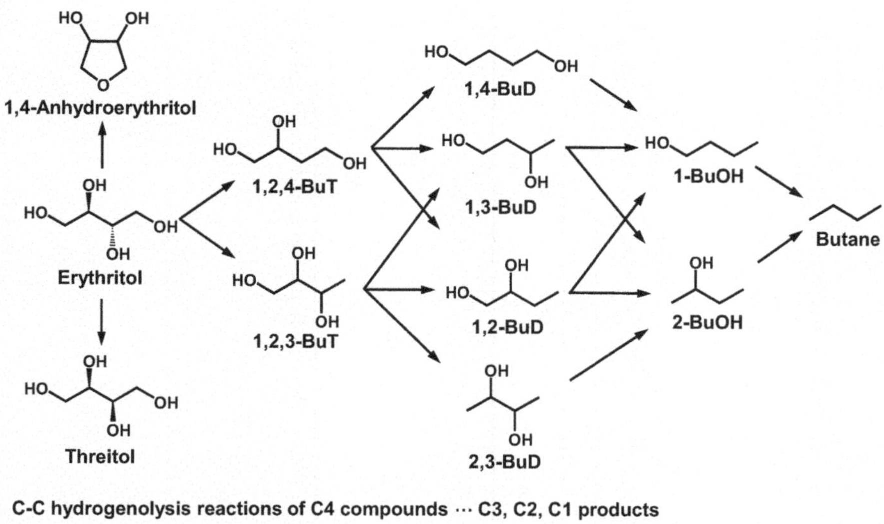

Fig. 19 Scheme of hydrogenolysis of erythritol and side reactions [58]. Reprinted from Wiley-VCH [58]

addition, a 56% yield of erythritol can be obtained from glycerol [61]. Based on these results, erythritol can be a candidate for a C4 building block in a future biomass refinery. However, valuable chemicals derived from erythritol are very limited. Therefore, the conversion of erythritol to 1,4- and 1,3-butanediols (BuD) was attempted by using Ir-ReO_x/SiO_2 and the direct hydrogenolysis mechanism.

Figure 19 shows the reaction scheme of erythritol hydrogenolysis including side reactions like dehydration and isomerization. The desired products can be produced by C–O hydrogenolysis reactions [58]. These reactions lead to two primary products – 1,2,4- and 1,2,3-butanetriol (BuT). The hydrogenolysis reactions of BuTs produce four BuDs, which are desired products as valuable chemicals. A problem is that over-hydrogenolysis of BuDs produces 1- and 2-butanols, and butane, which are less valuable. C–C hydrogenolysis reactions lead to many types of products such as 1,2-PrD, ethanol, and methane. The value of most of the products obtained by C–C hydrogenolysis reactions is not high except for that of 1,3-PrD. It has been reported that Ru/C and Raney Ni catalyzed the hydrogenolysis of erythritol to 1,2-BuD, 1,2,4-BuT glycerol, and so on [62].

One of the side reactions is the dehydration to 1,4-anhydroerythritol, and this reaction can proceed easily under acidic conditions [63]. It has been reported that Re-Pd/C + Nafion/SiO_2 gave 50% yield of tetrahydrofuran from erythritol via 1,4-anhydroerythritol in dioxane [64]. Another side reaction is the isomerization to threitol. The reports on the hydrogenolysis of erythritol are so limited that the selective hydrogenolysis to valuable BuDs is challenging.

Table 4 lists the erythritol hydrogenolysis on various catalysts including Ir-ReO_x/SiO_2 catalyst. It is found that Ir-ReO_x/SiO_2 catalyzed the erythritol

Table 4 Results of erythritol hydrogenolysis over various catalysts [58]

Catalyst	Conv. (%)	Selectivity (%)							
		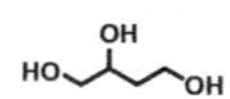	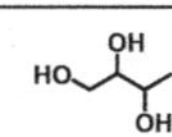	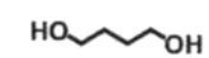	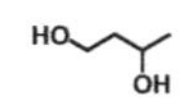	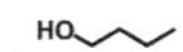	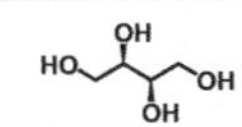	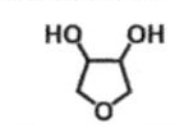	Others
Ir-ReO_x/SiO_2[a]	74	5	18	33	12	21	0	0	11
Ir/SiO_2[a,b]	3	4	18	0	0	0	0	70	9
ReO_x/SiO_2[a,b]	1	1	0	0	0	0	0	99	0
Ru/C	66	2	24	0	0	0	49	7	18
Raney Ni[b,c]	57	3	16	0	0	0	79	0	2
Rh/C	6	16	47	0	0	0	0	25	13

Reaction conditions: Erythritol 1 g, Water 4 g, $W_{cat} = 0.3$ g, $P(H_2) = 8$ MPa, $T = 373$ K, $t = 24$ h. [a]H_2SO_4 ($H^+/Ir = 1$). Reduction conditions: $T = 473$ K, $t = 1$ h, $P(H_2) = 8$ MPa. [b]$T = 413$ K. [c]$W_{cat} = 1.5$ g. Reprinted from Wiley-VCH [58]

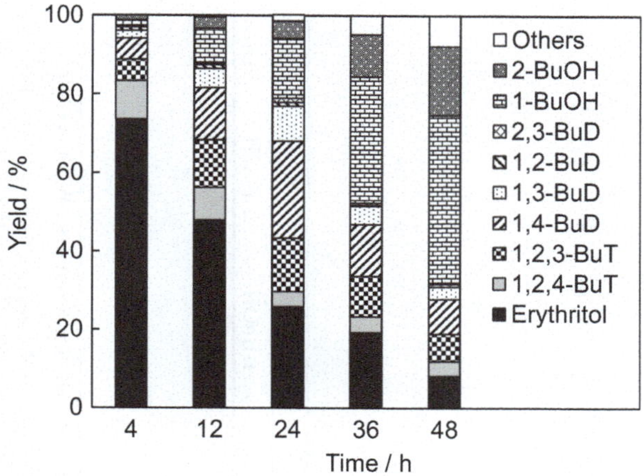

Fig. 20 Reaction time dependence in the erythritol hydrogenolysis over Ir-ReO$_x$/SiO$_2$[58]. Reaction conditions: erythritol 1 g, water 4 g, $W_{cat} = 0.3$ g, H_2SO_4 (H^+/Ir = 1), $P(H_2) = 8$ MPa, $T = 373$ K. *BuT* butanetriol, *BuD* butanediol, *BuOH* butanol. Reprinted from Wiley-VCH [58]

hydrogenolysis to 1,4- and 1,3-BuDs, and Ir/SiO$_2$ and ReO$_x$/SiO$_2$ had very low catalytic activity, as in the case of glycerol hydrogenolysis. In addition, Ru/C, Raney Ni, and Rh/C gave BuTs and/or threitol as a main product. Reaction time dependence in the erythritol hydrogenolysis over Ir-ReO$_x$/SiO$_2$ is shown in Fig. 20. At the initial stage, 1,2,4-BuTs was mainly produced, which is similar to the case of the glycerol hydrogenolysis to 1,3-PrD. The maximum yield of 1,4- and 1,3-BuD were 33% (24 h), and the yield of valuable products is not so high, and further improvement is necessary; however, the present performance was much higher than that of conventional hydrogenolysis catalysts such as Ru/C and Raney Ni.

The dehydration of erythritol to 1,4-anhydroerythritol proceeds much more easily than that of glycerol because of the stability of the five-membered ring [63]. The hydrogenolysis reaction via dehydration (indirect route) may thus also proceed in the case of erythritol. The reactivity of 1,4-anhydroerythritol on Ir-ReO$_x$/SiO$_2$ has been verified (Table 5). In the hydrogenolysis of 1,4-anhydroerythritol, 1,2,3-BuT and 1,3-BuD were mainly formed. The reaction scheme of the formation of 1,2,3-BuT and 1,3-BuD from 1,4-anhydroerythritol are described below. The different product distributions in the hydrogenolysis of erythritol and 1,4-anhydroerythritol indicate that 1,4-anhydroerythritol is not an intermediate in the hydrogenolysis of erythritol, supporting the direct reaction mechanism as in the case of glycerol. In the case of 1,4-anhydroerythritol, the formation route of 1,2,3-butanetriol and 1,3-butanetriol is shown in (5) and (6).

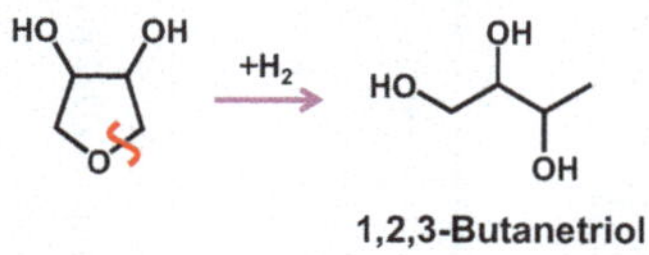

(5)

Table 5 Results of the hydrogenolysis of erythritol and 1,4-anhydroerythritol [58]

Reactant	Conv. (%)	Yield (%)							
		1,2,4-BuT	1,2,3-BuT	1,4-BuD	1,3-BuD	1,2-BuD	1-BuOH	2-BuOH	Others
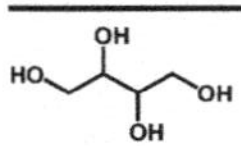	26.4	36.8	20.8	20.1	7.5	3.2	6.7	4.2	0.3
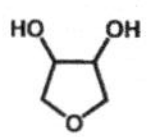	31.2	0.0	39.2	0.0	31.1	0.0	7.4	17.9	3.8

Reaction conditions: substrate 1 g, water 4 g, $W_{cat} = 0.3$ g, H_2SO_4 ($H^+/Ir = 1$), $P(H_2) = 8$ MPa, $T = 373$ K, $t = 4$ h. Reprinted from Wiley-VCH [58]
BuD butanediol, *BuOH* butanol, *BuT* butanetriol

Table 6 Reactivity trends of various polyols over Ir-ReO$_x$/SiO$_2$ [58]

Entry	Substrate	Conv. (%)	Products (selectivity (%))				
1	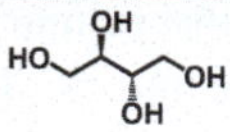	26	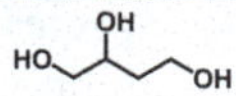 (37)	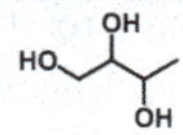 (21)	 (20)	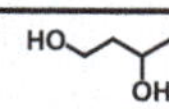 (8)	Others (14)
2	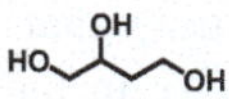	51	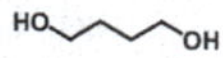 (73)	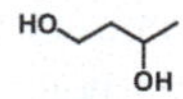 (12)	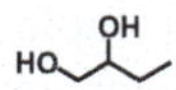 (7)	Others (7)	
3	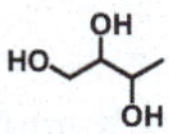	1	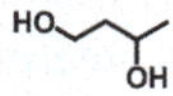 (47)	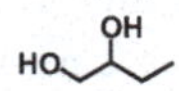 (11)	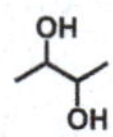 (20)	Others (23)	
4		12	 (97)	 (3)			
5	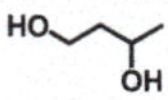	17	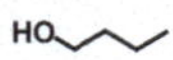 (51)	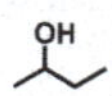 (47)	 (3)		
6	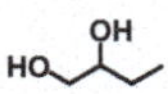	51	 (88)	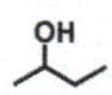 (10)	 (2)		
7	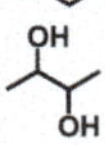	1	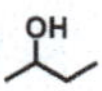 (71)	 (29)			
8		3	 (>99)				
9	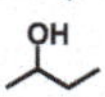	6	 (>99)				
10	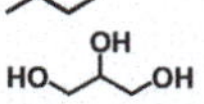	27	 (66)	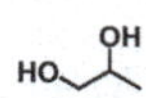 (7)	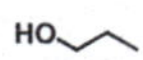 (19)	 (7)	Others (0)
11	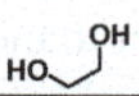	94	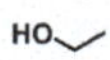 (98)	Others (2)			

Conditions: substrate 1 g, water 4 g, Ir-ReO$_x$/SiO$_2$ 0.3 g, H_2SO_4 (H^+/Ir = 1), $P(H_2)$ = 8 MPa, T = 373 K, t = 4 h. Reprinted from Wiley-VCH [58]

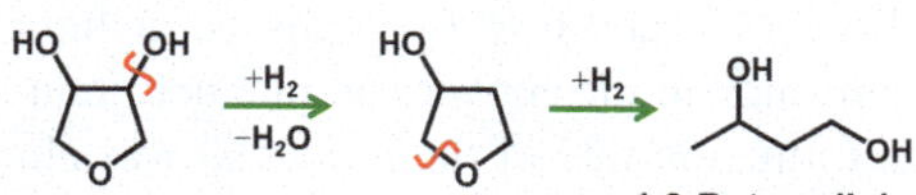

(6)

Table 6 lists the reactivity trends of various polyols on Ir-ReO$_x$/SiO$_2$ [58]. We also investigated the hydrogenolysis of the intermediates1,2,4-BuT and 1,2,3-BuT (Table 6, entries 2 and 3). The reactivity of 1,2,4-BuT was approximately equal to that of erythritol. The main product was 1,4-BuD, which was mainly obtained by the dissociation of the C–O bond neighboring the –CH_2OH group. However, 1,2,3-BuT showed very low activity (Table 6, entry 3), and its reactivity was much lower than that of glycerol (Table 6, entry 10). At present, the reason is not clear. Although the details are not shown here, the reactivity of threitol is much lower than that of the erythritol. The low reactivity of threitol and 1,2,3-BuT is another subject for the future. The reactivity of glycerol was comparable to that of erythritol. The hydrogenolysis of BuDs was also tested (Table 6, entries 4–7). The order of reactivity of the BuDs is as follows: 1,2-BuD $\gg$ 1,3-BuD > 1,4-BuD $\gg$ 2,3-BuD. The very low reactivity of 2,3-BuD is explained by no primary OH group, confirming that the interaction between the secondary OH group and the catalyst surface is weaker as mentioned earlier. Another important point is that the reactivity of BuDs with primary OH group is strongly dependent on the distance between the OH groups. This behavior is reflected in the probability of the hydride attack to the adsorbed diols attached to the Re species through the primary OH group. A greater distance means a lower probability of hydride attack. In the hydrogenolysis of glycols, the order of reactivity of ethylene glycol > 1,2-BuD > glycerol was observed (Table 6, entries 6, 10, and 11), confirming that steric hindrance around the breaking C–O bond may reduce the reactivity in the S_N2-like hydride attack. In addition, the high reactivity of ethylene glycol suggests that the reaction does not proceed via a carbocation intermediate. Low reactivity of mono-ols is also explained by the direct mechanism (Table 6, entries 8 and 9) in terms of the probability of the hydride attack.

8 Relation Between the Hydrogenolysis Selectivity and Catalyst Structure of Rh-ReO_x, Rh-MoO_x, and Ir-ReO_x

As explained above, Rh-ReO_x, Rh-MoO_x, and Ir-ReO_x catalysts had high C–O hydrogenolysis activity of THFA, glycerol, and so on. On the other hand, the selectivity in the glycerol hydrogenolysis was very different on these three catalysts. Relation between the catalyst structure and the selectivity, in particular, in the glycerol hydrogenolysis has been discussed. Figure 21 shows the formation rate of the products in the hydrogenolysis of THFA, 1,2-PrD, and glycerol over the optimized Rh-ReO_x, Rh-MoO_x, and Ir-ReO_x catalysts, including monometallic Rh and Ir catalysts. The tendency in the conversion rate per gram-catalyst of the three substrates (sum of the formation rate of the products) was almost same: Rh-ReO_x > Rh-MoO_x > Ir-ReO_x ≫ Rh > Ir. Based on the discussion on the reaction mechanism, the concentration of the active hydrogen species (hydride at the interface between the metal and the oxide species) can explain the rate tendency considering the first reaction order with respect to H_2 pressure. An interesting behavior is selectivity of products. In the case of the THFA hydrogenolysis, all the Rh-ReO_x, Rh-MoO_x, and Ir-ReO_x catalysts showed very high selectivity to 1,5-PeD (Fig. 21a). In the case of 1,2-PrD hydrogenolysis, 1-propanol (1-PrOH) was a main product and 2-PrOH a by-product. The selectivity to 1-PrOH on Ir-ReO_x was a little higher than that on Rh-ReO_x and Rh-MoO_x (Fig. 21b). In contrast, in the case of glycerol hydrogenolysis, the selectivity to 1,3-PrD on Rh-MoO_x was very low, the selectivity on Rh-ReO_x was a little higher, and the selectivity on Ir-ReO_x was much higher than Rh-MoO_x (Fig. 21c).

Figure 22 shows the hydrogenolysis selectivity trend on various substrates over Rh-ReO_x, Rh-MoO_x, and Ir-ReO_x, suggesting that the structure of MO_x (M = Re, Mo) can influence the selectivity in hydrogenolysis reactions. The tendency suggests that higher selectivity in hydrogenolysis of the substrates with larger numbers

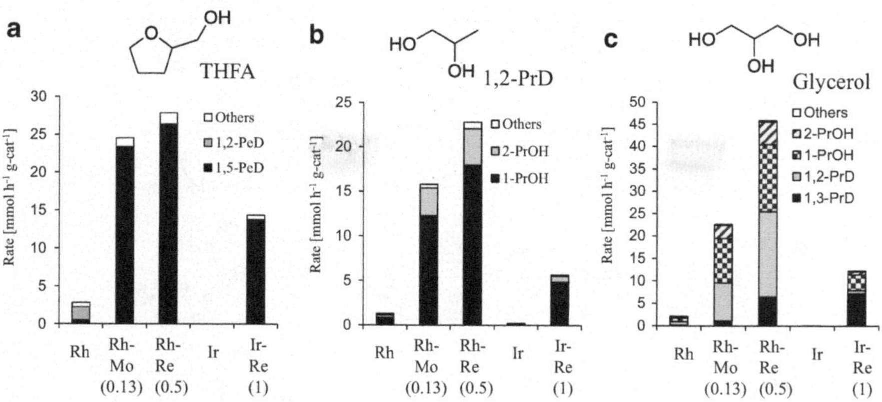

Fig. 21 Formation rate of products in the hydrogenolysis of (**a**) THFA, (**b**) 1,2-PrD, and (**c**) glycerol over various catalysts [33, 36, 38, 39, 47, 57]. Reaction conditions: (**a**) THFA 1 g (5 wt % eq.), $W_{cat} = 0.05$ g (for Ir-Re, 0.15 g), $P(H_2) = 8$ MPa, $H^+/Ir = 1$ (for only Ir-Re), 393 K. Time and conversion: Rh: 4 h, 5.7%; Rh-Mo: 4 h, 50.1%; Rh-Re: 4 h, 56.9%, Ir: 24 h, <0.1%; Ir-Re: 2 h, 43.9%. (**b**) 1,2-PrD 4 g (20 wt% eq.), $W_{cat} = 0.15$ g (for Ir, 0.3 g), $P(H_2) = 8$ MPa, $H^+/Ir = 1$ (for only Ir-Re), 393 K. Time and conversion: Rh: 24 h, 8.8%; Rh-Mo: 4 h, 18%; Rh-Re: 4 h, 26%; Ir: 48 h, 5.1%; Ir-Re, 24 h, 38.6%. (**c**) Glycerol 4 g (67 wt% eq.; for Ir-Re, 80 wt% eq.), $W_{cat} = 0.15$ g (for Ir, 0.6 g), $P(H_2) = 8$ MPa, $H^+/Ir = 1$ (for only Ir-Re), 393 K. Time and conversion: Rh: 5 h, 3.6%; Rh-Mo: 5 h, 38.9%; Rh-Re: 5 h, 79%; Ir: 240 h, 7.7%; Ir-Re: 12 h, 50.5%

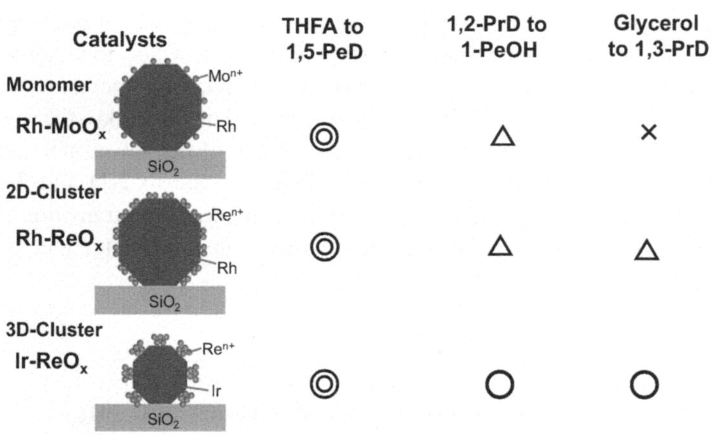

Fig. 22 Hydrogenolysis selectivity trend on various substrates over Rh-ReO_x, Rh-MoO_x, and Ir-ReO_x

of OH groups can be achieved by catalysts with more bulky MO_x (3-D cluster > 2-D cluster > isolated monomer). One interpretation is depicted in Fig. 23, showing the comparison of transition state in the glycerol hydrogenolysis based on the proposed model structures of Rh-ReO_x, Rh-MoO_x, and Ir-ReO_x, and the

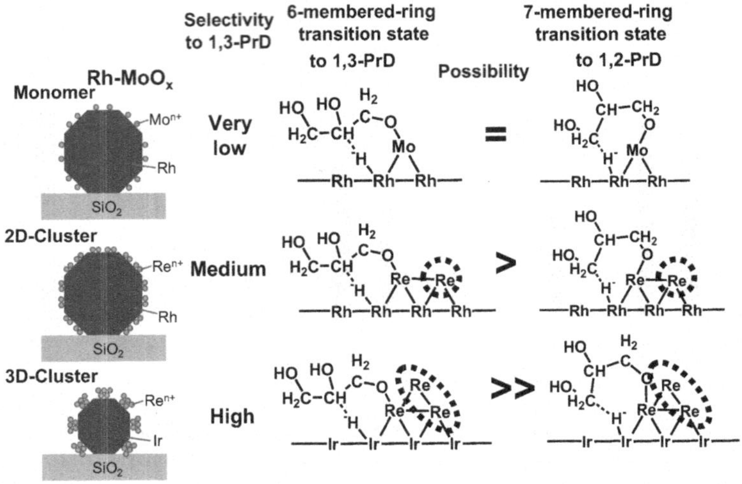

Fig. 23 Comparison of transition state in the glycerol hydrogenolysis based on the proposed model structures of Rh-ReO$_x$, Rh-MoO$_x$, and Ir-ReO$_x$

reaction mechanism. Primary OH groups have stronger interaction with secondary OH groups, and the 1-alkoxide species of glycerol is considered. The six-membered transition state can be described on the basis of the reaction mechanism for the formation of 1,3-PrD (Fig. 14). An important point is that 1,2-PrD was observed in the glycerol hydrogenolysis over Rh-ReO$_x$ and Rh-MoO$_x$, which is thought to be formed via a seven-membered transition state (Fig. 23). It is easy to imagine that the C–O–M bond in the 1-alkoxide species on isolated MoO$_x$ monomer can move much more freely than that on 3-D clusters. Free movement of 1-alkoxide species on isolated MoO$_x$ monomer can enhance the possibility of the seven-membered ring transition state. On the other hand, steric hindrance due to 2-D and 3-D ReO$_x$ clusters can decrease the possibility of a seven-membered ring transition state and relatively enhance that of a six-membered ring transition state, increasing the selectivity of 1,3-PrD formation.

9 Complete C–O Hydrogenolysis of Sugars and Sugar Alcohols to Alkanes Using Ir-ReO$_x$/SiO$_2$

In the above sections, Ir-ReO$_x$/SiO$_2$ catalyzed hydrogenolysis reactions using the interaction between OH groups and the catalyst surface, where the target products have two OH groups. In this case, the reactions are regarded as the partial C–O hydrogenolysis. In contrast, the target reactions in this chapter are complete hydrogenolysis, which means that all the C–O bonds are converted to C–H bonds by the hydrogenolysis reactions. Although the details are not shown in the previous

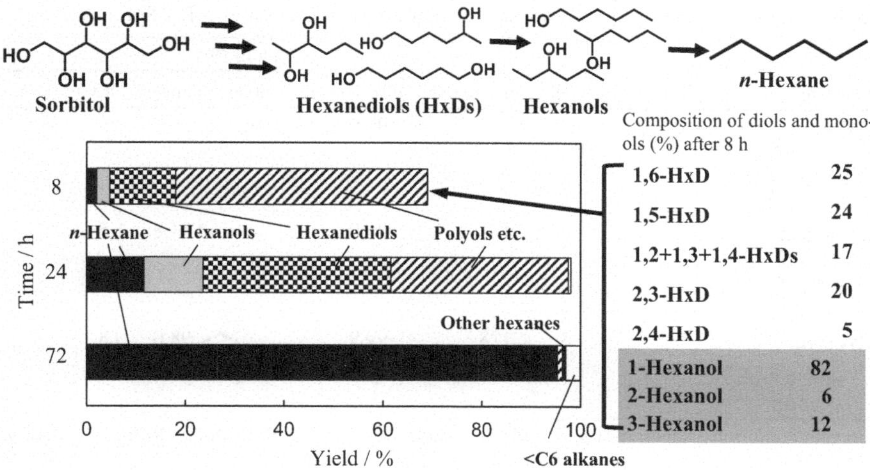

Fig. 24 Results of sorbitol hydrogenolysis over Ir-ReO$_x$/SiO$_2$ + H-ZSM-5 [68]. Reaction conditions: sorbitol 1 g, water 4 g, *n*-dodecane 4 mL, $P(H_2) = 8$ MPa, $T = 413$ K, $W_{cat} = 0.15$ g, H-ZSM-5 0.06 g. Reprinted from Wiley-VCH [68]

sections, the C–O hydrogenolysis reactions catalyzed by Ir-ReO$_x$/SiO$_2$ gave almost no degradation products formed by C–C hydrogenolysis reactions. It is concluded that Ir-ReO$_x$/SiO$_2$ catalyst showed very low activity of C–C hydrogenolysis reactions. Therefore Ir-ReO$_x$/SiO$_2$ catalyst is applied to the complete hydrogenolysis of sugars and sugar alcohols to alkanes without C–C hydrogenolysis. In this case, the reaction temperature for complete hydrogenolysis became a little higher than that for partial hydrogenolysis.

According to the previous reports on the conversion of sorbitol to alkanes, the yield of *n*-hexane is not so high, which is caused by C–C hydrogenolysis reactions [65–67], and low *n*-hexane yields were due to cracking and isomerization. In these studies, combination of metal catalyst and solid acid co-catalyst were applied. It is expected that Ir-ReO$_x$/SiO$_2$ can give high yields of alkanes because of high C–O and low C–C hydrogenolysis activity. Based on the previous reports, the combination of Ir-ReO$_x$/SiO$_2$ (Re/Ir $=$ 1) and H-ZSM-5 was applied to the conversion of sugars and sugar alcohols to corresponding alkanes [68].

Figure 24 shows the results of sorbitol hydrogenolysis using Ir-ReO$_x$/SiO$_2$ + H-ZSM-5 at various reaction times. In this experiment, *n*-dodecane was used to catch the product of *n*-hexane because of the decrease of loss to the gas phase, and it was verified that the effect of *n*-dodecane on the catalytic performance was very small. Regarding the result at 8 h, the composition of hexanediols and hexanols was also described. From these data it is found that the hydrogenolysis of inner C–O bonds proceeds preferentially, which can be explained by the behavior in the partial hydrogenolysis reaction on Ir-ReO$_x$/SiO$_2$ as mentioned in the above sections. In addition, at longer reaction time, 95% *n*-hexane yield was obtained and this yield was much higher than those reported previously. Although the details are not shown

Table 7 Results of hydrogenolysis of sugar alcohols over Ir-ReO_x/SiO_2 + H-ZSM-5 [68]

Substrate	Temperature (K)	Time (h)	Conv. (%)	Yield (%)	
				Target alkane	Other alkanes
Xylitol 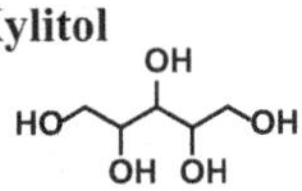	413	72	>99.9	 95.9	2.6
Erythritol 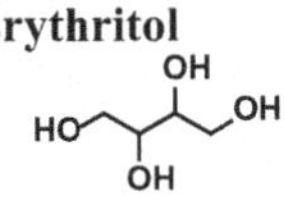	393	144	>99.9	 94.8	1.5
Glycerol 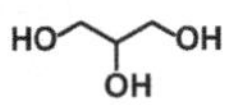	393	128	>99.9	 98.0	0.8

Reaction conditions: substrate 1 g, water 4 g, *n*-dodecane 4 mL, Ir-ReO_x/SiO_2 (Re/Ir = 1) 0.15 g, H-ZSM-5 0.06 g, $P(H_2)$ = 8 MPa. Reprinted from Wiley-VCH [68]

here, the highest yield of *n*-hexane on Rh-ReO_x/SiO_2 (Re/Rh = 0.5) + H-ZSM-5 was about 70% and this also demonstrates the superiority of Ir-ReO_x/SiO_2 + H-ZSM-5.

Table 7 lists results of hydrogenolysis of sugar alcohols over Ir-ReO_x/SiO_2 + H-ZSM-5. Ir-ReO_x/SiO_2 + H-ZSM-5 converted xylitol, erythritol, and glycerol to the corresponding linear alkanes with very high yield [68]. It should be noted that Ir-ReO_x/SiO_2 catalyst was very effective in the selective hydrogenation of unsaturated aldehyde to unsaturated alcohol. Considering that sugar alcohols are synthesized by the hydrogenation of sugars, the catalyst systems were applied to one-pot conversion of sugars to corresponding linear alkane by hydrogenation + hydrogenolysis reactions. Table 8 lists the results of the reaction of sugars over Ir-ReO_x/SiO_2 + H-ZSM-5, demonstrating that Ir-ReO_x/SiO_2 + H-ZSM-5 converted sugars to the corresponding alkanes with very high yield [68].

Figure 25 shows the results of gas chromatography (GC) analysis of the aqueous phase and *n*-dodecane for the reaction of cellobiose using Ir-ReO_x/SiO_2 + H-ZSM-5. In the phase of *n*-dodecane, *n*-hexane was observed as a main product and cracked products were also detected. In contrast, no products were detected in the aqueous phase, indicating the complete C–O hydrogenolysis of cellobiose. As mentioned above, under the reaction conditions for partial hydrogenolysis, the reactivities of mono-ols such as 1-PrOH and 2-PrOH were much lower than that of glycerol. H-ZSM-5 plays important roles in the complete hydrogenolysis.

Figure 26 shows the results of the hydrogenolysis of pentanols using Ir-ReO_x/SiO_2, H-ZSM-5, and Ir-ReO_x/SiO_2 + H-ZSM-5. It is clear that Ir-ReO_x/SiO_2 is not good at hydrogenolysis of secondary mono-ols. On the other hand, H-ZSM-5 catalyzed the dehydration of secondary mono-ols to alkenes efficiently. Therefore, it is thought that hydrogenolysis of 2- and 3-pentanols proceeds via dehydration to pentenes on H-ZSM-5 and hydrogenation of pentanes to *n*-pentane on Ir-ReO_x in

Table 8 Results of the reaction of sugars over Ir-ReO_x/SiO_2 + H-ZSM-5 [68]

			Yield (%)		
Substrate	Time (h)	Conv. (%)	*n*-Hexane	Other hexanes	Lower alkanes
Glucose 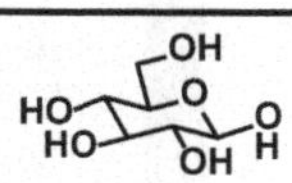	84	>99.9	94.4	0.8	3.0
Cellobiose 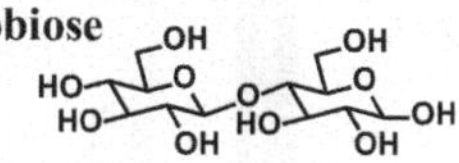	108	>99.9	94.8	0.8	3.3
Maltotriose 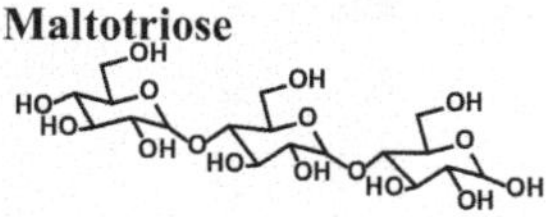	108	>99.9	94.0	0.9	3.6
Maltopentaose 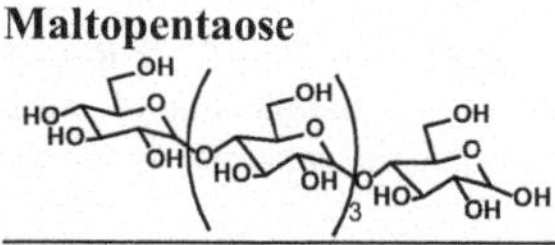	108	>99.9	94.0	0.9	3.6

Reaction conditions: substrate 1 g, water 4 g, *n*-dodecane 4 mL, Ir-ReO_x/SiO_2 (Re/Ir = 1) 0.15 g, H-ZSM-5 0.06 g, $P(H_2)$ = 8 MPa, T = 413 K. Reprinted from Wiley-VCH [68]

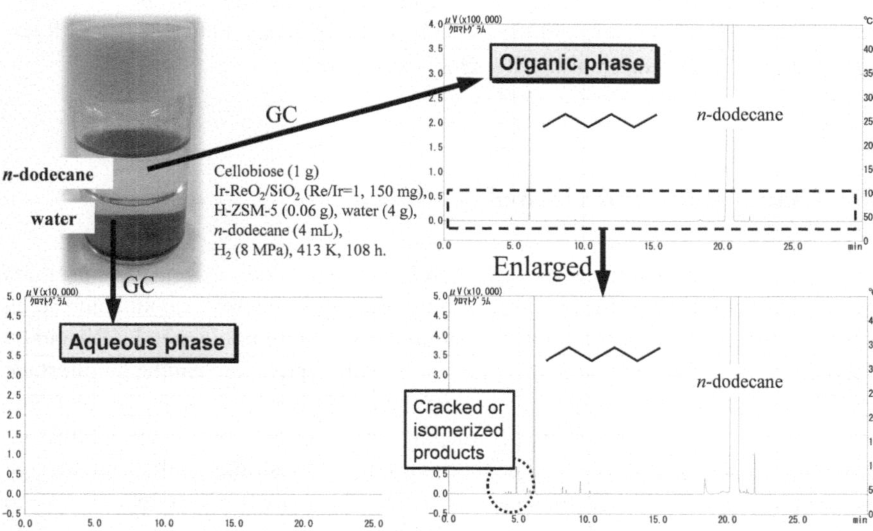

Fig. 25 Results of gas chromatography (GC) analysis of the aqueous phase and *n*-dodecane for the reaction of cellobiose using Ir-ReO_x/SiO_2 + H-ZSM-5 [68]. Reprinted from Wiley-VCH [68]

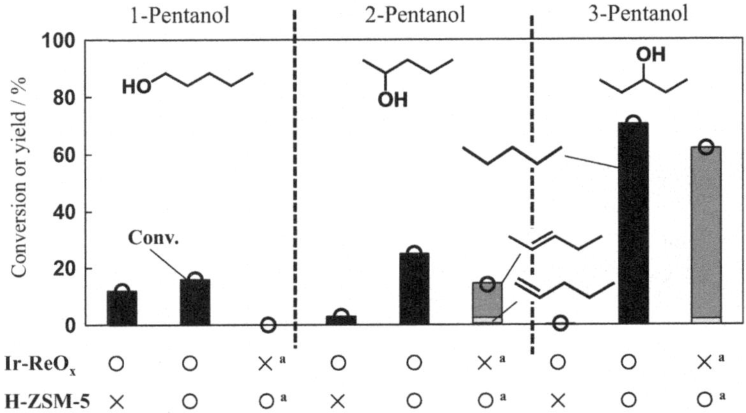

Fig. 26 Results of the hydrogenolysis of pentanols using Ir-ReO_x/SiO_2, H-ZSM-5, Ir-ReO_x/SiO_2 + H-ZSM-5 [68]. Reaction conditions: substrate 1 g, water 4 g, *n*-dodecane 4 mL, $P(H_2)$ = 8 MPa, T = 413 K, W_{cat} 0.15 g, H-ZSM-5 0.06 g. [a]The amount of substrate is 0.5 g. Reprinted from Wiley-VCH [68]

the dehydration + hydrogenation mechanism. In contrast, dehydration of 1-PeOH is not catalyzed on H-ZSM-5, and therefore the reaction of 1-PeOH to *n*-pentane proceeds on Ir-ReO_x/SiO_2. Here, the reaction mechanism can be the direct hydrogenolysis mechanism, and the hydride at the interface between Ir and ReO_x can attack the carbon atom at the Re–O–C– structure.

10 Conclusions and Outlook

The catalytic hydrogenolysis of C–O bonds in sugar alcohols and cyclic ethers including glycerol, erythritol, xylitol, sorbitol, tetrahydrofurfuryl alcohol, and so on proceeds selectively over the catalyst systems consisting of noble metals (Rh and Ir) and metal oxides (Re, Mo, and W). The catalytically active site can be an interface between the surface of the noble metal and the oxide species attached to the noble metal surface. The metal oxide species play an important role in the interaction between the catalyst surface and the OH groups of the substrates by the formation of the alkoxide species. The H_2 molecule dissociates heterolytically at the interface to give hydride and a proton. The hydride adsorbed on the interface noble metal atom can attack the carbon atom in ReO–C–C–O– structure, and this S_N2-like hydride attack dissociates the C–O bond in ReO–C–C–O– structure with the promotion of the desorption of H_2O instead of OH^- by the proton. This is a new reaction mechanism denoted as the direct mechanism. Applicability of the direct hydrogenolysis mechanism can be wider than that of conventional ones because of the simplicity. In particular, the catalysts giving the direct mechanism were highly active and can decrease the reaction temperature. This enables the conversion of thermally unstable substrates with complex structures. In addition, recently, Rh-ReO_x, Rh-MoO_x, and Ir-ReO_x catalysts have been applied to various reactions including selective hydrogenation of unsaturated aldehydes to unsaturated alcohols [69], selective dehydrogenation of vicinal alcohols to α-hydroxy ketones [70], one-pot conversion of 2,5-tetrahydrofuran-dimethanol to 1,6-hexanediol [71], one-pot conversion of furfural to 1,5-pentanediol utilizing metal-assisted catalysts [72, 73], the dehydration of fructose to HMF [74], and so on. In the future, the catalysts for direct hydrogenolysis will be used in wider applications, and the findings will be utilized to the development of new catalyst systems.

References

1. Schlaf M (2006) Selective deoxygenation of sugar polyols to α, ω-diols and other oxygen content reduced materials—a new challenge to homogeneous ionic hydrogenation and hydrogenolysis catalysis. Dalton Trans 4645–4653
2. Bozell JJ, Petersen GR (2010) Technology development for the production of biobased products from biorefinery carbohydrates—the US Department of Energy's "Top 10" revisited. Green Chem 12:539–554
3. Cantrell DG, Gillie LJ, Lee AF, Wilson K (2005) Structure-reactivity correlations in MgAl hydrotalcite catalysts for biodiesel synthesis. Appl Catal A 287:183–190
4. Ruppert AM, Weinberg K, Palkovits R (2012) Hydrogenolysis goes bio: from carbohydrates and sugar alcohols to platform chemicals. Angew Chem Int Ed 51:2564–2601
5. Nakagawa Y, Tomishige K (2011) Heterogeneous catalysis of the glycerol hydrogenolysis. Catal Sci Technol 1:179–190
6. Kusunoki Y, Miyazawa T, Kunimori K, Tomishige K (2005) Highly active metal-acid bifunctional catalyst system for hydrogenolysis of glycerol under mild reaction conditions. Catal Commun 6:645–649
7. Miyazawa T, Kusunoki Y, Kunimori K, Tomishige K (2006) Glycerol conversion in the aqueous solution under hydrogen over Ru/C + an ion-exchange resin and its reaction mechanism. J Catal 240:213–221
8. Miyazawa T, Koso S, Kunimori K, Tomishige K (2007) Catalyst development of Ru/C for glycerol hydrogenolysis in the combination with the ion exchange resin. Appl Catal A 318:244–251
9. Miyazawa T, Koso S, Kunimori K, Tomishige K (2007) Glycerol hydrogenolysis to propylene glycol catalyzed by a heat-resistant ion-exchange resin combined with Ru/C. Appl Catal A 329:30–35
10. Taher D, Thibault ME, Di Mondo D, Jennings M, Schlaf M (2009) Acid-, water- and high-temperature-stable ruthenium complexes for the total catalytic deoxygenation of glycerol to propane. Chem Eur J 15:10132–10143
11. Lao DB, Owens ACE, Heinekey M, Goldberg KI (2013) Partial deoxygenation of glycerol catalyzed by iridium pincer complexes. ACS Catal 3:2391–2396
12. Schlaf M, Ghosh P, Fagan PJ, Hauptman E, Bullock RM (2001) Metal-catalyzed selective deoxygenation of diols to alcohols. Angew Chem Int Ed 40:3887–3890
13. Ghosh P, Fagan PJ, Marshall WJ, Hauptman E, Bullock RM (2009) Synthesis of ruthenium carbonyl complexes with phosphine or substituted Cp ligands, and their activity in the catalytic deoxygenation of 1,2-propanediol. Inorg Chem 48:6490–6500
14. Foskey TJA, Heinekey DM, Goldberg KI (2012) Partial deoxygenation of 1,2-propanediol catalyzed by iridium pincer complexes. ACS Catal 2:1285–1289

15. Ueda N, Nakagawa Y, Tomishige K (2010) Conversion of glycerol to ethylene glycol over Pt-modified Ni catalyst. Chem Lett 39:506–507
16. Hirasawa S, Nakagawa Y, Tomishige K (2012) Selective oxidation of glycerol to dihydroxyacetone over Pd-Ag catalyst. Catal Sci Technol 2:1150–1152
17. Hirasawa S, Watanabe H, Kizuka T, Nakagawa Y, Tomishige K (2013) Performance, structure and mechanism of Pd-Ag alloy catalyst for selective oxidation of glycerol to dihydroxyacetone. J Catal 300:205–216
18. Murru S, Nicholas KM, Srivastava RS (2012) Ruthenium (II) sulfoxides-catalyzed hydrogenolysis of glycols and epoxides. J Mol Catal A 363–364:460–464
19. Stanowski S, Nicholas KM, Srivastava RS (2012) $[Cp^*Ru(CO)_2]_2$-catalyzed hydrodeoxygenation and hydrocracking of diols and epoxides. Organometallics 31:515–518
20. Shiramizu M, Toste FD (2012) Deoxygenation of biomass-derived feedstocks: oxorhenium-catalyzed deoxydehydration of sugars and sugar alcohols. Angew Chem Int Ed 51:8082–8086
21. Nakagawa Y, Nakazawa H, Watanabe H, Tomishige K (2012) Total hydrogenation of furfural over silica-supported nickel catalyst prepared by reduction of nickel nitrate precursor. ChemCatChem 4:1791–1797
22. Nakagawa Y, Tomishige K (2010) Total hydrogenation of furan derivatives over silica-supported Ni-Pd alloy catalyst. Catal Commun 12:154–156
23. Nakagawa Y, Tamura M, Tomishige K (2013) Catalytic reduction of biomass-derived furanic compounds with hydrogen. ACS Catal 3:2655–2668
24. Adkins H, Conner R (1931) The catalytic hydrogenation of organic compounds over copper chromite. J Am Chem Soc 53:1091–1095
25. Schniepp LE, Geller HH (1946) Preparation of dihydropyran, δ-hydroxyvaleraldehyde and 1,5-pentanediol from tetrahydrofurfuryl alcohol. J Am Chem Soc 68:1646–1648
26. Wang A, Zhang T (2013) One-pot conversion of cellulose to ethylene glycol with multifunctional tungsten-based catalysts. Acc Chem Res 46:1377–1386
27. Yabushita M, Kobayashi H, Fukuoka A (2014) Catalytic transformation of cellulose into platform chemicals. Appl Catal B 145:1–9
28. Liu Y, Luo C, Liu H (2012) Tungsten trioxide promoted selective conversion of cellulose into propylene glycol and ethylene glycol on a ruthenium catalyst. Angew Chem Int Ed 51:3249–3253
29. Furikado I, Miyazawa T, Koso S, Shimao A, Kunimori K, Tomishige K (2007) Catalytic performance of Rh/SiO_2 catalysts in the glycerol reaction under hydrogen. Green Chem 9:582–588
30. Yamagishi T, Ito S, Tomishige K, Kunimori K (2005) Selective formation of 1-propanol via ethylene hydroformylation over the catalyst originated from $RhVO_4$. Catal Commun 6:421–425
31. Tomishige K, Furikado I, Yamagishi T, Ito S, Kunimori K (2005) Promoting effect of Mo on alcohol formation in hydroformylation of propylene and ethylene on Mo-Rh/SiO_2. Catal Lett 103:15–21
32. Yamagishi T, Furikado I, Ito S, Miyao T, Naito S, Tomishige K, Kunimori K (2006) Catalyst performance and characterization of $RhVO_4/SiO_2$ for hydroformylation and CO hydrogenation. J Mol Catal A 244:201–212
33. Shinmi Y, Koso S, Kubota T, Nakagawa Y, Tomishige K (2010) Modification of Rh/SiO_2 catalyst for the hydrogenolysis of glycerol in water. Appl Catal B 94:318–326
34. Shimao A, Koso S, Ueda N, Shinmi Y, Furikado I, Tomishige K (2009) Promoting effect of Re addition to Rh/SiO_2 on glycerol hydrogenolysis. Chem Lett 38:540–541
35. Nakagawa Y, Tomishige K (2011) Catalyst development for the hydrogenolysis of biomass-derived chemicals to value-added ones. Catal Surv Asia 15:111–116
36. Amada Y, Koso S, Nakagawa Y, Tomishige K (2010) Hydrogenolysis of 1,2-propanediol for the production of biopropanols from glycerol. ChemSusChem 3:728–736
37. Takeda Y, Nakagawa Y, Tomishige K (2012) Selective hydrogenation of higher saturated carboxylic acids to alcohols using ReO_x-Pd/SiO_2 catalyst. Catal Sci Technol 2:2221–2223

38. Koso S, Furikado I, Shimao A, Miyazawa T, Kunimori K, Tomishige K (2009) Chemoselective hydrogenolysis of tetrahydrofurfuryl alcohol to 1,5-pentanediol. Chem Commun 2035–2037
39. Koso S, Ueda N, Shinmi Y, Okumura K, Kizuka T, Tomishige K (2009) Promoting effect of Mo on hydrogenolysis of tetrahydrofurfuryl alcohol to 1,5-pentanediol over Rh/SiO_2. J Catal 267:89–92
40. Chen K, Koso S, Kubota T, Nakagawa Y, Tomishige K (2010) Chemoselective hydrogenolysis of tetrahydropyran-2-methanol to 1,6-hexanediol over Re-modified carbon-supported Rh catalysts. ChemCatChem 2:547–555
41. Koso S, Nakagawa Y, Tomishige K (2011) Mechanism of the hydrogenolysis of ethers over silica-supported rhodium catalyst modified with rhenium oxide. J Catal 280:221–229
42. Koso S, Watanabe H, Okumura K, Nakagawa Y, Tomishige K (2012) Comparative study of Rh-MoO_x and Rh-ReO_x supported on SiO_2 for the hydrogenolysis of ethers and polyols. Appl Catal B 111–112:27–37
43. Koso S, Watanabe H, Okumura K, Nakagawa Y, Tomishige K (2012) Stable low-valence ReO_x cluster attached on Rh metal particles formed by hydrogen reduction and its formation mechanism. J Phys Chem C 116:3079–3090
44. Nakagawa Y, Tomishige K (2012) Production of 1,5-pentanediol from biomass via furfural and tetrahydrofurfuryl alcohol. Catal Today 195:136–143
45. Chia M, Pagan-Torres YJ, Hibbitts D, Tan Q, Pham HN, Datye AK, Neurock M, Davis RJ, Dumesic JA (2011) J Am Chem Soc 133:12675–12689
46. Amada Y, Shinmi Y, Koso S, Kubota T, Nakagawa Y, Tomishige K (2011) Reaction mechanism of the glycerol hydrogenolysis to 1,3-propandediol over Ir-ReO_x/SiO_2 catalyst. Appl Catal B 105:117–127
47. Nakagawa Y, Shinmi Y, Koso S, Tomishige K (2010) Direct hydrogenolysis of glycerol into 1,3-propanediol over rhenium-modified supported iridium catalyst. J Catal 272:191–194
48. Nakagawa Y, Ning X, Amada Y, Tomishige K (2012) Solid acid co-catalysts for the hydrogenolysis of glycerol to 1,3-propanediol over Ir-ReO_x/SiO_2. Appl Catal A: Gen 433–434:128–134
49. Amada Y, Watanabe H, Tamura M, Nakagawa Y, Okumura K, Tomishige K (2012) Structure of ReO_x clusters attached on Ir metal surface in Ir-ReO_x/SiO_2 for the hydrogenolysis reaction. J Phys Chem C 116:23503–23514
50. Nakagawa Y, Mori K, Chen K, Amada Y, Tamura M, Tomishige K (2013) C-O hydrogenolysis of Re-modified Ir catalyst in alkane solvent. Appl Catal A 468:418–425
51. Ueda R, Tomishige K, Fujimoto K (1999) Promoting effect of hydrogen spillover on pyridine migration adsorbed on Lewis acid site in USY zeolite. Catal Lett 57:145–149
52. Ueda R, Kusakari T, Tomishige K, Fujimoto K (2000) Nature of spilt-over hydrogen on acid sites in zeolites: the observation of the behavior of adsorbed pyridine on zeolite catalysts by means of FTIR. J Catal 14–22:194
53. Kusakari T, Tomishige K, Fujimoto K (2002) Hydrogen spillover effect on cumene cracking and n-pentane isomerization over catalysts prepared by physical mixture of H-β with Pt/SiO_2. Appl Catal A: Gen 224:219–228
54. Tomishige K, Okabe A, Fujimoto K (2000) Effect of hydrogen on n-butane isomerization over Pt/SO_4^{2-}–ZrO_2 and Pt/SiO_2+ SO_4^{2-}–ZrO_2. Appl Catal A Gen 194–195:383–393
55. Ishida Y, Ebashi T, Ito S, Kubota T, Kunimori K, Tomishige K (2009) Preferential CO oxidation in a H_2-rich stream promoted by ReO_x species attached on Pt metal particles. Chem Commun 5308–5310
56. Ebashi T, Ishida Y, Nakagawa Y, Ito S, Kubota T, Tomishige K (2010) Preferential CO oxidation in H_2-rich stream on Pt-ReO_x/SiO_2: catalyst structure and reaction mechanism. J Phys Chem C 114:6518–6526
57. Chen K, Mori K, Watanabe H, Nakagawa Y, Tomishige K (2012) C-O bond hydrogenolysis of cyclic ethers with OH groups over rhenium-modified supported iridium catalysts. J Catal 243:171–183

58. Amada Y, Watanabe H, Hirai Y, Kajikawa Y, Nakagawa Y, Tomishige K (2012) Production of biobutanediols by the hydrogenolysis of erythritol. ChemSusChem 5:1991–1999
59. Moon HJ, Jeya M, Kim IW, Lee JK (2010) Biotechnological production of erythritol and its applications. Appl Microbiol Biotechnol 86:1017–1025
60. Jeya M, Lee KM, Tiwari MK, Kim JS, Gunasekaran P, Kim SY, Kim IW, Lee JK (2009) Isolation of a novel high erythritol-producing Pseudozyma tsukubaensis and scale-up of erythritol fermentation to industrial level. Appl Microbiol Biotechnol 83:225–231
61. Rymowicz W, Rywińska A, Marcinkiewicz M (2009) High-yield production of erythritol from raw glycerol in fed-batch cultures of *Yarrowia lipolytica*. Biotechnol Lett 31:377–380
62. Montassier C, Ménézo JC, Hoang LC, Renaud C, Barbier J (1991) Aqueous polyol conversions on ruthenium and on sulfur-modified ruthenium. J Mol Catal 70:99–110
63. Yamaguchi A, Hiyoshi N, Sato O, Bando KK, Shirai M (2009) Enhancement of cyclic ether formation from polyalcohol compounds in high temperature liquid water by high pressure carbon dioxide. Green Chem 11:48–52
64. Manzer LE (2006) Hydrogenation of tetrahydroxybutane to tetrahydrofuran. US Patent, 7 019 155 B2
65. Huber GW, Cortright RD, Dumesic JA (2004) Renewable alkanes by aqueous-phase reforming of biomass-derived oxygenates. Angew Chem Int Ed 43:1549–1551
66. Zhang Q, Jiang T, Li B, Wang T, Zhang X, Zhang Q, Ma L (2012) Highly selective sorbitol hydrogenolysis to liquid alkanes over Ni/HZSM-5 catalysts modified with pure silica MCM-41. ChemCatChem 4:1084–1087
67. Kim YT, Dumesic JA, Huber GW (2013) Aqueous-phase hydrodeoxygenation of sorbitol: a comparative study of Pt/Zr phosphate and Pt-ReO_x/C. J Catal 304:72–85
68. Chen K, Tamura M, Yuan Z, Nakagawa Y, Tomishige K (2013) One pot conversion of sugar and sugar polyols to n-alkanes without C–C dissociation over Ir-ReO_x/SiO_2 catalyst combined with H-ZSM-5. ChemSusChem 6:613–621
69. Tamura M, Tokonami K, Nakagawa Y, Tomishige K (2013) Rapid synthesis of unsaturated alcohol in mild conditions by highly selective hydrogenation. Chem Commun 49:7034–7036
70. Sato H, Tamura M, Nakagawa Y, Tomishige K (2014) Synthesis of α-hydroxy ketones from vicinal diols by selective dehydrogenation over Ir-ReO_x/SiO_2 catalyst. Chem Lett 43:334–336
71. Buntara T, Noel S, Phua PH, Melin-Cabrera I, de Vries JG, Heeres HJ (2011) Caprolactam from renewable resources: catalytic conversion of 5-hydroxymethylfurfural into caprolactone. Angew Chem Int Ed 50:7083–7087
72. Liu S, Amada Y, Tamura M, Nakagawa Y, Tomishige K (2014) One-pot selective conversion of furfural into 1,5-pentanediol over Pd-added Ir-ReO_x/SiO_2 bifunctional catalyst. Green Chem 16:617–626
73. Tamura M, Amada Y, Liu S, Yuan Z, Nakagawa Y, Tomishige K (2014) Promoting effect of Ru on Ir-ReO_x/SiO_2 catalyst in hydrogenolysis of glycerol. J Mol Catal A Chem doi: org/10.1016/j.molcata.2013.09.006
74. Chia M, O'Neill BJ, Alamillo R, Dietrich PJ, Ribeiro FH, Miller JT, Dumesic JA (2013) Bimetallic RhRe/C catalysts for the production of biomass-derived chemicals. J Catal 308:226–236

Top Curr Chem (2014) 353: 163–184
DOI: 10.1007/128_2014_537

Published online: 23 April 2014

Deoxydehydration of Polyols

Camille Boucher-Jacobs and Kenneth M. Nicholas

Abstract The development of sustainable chemical processes for the conversion of highly oxygenated biomass feedstocks to chemical products requires efficient and selective processes for partial oxygen removal and refunctionalization. Here we review the development of the deoxydehydration (DODH) reaction, which converts vicinal diols (glycols) to olefins. Uncatalyzed deoxygenative eliminations were first established. The catalyzed DODH reactions have largely employed oxo-rhenium catalysts and a variety of reductants, including PR_3, dihydrogen, sulfite, and alcohols. A variety of glycol and biomass-derived polyol substrates undergo the DODH reaction in moderate to good efficiency, regioselectively, and stereoselectively. Observations regarding selectivity, mechanistic probes, and computational studies support the general operation of a catalytic process involving three basic stages: glycol condensation to an M-glycolate, reduction of the oxo-metal, glycol condensation to produce a metal-glycolate, and alkene extrusion from the reduced metal-glycolate. Recent practical developments include the discovery of non-precious V- and Mo-oxo DODH catalysis.

Keywords Deoxydehydration · Oxo-metal complexes · Polyols · Reductants · Rhenium

Contents

C. Boucher-Jacobs and K.M. Nicholas (✉)
Department of Chemistry and Biochemistry, University of Oklahoma, 101 Stephenson Parkway, Norman, OK 73019, USA
e-mail: knicholas@ou.edu

Abbreviations

APR	Ammonium perrhenate
Bn	Benzyl
Cp	Cyclopentadienyl
DFT	Density functional theory
DODH	Deoxydehydration
e.u.	entropy
MTO	Methyltrioxorhenium
TPB	tris-Pyrazolylborate
TsOH	*para*-Toluenesulfonic acid

1 Introduction

The highly oxygenated nature of biomass-derived feedstocks has stimulated the discovery and development of chemical processes for the efficient and selective deoxygenation and refunctionalization of these renewable chemical resources [1–4]. Cellulose, carbohydrates, and hydrolyzed plant oils are polyhydroxylated materials (polyols) and methods for their partial deoxygenation have largely focused on their dehydration reactions (Scheme 1) [5–9]. More recently, reductive processes for oxygen removal have garnered increasing attention. These approaches include catalytic hydrodeoxygenation, which replaces C–O bonds with C–H bonds [10–13], and deoxydehydration (DODH), which effects vicinal OH removal with the formation of C–C unsaturation. It produces value-added unsaturated products that have widespread use as laboratory and industrial intermediates and end-products. The development of the DODH reaction is thoroughly reviewed in this chapter (a mini-review of catalytic DODH was published recently [14]).

2 Uncatalyzed Didehydroxylation of Glycols

The earliest reported methods for vicinal dehydroxylation of glycols involved a two-stage process beginning with the acid-catalyzed condensation with ortho-esters (Scheme 2) to form dioxolanes that are subsequently thermolyzed (180–200°C) [15] or heated with acetic anhydride (140°C) [16] to produce olefins. These reactions provide moderate to good yields of olefins (40–95%) for a variety of glycols. The reactions are highly *regioselective*, locating the C–C unsaturation between the hydroxyl-bearing carbons (i.e., no double bond isomerization) and *stereospecific*, resulting from a *syn*-elimination process. Co-products of these processes include carbon dioxide, alcohols, and, in the latter case, acetic acid; hence the process is neither atom-economical nor carbon-neutral.

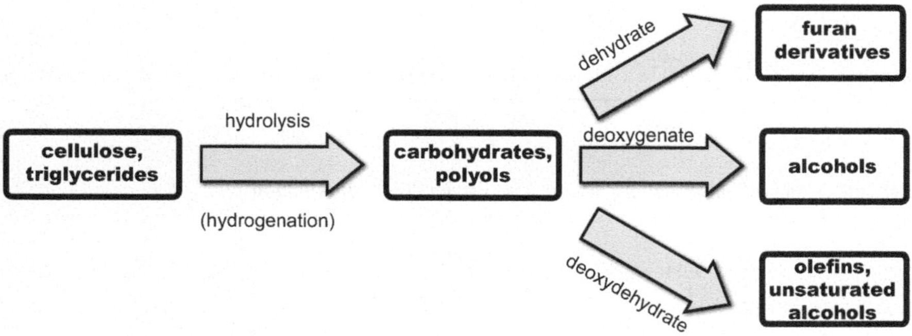

Scheme 1 Hydrolysis/deoxygenation processes for biomass oxygenates

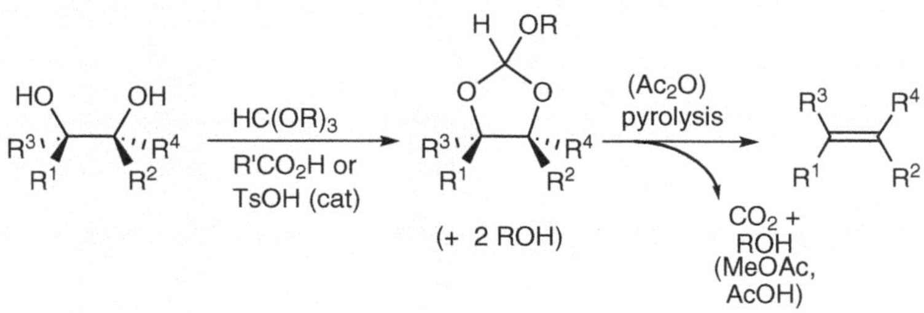

Scheme 2 Non-catalytic glycol dihydroxy elimination

A renewed interest in the glycol-to-olefin conversion was highlighted by the Bergman–Ellman team who reported a single-stage process involving high temperature reactions between glycols and polyols with formic acid as a reductant (Scheme 3) [17–19]. Water and carbon dioxide are the co-products of these transformations; hence the process is not carbon-neutral. When heated at 230–240°C under a stream of nitrogen, unsaturated products were produced in high yields if distilled as they are formed; 80–90% yields were obtained with simple glycols and glycerol. Erythritol was converted to 2,5-dihydrofuran in moderate yield, presumably via dehydration to 3,4-dihydroxyfuran followed by didehydroxylation (Scheme 4). In their patent several other natural polyols were claimed as substrates; the cyclohexane derivatives, quinic and shikimic acids, were both largely converted to benzoic acid, the result of exhaustive dehydration/dehydroxylation. Reactivity studies with diastereomeric decane-3,4-diols demonstrated a stereospecific *syn*-elimination process. This feature, together with isotopic labeling experiments, supports a proposed mechanism involving an intermediate orthoester-carbocation.

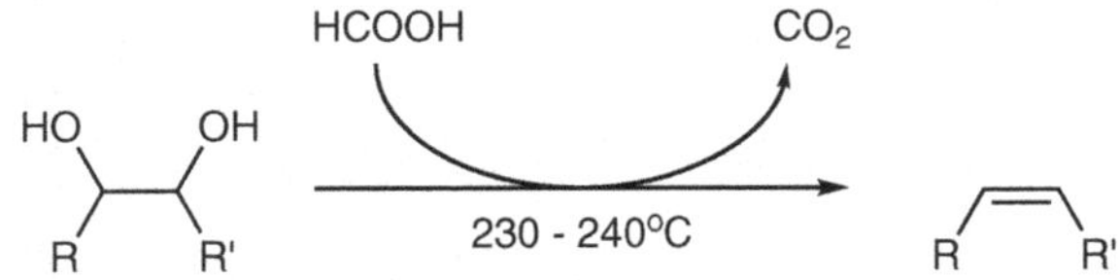

Scheme 3 Polyol substrates for formic acid-driven didehydroxylation

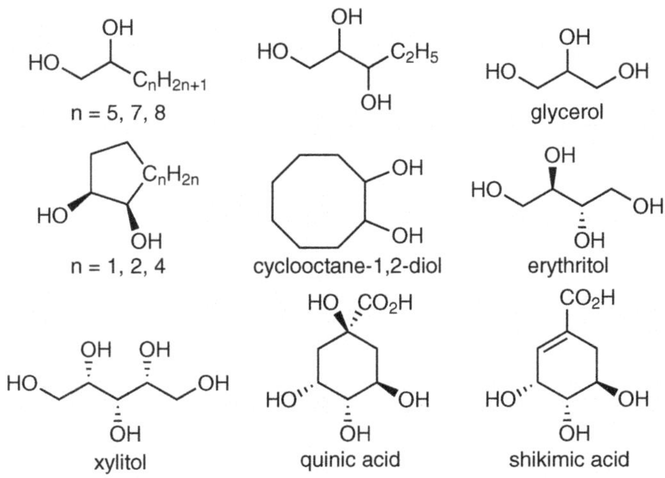

Scheme 4 Formic acid-driven didehydroxylation

3 Oxo-Metal Promoted and Catalyzed Deoxydehydration

3.1 ReO$_x$-Based Catalysts

3.1.1 Phosphine Reductants

The first transition metal-catalyzed deoxydehydration (DODH) reaction was reported by Cook and Andrews [20] employing Cp*ReO_3 (**1**, Cp*=pentamethylcyclopentadienyl) as the catalyst with PPh_3 as the stoichiometric reductant (Scheme 5). Typically, reactions were conducted with 2 mol% **1** at 90–100°C in chlorobenzene as solvent for 1–2 days. Among the substrates evaluated were 1-phenyl-1,2-ethanediol (styrene diol), a 1,2:5,6-diketal of mannitol, glycerol, and erythritol. The first two provided the corresponding alkenes in 80–95% yield, while the latter largely produced butadiene (ca. 80%) with smaller amounts of the butene-1,4- and 1,2-diols (6:1). Coordinating solvents were found to inhibit the reaction, while TsOH was found to be a promoter. Catalyst deactivation, ascribed to over-reduction to Re^{III} was noted. A three-stage catalytic cycle (Scheme 6) was proposed involving (1) reduction of **1** to Cp*ReO_2 (**2**), (2) condensation with the diol

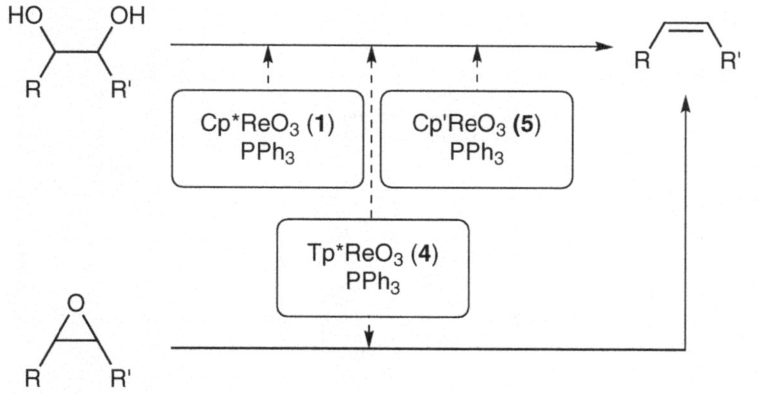

Scheme 5 Putative catalytic cycle for Cp^*ReO_3-catalyzed DODH by phosphines

Scheme 6 Phosphine-driven DODH and deoxygenation

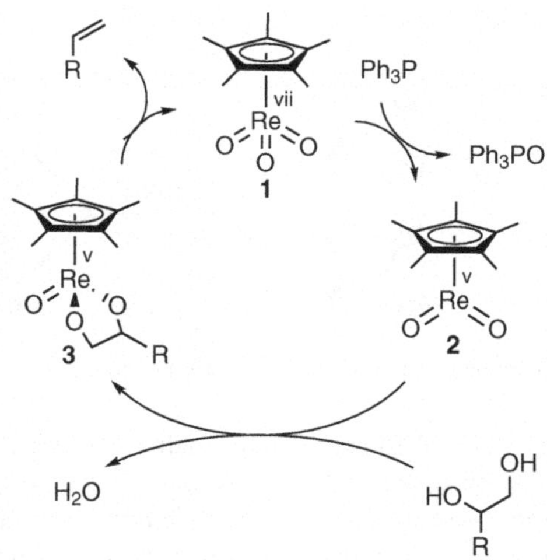

with **2** to form the Re-glycolate **3**, and (3) extrusion (retrocyclization) of the alkene from **3**. The rate of styrene formation from its glycol (**1**/TsOH co-catalyzed) was independent of the concentrations of both reactants and comparable to its rate of formation from the isolated Cp*ReO(glycolate) (see below), consistent with the Re-glycolate retrocyclization being rate-limiting.

Several years after the initial catalytic DODH report by Andrews, the Gable group reported that phosphine-driven glycol DODH and epoxide deoxygenation is catalyzed by (tris-dimethylpyrazolylborate)ReO_3 (Tp^*ReO_3) (**4**) (Scheme 7) [21]. The sterically hindered and stronger donor TPB ligand provides a more robust but somewhat less active catalyst. With the substrates styrene diol, glycerol, erythritol, and threitol, reactions catalyzed by **4** (5 mol%) proceed at 120°C over 1–5 days in toluene. Yields were not reported for the diol or triol, but the tetrols were converted in 20–40% yields to unsaturated products. For erythritol, with

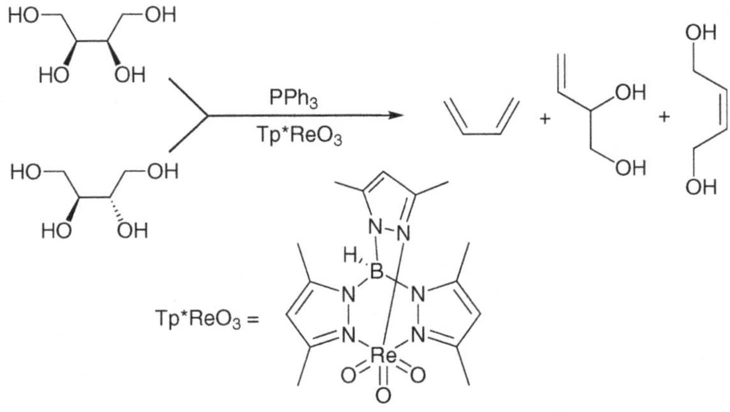

4

Scheme 7 (Tp)ReO_3-catalyzed DODH

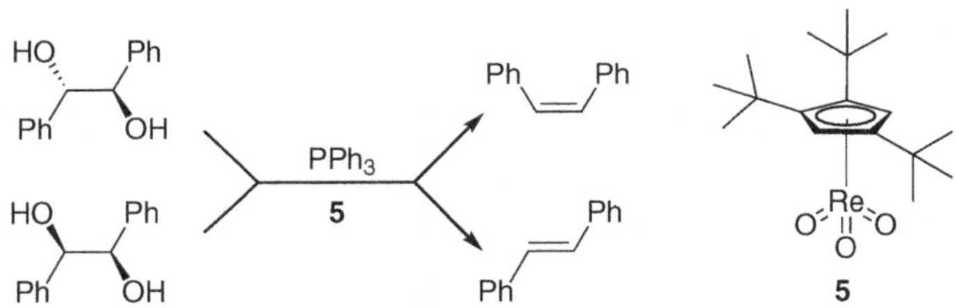

5

Scheme 8 Cp′ReO_3-catalyzed DODH reactions

1 equiv. of PPh_3, moderate selectivity for the terminal ene-diol (7.2:1 vs internal) was observed at 40% conversion, whereas threitol produced the terminal ene-diol more selectively along with relatively more butadiene. These selectivities were explained in terms of the distribution of intermediate Re-diolates produced from each polyol. Kinetic studies again indicated that alkene extrusion is turnover-limiting in this system.

A very recent report has revisited the Cp*ReO_3/PPh_3 DODH system, focusing on the use of the sterically bulky (1,2,4-tri-*tert*-butyl-cyclopentadienyl)ReO_3 (Cp′ReO_3) complex **5** as catalyst with a range of glycols and polyols (Scheme 8) [22]. The DODH reactions with **5** (0.05–2 mol%) are effective at 135–180°C (chlorobenzene), producing alkenes from acyclic glycols in generally good yields (80–95%), albeit with small amounts (2–8%) of isomerized olefins being detected. The high turnover number achieved at low catalyst loading (ca. 1,600) indicates a more robust catalyst than the original Cp*ReO_3. *cis*-Cyclic diols gave relatively poor yields (10–50%), perhaps a result of limited access to the sterically hindered Cp′-catalyst. DODH reactions with *meso- and d,l*-dihydrobenzoin were stereospecific, providing the *cis*- and *trans*-stilbenes, respectively, indicative of a *syn*-elimination process via a stereoselective alkene extrusion from an Re-glycolate

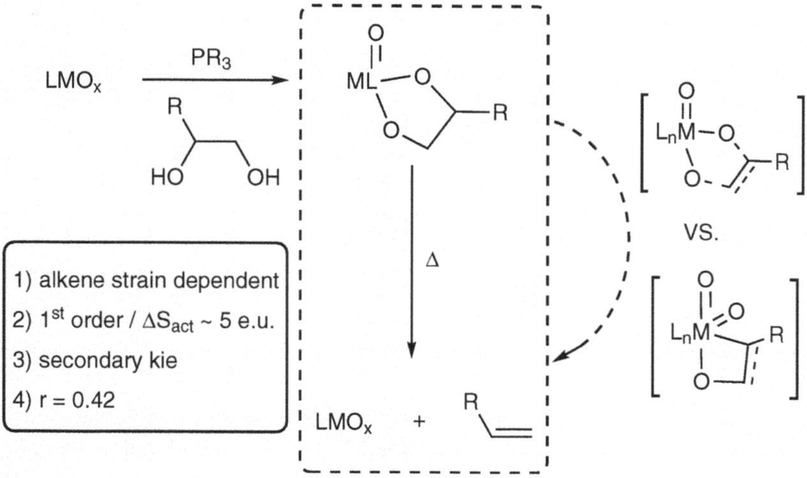

Scheme 9 Mechanistic probes of alkene extrusion from LReO(glycolate)

intermediate. Glycerol was efficiently converted to allyl alcohol (91%) but the reaction with erythritol gave a mixture of butadiene and butene-diols in modest yield. Although reactions with alternative reductants (other phosphines, H_2, *sec*-alcohols, Na_2SO_3) were generally much less efficient than with PPh_3, with 3-octanol as solvent erythritol gave butadiene selectively in 67% yield.

Gable and co-workers have made valuable contributions to elucidating the mechanism of oxo-Re-promoted DODH reactions, particularly regarding features of the alkene extrusion from LRe^V(O)-glycolate species. It is interesting from a historical perspective that their work on the Cp*Re^V(O)(glycolate) (**6a**) reactions actually preceded the report by Andrews and Cook of the Re-catalyzed DODH reaction. Producing **6a** by glycol condensation/PPh_3-reduction of Cp*ReO_3 (Scheme 9), the Gable group established: that (1) the addition/extrusion equilibrium is dependent on the strain present in the alkene; (2) the rate dependence for alkene extrusion is first-order in [**6a**] with a small ΔS_{act} (±5 e.u.); (3) there is a significant normal secondary kinetic isotope effect (1.3) with C–H(D)-labeled glycols; and (4) the reaction rate for a series of Re-glycolates derived from 4-*Z*-ArCH(OH) CH_2OH correlated with σ^- and gave a reaction parameter $\rho = 0.42$ [23–25]. When taken together with extended Huckel MO calculations, these results were interpreted in favor of an early (reactant-like), slightly polar transition state with asynchronous character with respect to the degree of C–O bond-breaking, favoring rate-limiting formation of a metallaoxetane intermediate (**7**) over a concerted [3+2] transition state (**8**).

Subsequently, the Gable group investigated the alkene extrusion process from (tris-pyrazolylborate)Re(*O*)(glycolate) derivatives **6b**. The probes and findings from these studies were generally quite similar to those from their investigation of the Cp*Re-derivatives. The Tpb-derivatives react with a slightly higher ΔH_{act} (2–4 kcal/mol) and lower rate (Scheme 9) [26, 27]. A later study focused on the

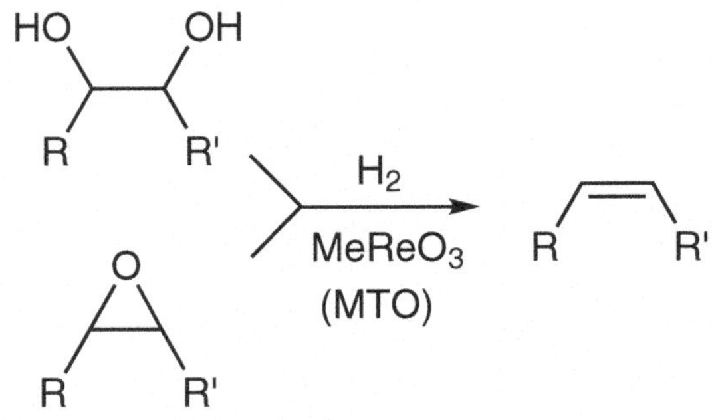

Scheme 10 Hydrogen-driven, MTO-catalyzed DODH and deoxygenation

kinetic isotope effects for derivatives separately labeled on the glycolate carbons, substituent effects for a series of Z-ArCHOHCH$_2$OH-derived glycolate complexes, and DFT computational analysis, including transition state modeling. The differences in the measured kinetic isotope effects were small, but significant, the Hammett substituent effect study showed dichotomous behavior, and the DFT-calculated transition state showed significantly different C–O bond lengths, consistent with an asynchronous extrusion process. On this issue we note that fragmentation of M-glycolates derived from symmetrical or aliphatic glycols may be more prone to concerted [3+2] cycloreversions (see below). This mechanistic subtlety notwithstanding, the examples of stereospecific *syn*-eliminations observed to date are indicative of a net *syn*-metalloglycolate cycloreversion, whether concerted or step-wise.

3.1.2 H$_2$ Reductant

The discovery of more practical, non-phosphine reductants for Re-catalyzed DODH began with a report by the Abu-Omar group employing hydrogen with the commercially available MeReO$_3$ as catalyst (Scheme 10) [28]. Reactions were conducted with epoxides and glycols at 150°C in THF at 5–20 atm H$_2$ and 5–10 mol% MeReO$_3$ over 1–16 h. Representative epoxides and glycols are converted moderately efficiently to olefins with high selectivity at lower pressures and shorter reaction times; alkanes, from over-reduction, were favored at higher pressures and longer times. The deoxygenation of *cis*- and *trans*-stilbene oxides, exhibit good but not complete stereoretentive selectivity; *cis*-cyclic glycols eliminate much more effectively than the *trans*-glycols, consistent with the intervention and concerted fragmentation of MeRe^{V}O(glycolate). A catalytic process for deoxygenation and DODH was proposed involving initial hydrogenative reduction of MeReO$_3$ to MeReO$_2$L (L=THF or H$_2$O), condensation of the latter with the glycol to form the Re^V-glycolate, and subsequent alkene extrusion.

Lin and coworkers reported a DFT-level computational study of the hydrogen-driven epoxide deoxygenation and glycol DODH catalyzed by MeReO$_3$ (MTO) [29]. Very high enthalpic barriers were found for both the [2+3] and the [2+2]

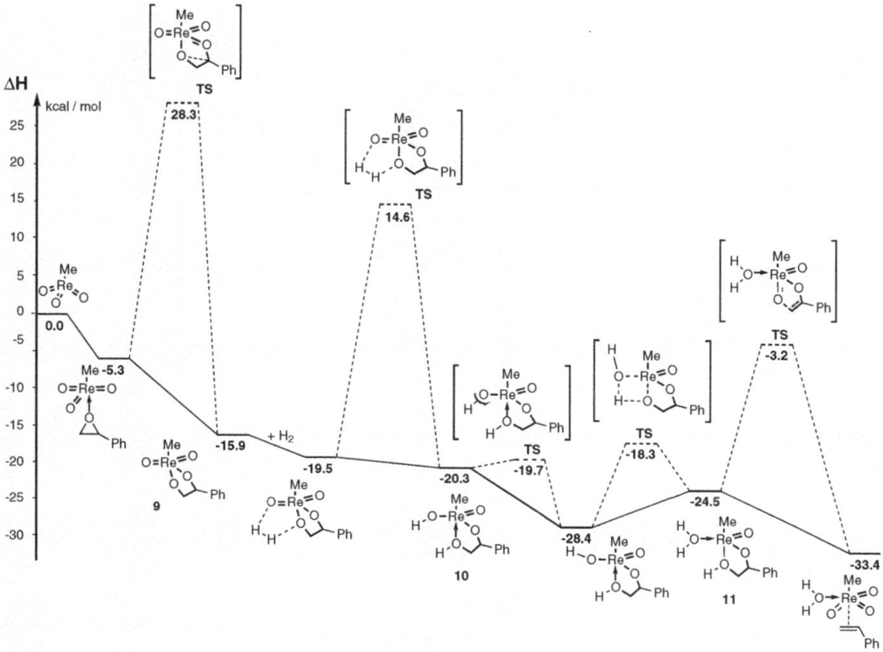

Scheme 11 Calculated lowest energy profile for H_2-driven, MTO-catalyzed DODH

addition of hydrogen to MTO (ΔH_{act} 61 and 43 kcal respectively), and was judged not to be viable as a step in the catalytic process. Instead, reaction of MTO with the epoxide or the glycol (not shown) to produce $MeRe^{VII}O_2$(glycolate) (**9**) is more favorable (Scheme 11), followed by [2+3] addition of H_2 to **9** to give the Re^{V}-glycol derivative **10** (ΔH_{act} 33 kcal). Extrusion of alkene from Re^{V}-glycolate **11** was calculated to proceed via a concerted [2+3] process with a barrier of 21 kcal. Thus, in the H_2-driven MTO reactions it appears that reduction of the glycolate **9** is likely turnover-limiting.

3.1.3 Sulfite Reductant

Soon thereafter the Nicholas group reported DODH and epoxide deoxygenation reactions driven by sulfite salts and catalyzed by $LReO_{3,4}$ derivatives (Scheme 12). Sulfite is an inexpensive and thermodynamically strong reductant, comparable to H_2 and CO [30]. Several glycol substrates are efficiently transformed into olefins (50–80% yields) with this system, operating most effectively with $MeReO_3$ or $Bu_4N^+ReO_4^-$ (2–10 mol%) in aromatic solvents at 150°C; yields and reaction rates are improved by the inclusion of the polyether, 15-crown-5, presumably by increasing the solubility of the sulfite salt reductant. Selective *syn*-elimination of the glycol was demonstrated by the preferential DODH of *cis*-1,2-cyclohexanediol

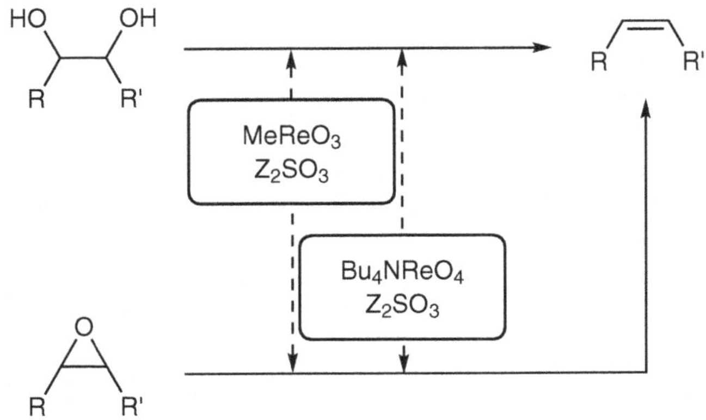

Scheme 12 Sulfite-driven Re-catalyzed DODH and epoxide deoxygenation

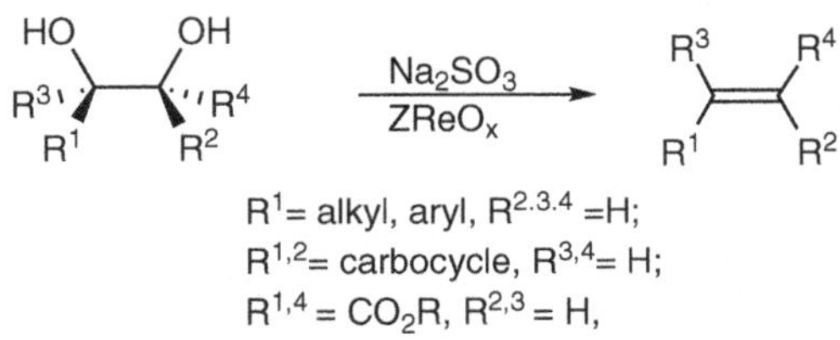

Scheme 13 Sulfite-driven Re-catalyzed DODH

and the stereoselective conversion of *l*-diethyl tartrate to (*trans*)-diethyl fumarate (Scheme 13). In a limited study, erythritol was converted by Bu_4NReO_4/Na_2SO_3 largely to 1,3-butadiene along with 2,5-dihydrofuran and 2-butene-1,4-diol (5:1:0.5). Representative epoxides were deoxygenated by MTO/Na_2SO_3 in moderate yields. Lewis base additives (e.g., amines) and potentially coordinating solvents (e.g., ethers, nitriles) were found to inhibit the MTO-based reactions, but had little effect on the yields/conversions of those catalyzed by $Bu_4N^+ReO_4^-$. This may be the result of the high Lewis acidity of $MeReO_3$ [31–33], by which coordination of donor ligands can inhibit both glycol association and alkene extrusion from the metalloglycolate (see below).

A stoichiometric reactivity study with styrene-1,2-diol/$MeReO_3$ demonstrated the viability of a condensation/reduction/alkene extrusion pathway (Scheme 14) by spectroscopically detecting [34]: (1) that $MeReO_3$ reversibly forms $MeRe^{VII}$(O)(glycolate) (**9**) from the glycol (K_{eq} ca. 0.2) or more favorably from styrene oxide at 20°C; (2) the Re^{VII}-glycolate **9** is reduced at 20°C by PPh_3 or $(Bu_4N)_2SO_3$ to a mixture of Re^V-glycolates **12/12a**; and (3) the reduced Re-glycolates cleanly produce styrene and $MeReO_3$ upon heating at 60–100°C. These observations indicate that the latter step is likely turnover-limiting.

The mechanism of the $MeReO_3$-catalyzed deoxydehydration of glycols to olefins by sulfite salts has been probed with Density Functional Theory (DFT)

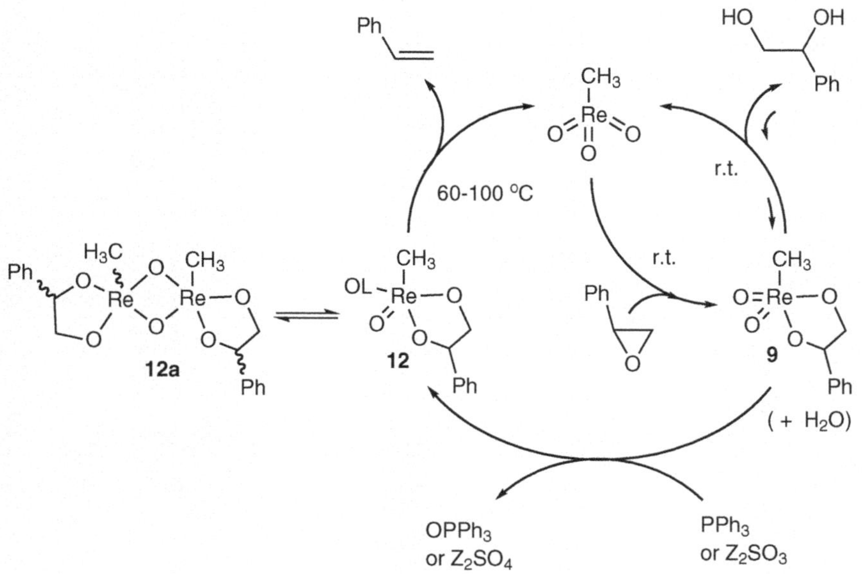

Scheme 14 Intermediates in MTO-promoted DODH/deoxygenation

calculations [35]. Potential intermediates and transition states were evaluated for the three stages of reaction: (1) dehydration of the glycol by an oxo-rhenium complex to form an Re-(O,O-glycolate); (2) sulfite-induced O-transfer (to sulfate and a reduced oxo-Re); and (3) fragmentation of the Re^V-glycolate to give the olefin and to regenerate $MeReO_3$. Various sulfite, sulfate, and Na-sulfite/sulfate species were evaluated as reactants/products and as ligands, and solvation was taken into account. Transition states and activation energies were calculated for several of the key transformations, including the H-transfer glycol dehydration, the $LMeRe^VO$(glycolate) fragmentations (L=H_2O, $NaSO_3^-$, $NaSO_4^-$), and $NaSO_3^-$ attack on oxo-Re^{VII} species. The lowest energy catalytic pathway identified involves (Scheme 15) $NaSO_3^-$ attack on an oxo-oxygen of $MeReO_3$ to produce $MeRe^VO_2(OSO_3Na)^-$ (**13**), glycol coordination by **13** followed by a series of H-transfer steps to LRe=O and/or LRe-OSO_3Na^- to give $MeRe^VO$(glycolate)$(OSO_3Na)(H_2O)^-$ (**14**), concerted extrusion of olefin from the Re^V-glycolate **14**, and dissociation of $NaSO_4^-$ from $MeReO_3(OSO_3Na)$ (**15**) to regenerate $MeReO_3$. Fragmentation of the Re-glycolate **14** is turnover-limiting with a calculated activation free energy of 35 kcal, a value consistent with the typical operating temperatures of these reactions. Coordination of H_2O, the reductant or its oxidized form to the Re(V)-glycolate is calculated to greatly affect the facility of the olefin extrusion. The nature of the reductant, e.g., an O-transfer agent like PPh_3 or SO_3^{2-} vs the H-transfer agents, H_2, or *sec*-alcohols, thus likely affects both the thermodynamics and the kinetics of the reduction step.

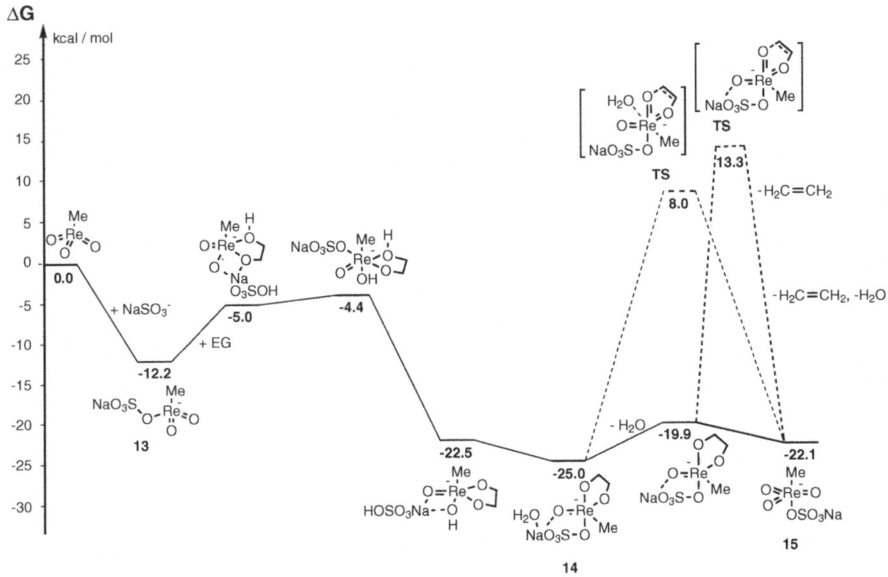

Scheme 15 Calculated lowest energy profile for sulfite-driven, MTO-catalyzed DODH

3.1.4 Alcohol Reductants

The use of secondary alcohols as reductants for DODH was first reported by Ellman, Bergman, and coworkers, who employed Re-carbonyl compounds, e.g., $Re_2(CO)_{10}$, as pre-catalysts under aerobic conditions (Scheme 16) [36]. Optimized conditions used the glycol substrate with the mono-alcohol as the solvent, e.g., 3-octanol, at 150–175°C, with 1–2.5 mol% $Re_2(CO)_{10}$ and TsOH as a co-catalyst (2–5 mol%). Good yields of the olefin (50–84%) were obtained with representative glycols. The *syn*-3,4-decanediol was converted highly selectively to *trans*-3-decene, implicating a *syn*-elimination process in the diol to olefin conversion (Scheme 17). Erythritol was converted moderately efficiently to 2,5-dihydrofuran (62% yield), presumably the result of initial 1,4-diol dehydration followed by DODH of the THF-diol intermediate. The nature of the active catalyst was unknown at the time, but was speculated to be an oxidized Re species.

The Abu-Omar group demonstrated that, in the absence of an added reductant, MeReO (MTO) catalyzes redox disproportionation of glycols. Glycerol was shown to serve as both substrate and reductant (165°C, 2 mol% MTO) in a reactive distillation process, producing moderate yields of volatile allyl alcohol, acrolein, and propanal (1.0: 0.22: 0.15), and non-volatile, reactive 1,3-dihydroxyacetone (Scheme 18). Redox disproportionation of *cis*-cyclohexanediol catalyzed by MTO was also efficient producing cyclohexene and 1,2-cyclohexanedione [37]. The rate of the glycerol reaction was determined to be first-order in the diol and first-order in MTO and a kinetic isotope effect of 2.4 was measured for

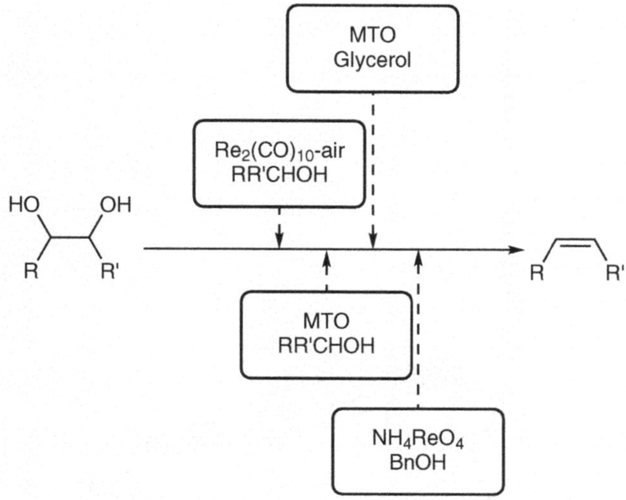

Scheme 16 Systems for alcohol-driven Re-catalyzed DODH

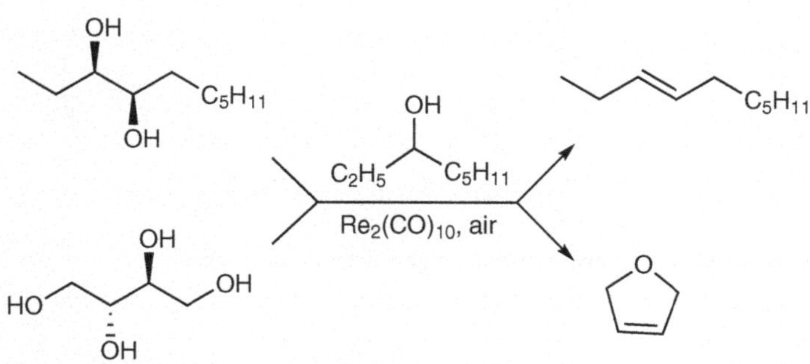

Scheme 17 Alcohol-driven, Re-catalyzed aerobic DODH

d_5-glycerol$(OH)_3$, indicating turnover-limiting C–H(D) bond breaking. This was interpreted in terms of a catalytic cycle in which the Re^{VII}-glycolate intermediate undergoes rate-limiting H(D)-transfer redox reaction by glycerol.

Other researchers have expanded the scope of alcohol-driven DODH reactions. Shiramizu and Toste demonstrated that $MeReO_3$ (MTO) was an effective catalyst for the DODH using *sec*-alcohol reductants and they extended the substrate scope to more complex polyols [38]. The alcohol reductant, preferably 3-octanol, was generally used as the solvent with 2.5 mol% of MTO at 155–200°C in air; at the higher temperatures the DODH reactions of glycols and polyols were complete in 1–3 h. The polyols glycerol, erythritol, and threitol were converted in high yields (80–90%) to allyl alcohol and 1,3-butadiene respectively, with 11–13% of dihydrofuran derivatives as minor products from the tetraols. The C_5-sugar alcohols xylitol, arabinatol, and ribitol were converted in moderate (33–61%) yields to the

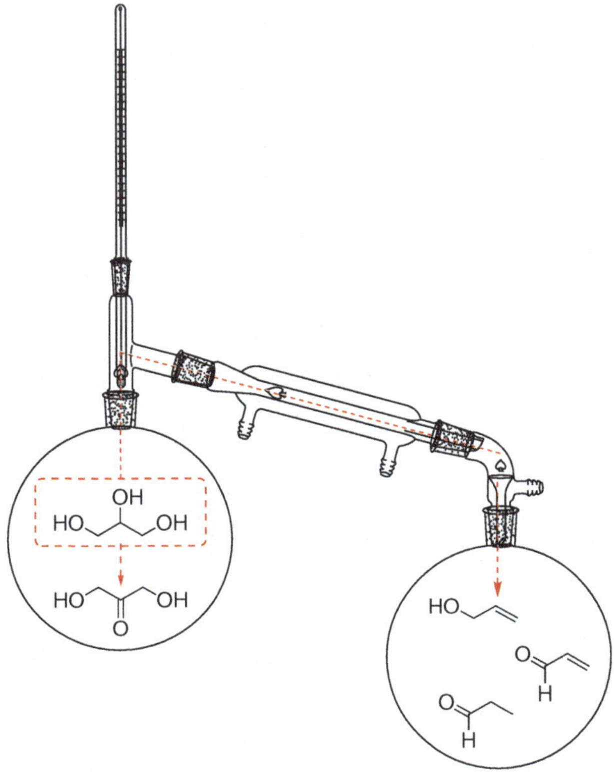

Scheme 18 MTO-catalyzed, distillative disproportionation of glycerol

same *E*-2,4-pentadienyl 3-pentyl ether derived from reaction with 3-pentanol as solvent. Similarly, the C_6-sugar alcohols sorbitol, and mannitol were completely deoxygenated to the same *E*-hexatriene (54% yield). A series of stereoisomeric inositols were converted in low to moderate yields to mixtures of benzene (from exhaustive DODH) and phenol (from DODH and dehydration), with the former generally predominating (ca. 1:1 to 3:1). Application of the alcohol/MTO protocol on sugars provided moderate yields of furan (47–60%) from erythrose and threose, while pentoses gave low yields of 2-alkoxymethylfuran, resulting from a combination of dehydration and DODH steps. Regarding the mechanism of the catalytic process, it was found that reactions run in the presence of 3-hexyne resulted in formation of isolable $MeRe^{V}O_2$(alkyne), which reacted at room temperature with glycol to form $MeRe^{V}O$(glycolate), and had comparable catalytic activity to MTO, supporting $MeRe^{V}O_2$ as the catalytically relevant species.

Shiramizu and Toste recently expanded the scope and utility of alcohol-driven, MTO-catalyzed DODH by: (1) using MTO's ability to promote allylic alcohol 1,3-transposition to enable the conversion of 1,4-unsaturated alcohols to 1,3-dienes; (2) including new polyfunctional natural substrates in DODH; and (3) carrying out a

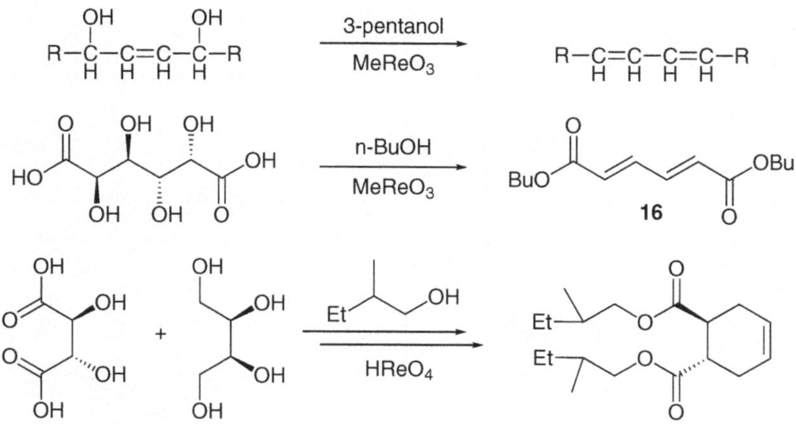

Scheme 19 Alcohol-driven, MTO-catalyzed DODH of polyols

one-pot tandem DODH-Diels–Alder reaction sequence (Scheme 19) [39]. Representative acyclic and cyclic ene-1,4-diols were converted in moderate to good yields to 1,3-dienes (18–70%) with α,β-unsaturated ketones as minor side-products. The C_6 sugar acid, mucic acid, and its ester could be efficiently converted to the dienic muconic ester **16** (71%) with the primary alcohol 1-butanol as solvent/reductant and acidic $HReO_4$ as the catalyst. Similarly the C_6 gluconic acid was efficiently converted to the *E,E*-dienic ester **16** (47%). The tandem DODH/Diels–Alder process was demonstrated with tartaric acid and erythritol as the dienophile and diene precursors, respectively. These were first heated with $HReO_4$/2-methyl-1-butanol (170°C, 4 h) to generate butadiene and the fumarate ester; continued heating at 120°C (42 h) provided the Diels–Alder adduct in 70% yield from tartaric acid.

Although secondary alcohols have generally been found to be more effective reductants in DODH reactions, Boucher-Jacobs and Nicholas showed that the reactive primary alcohol, benzyl alcohol was also an efficient reaction partner for DODH of representative glycols and polyols with less expensive NH_4ReO_4 (APR) as the catalyst (Scheme 20) [40]. This system has practical advantages for facilitating separation of the alkene and aldehyde co-products via the insoluble bisulfite adduct of the aldehyde (not effective with ketones) and also enabling efficient recovery/re-use of the insoluble catalyst. Operating at 150°C in aromatic solvents with 2.5 mol% APR, several glycols were converted to corresponding alkenes in moderate to excellent yields (50–95%).

An experimental mechanistic study of the MTO-catalyzed, alcohol driven DODH of hydrobenzoin to *trans*-stilbene was recently reported by Abu-Omar and coworkers (Scheme 21; R=Ph) [41]. Kinetic studies of the catalytic reaction in excess 3-octanol (solvent) at 140°C revealed an induction period, a zeroth-order dependence on the glycol and half order behavior in MTO; the rate dependence on the alcohol reductant was not determined. The half-order catalyst dependence suggests the involvement of a monomer–dimer equilibrium of Re-complexes. Stoichiometric reactivity experiments with NMR monitoring showed that the $MeRe^{V}O$(glycolate) intermediate could be

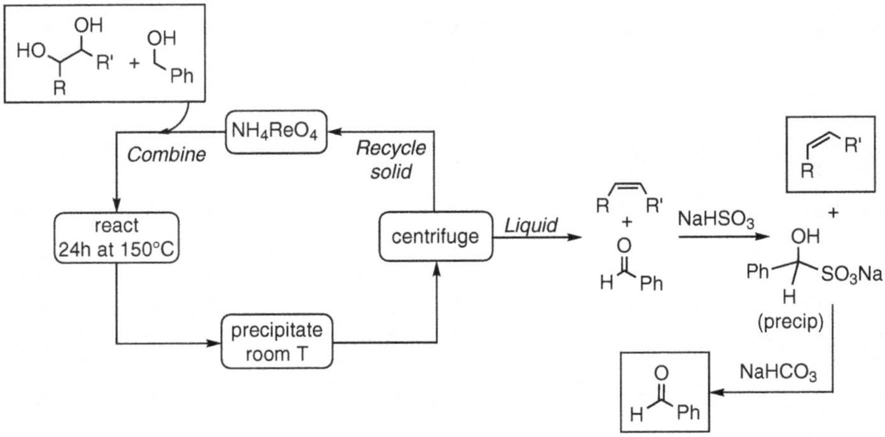

Scheme 20 Benzyl alcohol-driven, ammonium perrhenate-catalyzed DODH

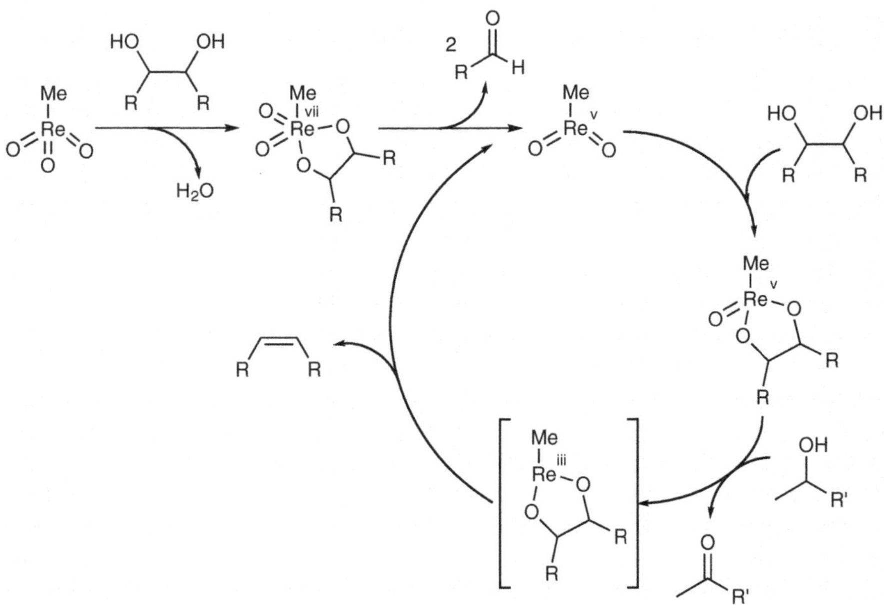

Scheme 21 Proposed catalytic cycle for alcohol-driven, MTO-catalyzed DODH

detected throughout the reaction and hence its conversion to product is probably rate-limiting. The authors claim that this conversion requires further reaction with the octanol to produce alkene and suggested that reduction to an Re^{III}-diolate precedes alkene extrusion. A small primary kinetic isotope effect (1.4) with 3-D-octanol and a large negative activation entropy (−37 e.u.) was taken as support for H(D)-transfer from the alcohol to the Re(diolate) as the rate-limiting step.

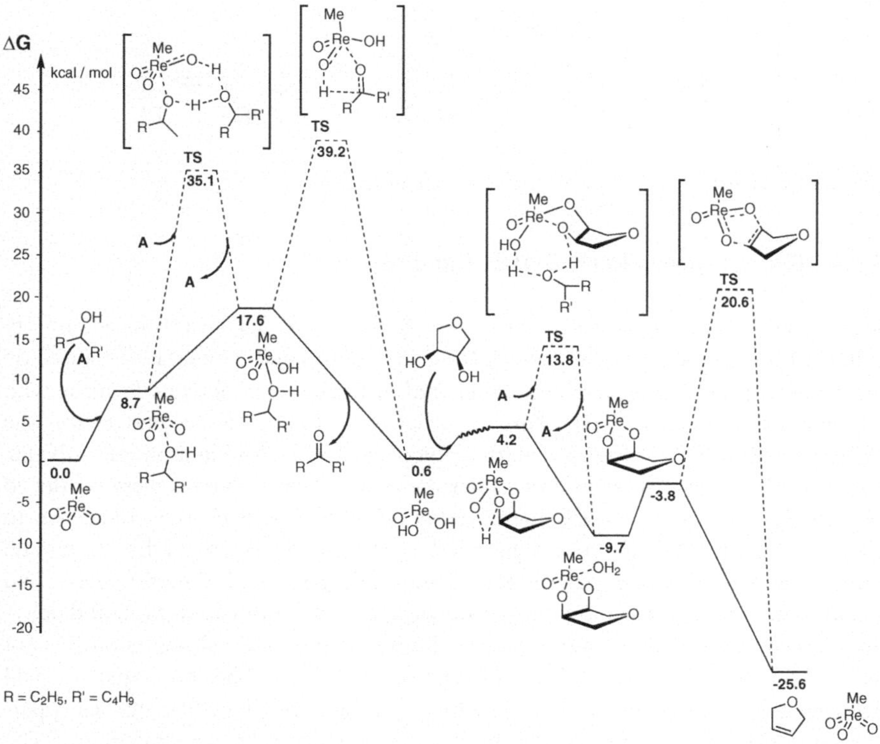

Scheme 22 Calculated lowest energy profile for alcohol-driven, MTO-catalyzed DODH

A recent computational study by Wang and coworkers analyzed the energetics for various pathways in the alcohol-mediated DODH of glycols catalyzed by MTO [42]. Free energies and enthalpies were calculated using DFT methods with the MO6 functional and corrected for temperature and solvent. Three pathways A, B, and C were compared, differing in the timing of the $Re^{VII/V}$ reduction and the glycol condensation stage, and in the condensation intermediates. In pathway A the MTO reduction precedes the condensation and the highest barrier (45/33 kcal for $\Delta G_{act}/\Delta H_{act}$) was found for intramolecular H-transfer reduction of an Re^{VII}- alkoxide-OH species. Pathway B, involving initial condensation followed by reduction, was found to have its highest barrier (53/38 kcal) in the H-transfer reduction of an Re^{VII}-H-glycolate by the alcohol. The lowest energy pathway, C (Scheme 22), like A, involves initial MTO reduction followed by condensation, but differed from A in finding a somewhat lower barrier H-transfer step (39/27 kcal) for the e^{VII}-alkoxide to form $MeReO(OH)_2$. The possibility that these reactions proceed through a Re^{III}-glycolate, as proposed by Abu-Omar, was not considered.

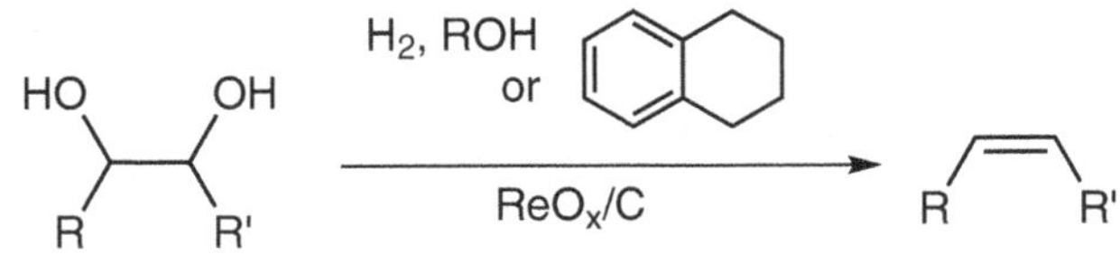

Scheme 23 DODH catalyzed by ReO_x/C with various reductants

3.1.5 Heterogeneous ReO_x DODH Catalysts

In an initial effort to develop *heterogeneous* (supported) DODH catalysts potentially suitable for industrial scale processes, Nicholas, Jentoft and co-workers reported on the catalytic properties of a material prepared by treatment of activated carbon with ammonium perrhenate [43]. This material was found to be active for the hydrogenative DODH of representative glycols at 150–175°C in aromatic solvents under 6–12 atm H_2 (Scheme 23). Under these conditions, alkenes were produced selectively in moderate to excellent yields (40–90%) with no over-reduction to alkanes; *l*-diethyl tartrate was converted stereoselectively to diethyl fumarate. From a preparative scale experiment, corresponding ketones, dimeric ethers, and acetals were identified as minor by-products, apparently from acid-promoted dehydration processes. Catalyst recovery and filtrate activity tests show partial loss of activity by the recovered catalyst and suggest catalysis by both homogeneous and heterogeneous components. Partial leaching of a catalytically active species apparently occurs under operating conditions that is re-adsorbed at room temperature. The ReO_x/C material also catalyzes moderately efficient DODH reactions (40–52%) with hydrogen transfer reductants, including 3-hexanol, benzyl alcohol, and tetralin.

3.2 Non-precious Metal Catalyzed DODH

Until recently, all the reported metal-catalyzed DODH systems have utilized oxo-rhenium catalysts. The low natural abundance and high cost of rhenium and its derivatives [44] provides an incentive for the discovery of non-precious metal catalysts for deoxydehydration that could be practically applied to large-scale biomass conversion processes.

Three brief notes of modest Mo-based DODH activity have appeared (Scheme 24), the first with styrene oxide and styrene diol promoted by (dithiocarbamate)$_2MoO_2/Na_2SO_3$ (**17**, 10–30% yield); two examples employed (acylpyrazonolate)MoO_2/PPh_3 (**18**; 10, 55% yields); [45] and a survey of several oxo-Mo-complexes with no reductant gave 35–45% olefin yields from tetradecanediol, with $(NH_4)_6Mo_7O_{24}$ [46] and $MoO_2(DMF)_2Cl_2$ (Maradur S, Nicholas KM (2012) unpublished results) being the best. Molybdenum's relative abundance [47] and lower cost should stimulate more thorough investigations to evaluate the scope and efficacy of Mo-catalyzed DODH reactions.

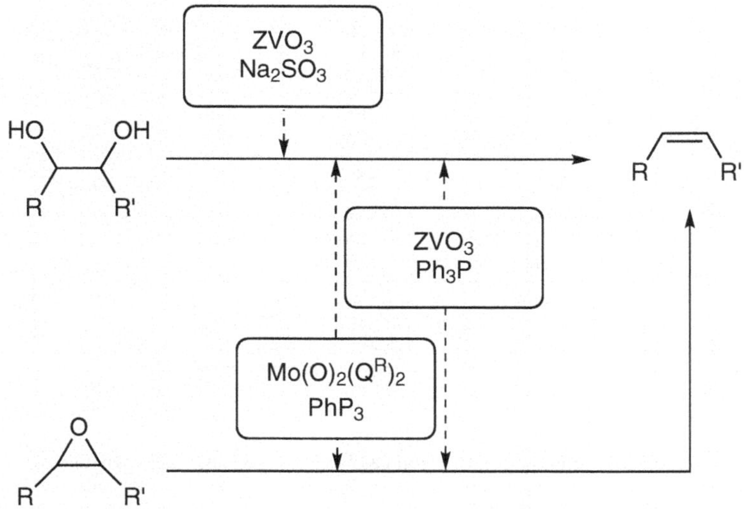

Scheme 24 Non-precious metal-catalyzed DODH and deoxygenation

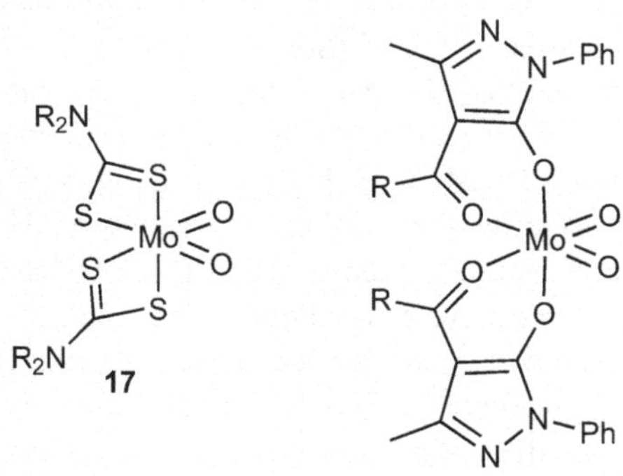

17

18

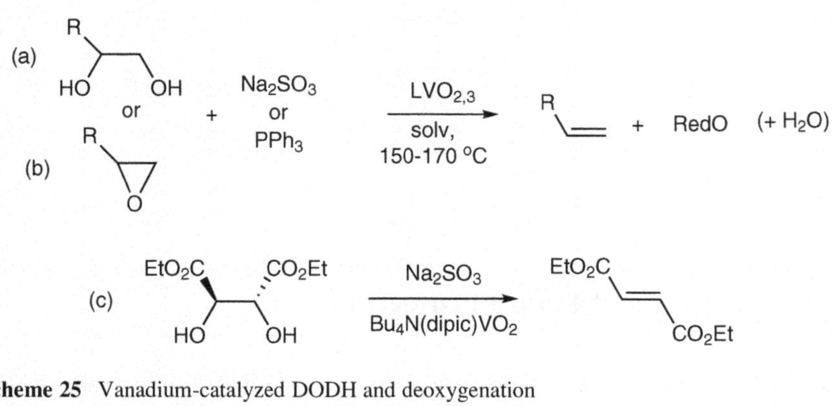

Scheme 25 Vanadium-catalyzed DODH and deoxygenation

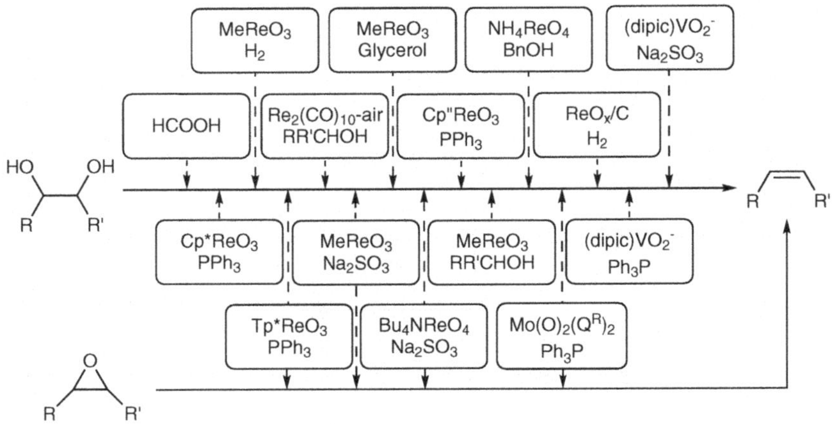

Scheme 26 Graphical summary of oxo-metal catalyzed DODH and deoxygenation

Recently, a more thorough study of DODH reactions catalyzed by non-precious vanadium complexes [48] was reported by Nicholas and coworkers. Several inexpensive and readily available oxo-vanadium compounds, including metavanadate salts ($Z^+VO_3^-$) and dioxo-vanadium complexes, were evaluated for their catalytic activity with representative glycols (Scheme 25) [49]. Among these, Bu_4N [(2,6-pyridinecarboxylate)VO_2] (**19**) was found to be the most effective with either PPh_3 or Na_2SO_4 serving as the reductant. Under optimized conditions (150–170°C, aromatic solvent, 10 mol% **19**, 24–48 h), high conversions and good yields of alkene were achieved. Highly selective *syn*-elimination was observed in the conversion of *l*-diethyl tartrate to diethyl fumarate catalyzed by PPh_3/**19** (Scheme 25c). This result supports a proposed catalytic cycle involving a reduction/condensation sequence (in either order) followed by stereoselective olefin extrusion from a V^{III}-glycolate intermediate, (dipic)V(glycolate)$^-$. The Na_2SO_3/**19** combination is also highly efficient for the deoxygenation of epoxides to olefins (Scheme 25b) [50].

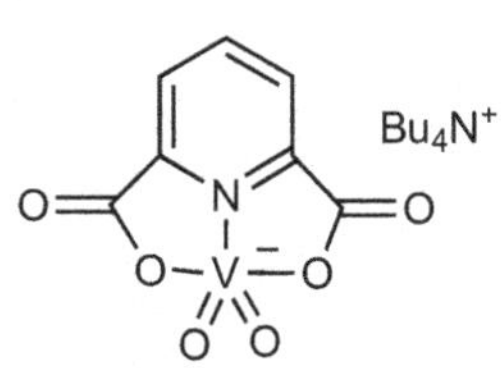

19

4 Conclusions and Future Prospects

A graphical summary of the catalytic glycol deoxydehydration and epoxide deoxygenation reactions reported to date is given in Scheme 26.

Since its discovery, the catalytic deoxydehydration reaction has seen rapid development. A variety of oxo-metal catalysts (mostly of rhenium) and reductants, including phosphines, hydrogen, sulfites, and alcohols, have been shown to be effective. Numerous glycols and a growing set of biomass-derived polyol substrates undergo the reaction with good efficiency. The reactions are typically regiospecific and highly stereoselective. Results from experimental and computational mechanistic studies suggest the general operation of a catalytic process involving three basic stages: glycol condensation to an M-glycolate, reduction of the oxo-metal, and alkene extrusion from the reduced metal-glycolate. The preferred sequence of the condensation and reduction steps and which step of the catalytic pathway is turnover-limiting depend on the catalyst and the reductant. Recent practical DODH developments include the discovery of non-precious V- and Mo-oxo DODH catalytic systems and supported oxo-rhenium catalysts.

There remain important needs for new, more active, economical deoxygenation catalysts, reductants, and practical reaction media, to improve the efficiency and extend the substrate scope to higher polyols. The development of reagent/catalyst systems that achieve high chemo-, regio-, and stereoselectivity is another important goal. Additional experimental and computational mechanistic studies are needed to provide deeper insights into the catalytic pathways and the factors correlating structure/reactivity. The development of other selective deoxygenation and refunctionalization processes may also be anticipated. We look forward with much anticipation to exciting developments in these and other, unforeseen aspects of deoxygenation reactions of biomass-derived substrates.

Acknowledgments We are grateful for financial support of our research on catalytic deoxydehydration provided by the U.S. Department of Energy (Basic Energy Sciences) and by the National Science Foundation.

References

1. Corma A, Iborra S, Velty A (2007) Chem Rev 107:2411–2502
2. Huber GW, Iborra S, Corma A (2006) Chem Rev 106:4044–4098
3. Chheda JN, Huber GW, Dumesic JA (2007) Angew Chem Int Ed 46:7164–7183
4. Aden A, Bozell J, Holladay J, White J, Manheim A (2004) Top value-added chemicals from biomass. In: Werpy T, Peterson G (eds) U.S. D.O.E. Report. See also the review by M. Mascal, M. Dusselier and B. Sels in this Topics in Current Chemistry volume
5. Agirrezabal-Telleria I, Gandarias I, Arias PL (in press) Catal Today. Submitted for publication dx.doi.org/10.1016/j.cattod.2013.11.027
6. Mascal M, Nikitin EB (2009) ChemSusChem 2:859–861
7. Mascal M, Nikitin EB (2009) ChemSusChem 2:423–426
8. Yang W, Sen A (2010) ChemSusChem 3:597–603
9. Binder JB, Raines RT (2009) J Am Chem Soc 131:1979–1985. See also the review by M. Mascal in this Topics in Current Chemistry volume
10. Ruppert AM, Weinberg K, Palkovits R (2012) Angew Chem Int Ed 51:2564–2601
11. Deng W, Tan X, Fang W, Zhang Q, Wang Y (2009) Catal Lett 133:167
12. Yan N, Zhao C, Luo C, Dyson PJ, Liu H, Kou Y (2006) J Am Chem Soc 128:8714–8715

13. Schlaf M, Ghosh P, Fagan PJ, Hauptman E, Bullock RM (2009) Adv Synth Catal 351:789. See also the review by K. Tomishige et al. in this Topics in Current Chemistry volume
14. Metzger JO (2013) ChemCatChem 5:680–682
15. Crank G, Eastwood FW (1964) Aust J Chem 17:1392–1398
16. Ando M, Ohhara H, Takase T (1986) J Chem Soc Jpn 879–882
17. Bergman R, Ellman J, Arceo E (2008) WO 2008/092,115
18. Bergman R, Ellman J, Arceo E, Marsden P (2009) US 2009/0,287,004
19. Arceo E, Marsden P, Bergman R, Ellman J (2009) Chem Commun 3357–3359
20. Cook GK, Andrews MA (1996) J Am Chem Soc 118:9448–9449
21. Gable KP, Ross B (2006) ACS Symposium Series (Feedstocks for the Future) 921:143–155
22. Raju S, Jastrzebski JTBH, Lutz M, Klein Gebbink RJM (2013) ChemSusChem 1673–1680
23. Gable KP, Phan TN (1994) J Am Chem Soc 116:833–839
24. Gable KP, Juliette JJJ (1995) J Am Chem Soc 117:955–962
25. Gable KP, Juliette JJJ (1996) J Am Chem Soc 118:2625–2633
26. Gable KP, AbuBaker A, Zientara K, Wainwright AM (1999) Organometallics 18:173–179
27. Gable KP, Zhuravlev FA (2002) J Am Chem Soc 124:3970–3979
28. Ziegler JE, Zdilla MJ, Evans AJ, Abu-Omar MM (2009) Inorg Chem 48:9998–10000
29. Bi S, Wang J, Liu L, Li P, Lin Z (2012) Organometallics 31:6139–6147
30. Vkuturi S, Chapman G, Ahmad I, Nicholas KM (2010) Inorg Chem 49:4744–4746
31. Shatnawi MY, Al-Ajlouni AM (2009) Jordan J Chem 4:119–130
32. Xu Z, Zhou M-D, Drees M, Chaffey-Millar H, Herdtweck E, Herrmann WA, Kuhn FE (2009) Inorg Chem 48:6812–6822
33. Rietveld MHP, Nagelholt L, Grove DM, Veldman N, Spek AL, Rauch MU, Herrmann WA, van Koten G (1997) J Organometal Chem 530:159–167
34. Ahmad I, Chapman G, Nicholas KM (2011) Organometallics 30:2810–2818
35. Liu P, Nicholas KM (2013) Organometallics 1821–1831
36. Arceo E, Ellman JA, Bergman RG (2010) J Am Chem Soc 132:11408–11409
37. Yi J, Liu S, Abu-Omar MM (2012) ChemSusChem 5:1401–1404
38. Shiramizu M, Toste FD (2012) Angew Chem Int Ed 51:8082–8086
39. Shiramizu M, Toste FD (2013) Angew Chem Int Ed 52:12905–12909
40. Boucher-Jacobs C, Nicholas KM (2013) ChemSusChem 6:597–599
41. Liu S, Senocak A, Smeltz JL, Yang L, Wegenhart B, Yi J, Kenttamaa HI, Ison EA, Abu-Omar MM (2013) Organometallics 32:3210–3219
42. Qu S, Dang Y, Wen M, Wang Z (2013) Chem Eur J 19:3827–3832
43. Denning L, Dang H, Liu Z, Nicholas KM, Jentoft FC (2013) ChemCatChem 5:3567–3570
44. Emsley J (2001) Rhenium is one of the rarest elements in the Earth's crust with an average concentration of ca. 7 parts per billion by weight, making it the 77th most abundant element in Earth's crust. In: Nature's building blocks: an A-Z guide to the elements. Oxford University Press, Oxford, pp 358–360
45. Hills L, Moyano R, Montilla F, Pastor A, Galindo A, Alvarez E, Marchetti F, Pettinari C (2013) Eur J Inorg Chem 3352–3361
46. Dethlefsen JR, Lupp D, Oh B-C, Fristrup P (2014) ChemSusChem. doi:10.1002/cssc.201300945
47. Eagleson M, de Gruyter W (1994) Molybdenum's abundance in the Earth's crust is 1.2 parts per million by weight. Concise Encyclopedia Chem 662
48. Emsley J (2001) Vanadium's abundance in the earth's crust is 120 parts per million by weight. In: Nature's building blocks: an A–Z guide to the elements. Oxford University Press, Oxford
49. Chapman G Jr, Nicholas KM (2013) ChemComm 49:8199–8201
50. Chapman G Jr, Nicholas KM (2014) Submitted for publication

Top Curr Chem (2014) 353: 185–228
DOI: 10.1007/128_2014_539

Published online: 4 April 2014

Polymerization of Nonfood Biomass-Derived Monomers to Sustainable Polymers

Yuetao Zhang and Eugene Y-X Chen

Abstract The development of sustainable routes to fine chemicals, liquid fuels, and polymeric materials from natural resources has attracted significant attention from academia, industry, the general public, and governments owing to dwindling fossil resources, surging energy demand, global warming concerns, and other environmental problems. Cellulosic material, such as grasses, trees, corn stover, or wheat straw, is the most abundant nonfood renewable biomass resources on earth. Such annually renewable material can potentially meet our future needs with a low carbon footprint if it can be efficiently converted into fuels, value added chemicals, or polymeric materials. This chapter focuses on various renewable monomers derived directly from cellulose or cellulose platforms and corresponding sustainable polymers or copolymers produced therefrom. Recent advances related to the polymerization processes and the properties of novel biomass-derived polymers are also reviewed and discussed.

Keywords Biomass · Cellulose · Polymerization · Renewable monomer · Sustainable polymer

Contents

Y. Zhang (✉)
State Key Laboratory of Supramolecular Structure and Materials, College of Chemistry, Jilin University, 2699 Qianjin Street, Changchun, Jilin 130012, People's Republic of China

Department of Chemistry, Colorado State University, Fort Collins, CO 80523-1872, USA
e-mail: ytzhang2009@jlu.edu.cn

E.Y-X. Chen
Department of Chemistry, Colorado State University, Fort Collins, CO 80523-1872, USA

Abbreviations

$_{\beta}$MMBL	β-Methyl-α-methylene-γ-butyrolactone
$_{\gamma}$MMBL	γ-Methyl-α-methylene-γ-butyrolactone
ATRP	Atom transfer radical polymerization
^{n}BA	*n*-Butyl acrylate
BM	Bismaleimide
CGC	$Me_2Si(\eta^5$-$(Me_4C_5)(^tBuN)$
CL	Caprolactone
CLP	Classical Lewis pair
Cp	η^5-Cyclopentadienyl
DA	Diels–Alder
DIOP	Diisooctyl phthalate
DMAP	4-Dimethylaminopyridine
DMF	*N,N*-Dimethylformamide
DOE	Department of energy
EBDMI	$C_2H_4(\eta^5$-4,7-dimethylindenyl$)_2$
EBI	$C_2H_4(\eta^5$-indenyl$)_2$
FLP	Frustrated Lewis pair
Flu	η^5- or η^3-Fluorenyl
FA	Furfuryl alcohol
FMA	Furfuryl methacrylate
GHG	Greenhouse gas
GPC	Gel permeation chromatography
GTP	Group transfer polymerization
HMF	5-Hydroxymethylfurfural
I^{t}Bu	1,3-Di-*tert*-butylimidazol-2-ylidene
ICD	β-Isocupreidine
ICD	β-Isocupreidine
IMes	1,3-Bis(2,4,6-trimethylphenyl)imidazol-2-ylidene
it	Isotactic (*mm*)
LA	Lactic acid or lactide
LPP	Lewis pair polymerization
MBL	α-Methylene-γ-butyrolactone
MEP	1,8-Bis(maleimido)-1-ethylpropane
MIMA	Maleimide methacrylate
MMA	Methyl methacrylate
M_n (M_w)	Number (weight) average molecular weight

MW	Molecular weight
MWD	Molecular weight distribution
NHC	*N*-Heterocyclic carbene
OSA	Oligo(isosorbide adipate)
OSS	Oligo(isosorbide suberate)
PASA	Poly(aspartic acid)
PDI	Polydispersity index
PET	Poly(ethylene terephthalate)
PFS	Poly(2,5-furandimethylene succinate)
PGA	Poly(glutamic acid)
PHBHV	Bis-hydroxylated poly(3-hydroxybutyrate-*co*-3-hydroxyvalerate) oligomers
PHUs	Polyhydroxyurethanes
PLA	Polylactide
PMMA	Poly(methyl methacrylate)
PP	Polypropylene
REM	Rare earth metal
ROP	Ring-opening polymerization
^{R}SKA	Trialkylsilyl methyl dimethylketene acetal
RT	Room temperature
st	Syndiotactic (*rr*)
SA	Succinic acid
SDH	Isosorbide dihexanoate
SEM	Scanning electron microscopy
T_g	Glass transition temperature
THF	Tetrahydrofuran
TOF	Turnover frequency
TPT	1,3,4-Triphenyl-4,5-dihydro-1*H*-1,2,4-triazol-5-ylidene

1 Introduction

The development and use of materials from renewable sources is not a new concept. Besides providing food, feed, clothes, shelter, and energy, biomass has been employed since ancient times to extract valuable products such as medicinal drugs, flavors, and fragrances. With the development of civilization of human society, in the nineteenth century various biomass resources were employed for the large-scale industrial production of chemicals and durable materials, such as cellulose esters (nitrate and acetate), oxidized linseed oil (linoleum), vulcanized rubber, adhesives from starches, and so on. However, the widespread use of such renewable materials diminished in the twentieth century since the development of fossil fuel derivatives, leading to the polymer renaissance. Today commodity polymers such as polyolefins are ubiquitous in our societies because they represent the optimal choice based on several factors, including monomer cost and availability, high polymer production efficiency, and excellent properties. Nowadays, coal and petroleum-based polymers can be found in nearly every item we touch, including clothing, packaging, paints, adhesives, and plastics. The world's primary sources of energy for the transportation and production of chemicals are also fossil fuels. World demand is approximately 84 million barrels per day and is projected to increase to about 116 million barrels per day by 2030 [1]. However, fossil resources are finite and will begin to dwindle in the future. Many studies surmise that all fossil resources will be depleted within a few centuries ([2], and at the present utilization rate the first fossil resource anticipated to be depleted is oil, followed by natural gas and finally coal, which is estimated to last for about another 200 years from now: [3, 4]). In addition, the emissions of greenhouse gases (GHG) caused by usage and consumption of fossil resources are perturbing the Earth's climate [5]. Lastly, as the world begins to become much more aware of the need for a sustainable future, there will be increasing pressure to search for sustainable materials. With these concerns there is now a growing shift back to polymeric materials derived from renewable sources. Intensive studies are ongoing to develop new or improved products and processes based on sustainability.

Sustainability has many definitions. One way to think of it is "meeting the needs of the present without compromising the ability of future generations to meet their needs" (defined by the World Commission on Environment and Development held by the United Nations in 1983). The concept of sustainability is that we should synchronize our consumption of natural resources with the Earth's production – in other words, using up natural resources at the same rate at which they are produced. Compared to traditional polymers typically made from petroleum and other fossil resources such as natural gas, sustainable polymers are fully or partially biobased and/or biodegradable or compostable. They are bioplastics made from renewable resources (biomass) and can be broken down faster than traditional plastics. Sustainable polymers could also protect our Earth by offering a reduced carbon footprint, a reduced use of fossil resources, and improved end-of-life options.

Biomass is most likely to be the only viable alternative to fossil resources for production of transportation fuels and fine chemicals, since it is the only carbon-rich material source available on Earth besides fossil resources. There are a large number of biomass sources that could be converted into fuels, fine chemicals, and renewable monomers for production of sustainable polymers, including simple sugars, starch, lignocelluloses, plant oils, and so on. Considering the fact that edible biomass is extensively used in the food and feed industries, the nonfood plant biomass, produced via photosynthesis utilizing solar energy, which exists in the form of lignocellulosic materials such as grasses, trees, corn stover, or wheat straw, will definitely provide ample biorenewable resources for production of fuels and chemicals. The majority (60–90 wt%) of plant biomass are the biopolymer carbohydrates stored in the form of cellulose and hemicelluloses. As cellulosic material is the most abundant renewable biomass resource on Earth, it can potentially meet our future needs if it can be efficiently converted into fuels and value added chemicals. The cellulose can be depolymerized by hydrolysis of the β-1,4-glycosidic bond to glucose. This C_6 saccharide is the starting material for a large variety of fine chemicals and potential monomers. Accordingly, this review chapter focuses on the recent advances on various chemical platforms derived from these nonfood biomass resources, including ethanol, lactic acid or lactide, furfural, 5-hydroxymethylfurfural (HMF), levulinic acid, sorbitol, and several dicarboxylic acids containing different numbers of carbons. These platforms can be further employed to produce renewable monomers for sustainable polymer production.

2 Sustainable Polymers Based on Cellulose

2.1 Ethanol Platform

Currently, ethanol from renewable resources is mainly obtained by fermentation of glucose derived from sugar cane or sugar beet with yeast. In the long run, the nonfood biomass, cellulose, should be the best candidate for production of ethanol. Cellulose is a major constituent of plants and may soon become an important component in our mix of energy sources. Cellulosic ethanol is made from biomass, including wood, grasses, agricultural residues, and municipal solid waste. By producing ethanol, we can reduce our demands for petroleum, lower transportation costs, reduce greenhouse gas emissions, and provide more economic stability. The federal Energy Independence and Security Act of 2007 mandates that by 2022 the U.S. produce 36 billion gallons of biofuel per year, of which 16 billion gallons (nearly 10% of the total U.S. transportation fuel supply) must be cellulosic ethanol (http://www.epa.gov/oms/renewablefuels/420f10007.htm).

Besides serving as transportation fuel, by means of dehydration ethanol can be converted to ethylene which is mainly produced in the petrochemical industry by steam cracking processes and is used in the production of polyethylene, polyethylene oxide, polyvinylchloride, and polystyrene. Vapor phase dehydration of ethanol at 400°C affords 99.9% selectivity to ethylene at 99.5% conversion [2, 6]. Recently, both the low price of sugar cane in Brazil and the increasing crude oil cost has spurred the renewed interest in ethanol dehydration. Dow (http://news.dow.com/dow_news/prodbus/2007/20070719a.htm), Braskem (http://www.reuters.com/article/pressRelease/idUS246273+05-Jun-2008+PRN20080605) (Brazil's largest plastics producer), and Solvay (http://www.solvinpvc.com/static/wma/pdf/1/2/1/0/0/Press_release_Brasilian_SolVinPVC_EN_141207.pdf) have announced separate projects for large-scale production of ethylene from renewable ethanol based on sugar cane. Dow and Braskem will ultimately manufacture "green" polyethylene while Solvay will use ethylene to supply its polyvinylchloride capacity. The projects of Braskem and Solvay, with production of 180,000 and 55,000 tons/year, respectively, are underway. However, Dow (estimated polyethylene capacity of 320,000 tons/year) later announced a delay in their construction of their plant. Today, Braskem is the world's leading producer of green polyethylene (Green PE) producing 200 ktons of ethylene per year from sugar cane ethanol. In addition to using renewable raw materials, the production process adopted for the Green PE also has a lower environmental impact, because every metric ton produced removes up to 2.5 tons of CO_2 from the atmosphere. After inaugurating the world's first industrial-scale green ethylene plant, Braskem is now focusing on the development of an innovative and economically competitive process for the production of green polypropylene. In 2008, Braskem produced green polypropylene in its laboratories. The samples were obtained at Braskem's Technology and Innovation Center on a pilot scale, where homopolymers and copolymers were produced and certified by the U.S. laboratory Beta Analytic Inc., the world leader in carbon isotope analysis, as being 100% from renewable raw materials. Now Braskem has been intensifying its research to improve competitiveness on an industrial scale (http://www.braskem.com.br/plasticoverde/eng/braskem.html, accessed Dec 23, 2013).

2.2 Lactic Acid Platform

Lactic acid (LA), or 2-hydroxy-propanoic acid, is one of the high potential and versatile biomass-derived platform chemicals [7, 8], and has been widely utilized in the food, cosmetics, pharmaceutical, and chemical industries [9–12]. LA could serve as an active precursor and renewable feedstock for the production of a wide range of useful intermediates[7, 13–15] and various polymers; through dehydration (combined with other reactions) resulting in acetaldehyde, acrylic acid (to polyacrylates), 2,3-pentanedione, and propionic acid; through reduction and oxidation yielding propylene glycol (to polyester, polycarbonates, polyurethanes, and polypropylene oxide) or pyruvic acid; through catalytic upgrading of LA via propionic acid and acetaldehyde affording C_5–C_7 ketones; and through esterification providing the synthesis of alkyl lactates, lactide (to polylactide or poly(lactic acid), PLA) (Scheme 1). Among these, PLA has received the most attention due to its intriguing physico-chemical and mechanical properties. PLA exhibits mechanical similarities to poly(ethylene terephthalate) (PET) and poly(propylene) (PP). It can be processed in most polymer processing equipment, which is a critical factor for industrial use. However, more importantly, it appears to be the polymer with the broadest range of applications, because of its ability to be stressed or thermally crystallized, filled, and copolymerized [16, 17]. In addition, as a renewable polymer, PLA has the potential to replace fossil-derived plastics in particular applications, and, based on its life cycle analyses, it has a more positive impact on the environment [11]. Therefore, PLA is the most well-known biocompatible and biodegradable polymer and is one of the most widely used bioplastics in packaging and medical applications. PLA can be obtained either by direct polycondensation [18, 19] of lactic acid or by ring-opening polymerization (ROP) [20] of lactide, a cyclic dimer of lactic acid. With the polycondensation method, it is difficult to obtain high molecular weight PLA due to the equilibrium between lactic acid, water, and lactoyl oligomers, and the increasing viscosity during polymerization. Currently, the most convenient and efficient production of PLA is mainly carried

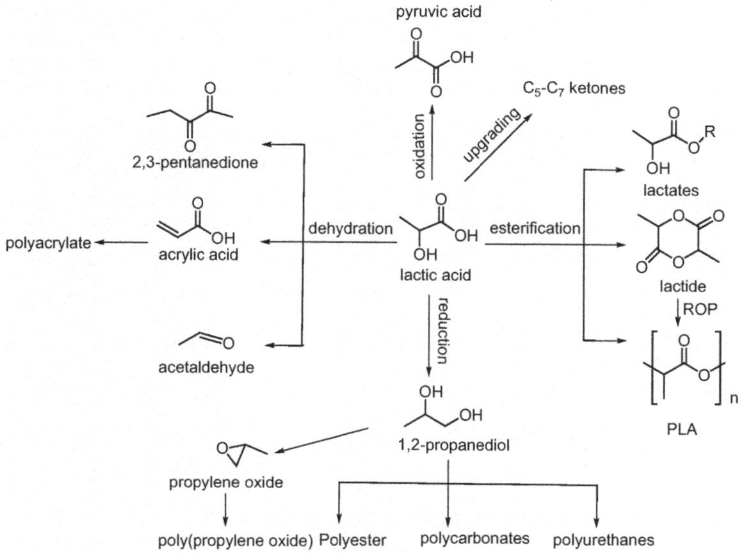

Scheme 1 Chemical intermediates and polymers from lactic acid

out by ROP of lactides, which involves multiple steps. First, lactic acid is produced by sugar fermentation or by non-fermentative chemical synthesis starting from various biomass resources such as triose [21–29], hexose [30–40], and even cellulose [41–46], the most abundant and non-edible biomass on the earth, then followed by preparation of lactide from lactic acid. After purification, lactide can produce PLA via the ROP process.

Within the last decade, several ROP processes have been developed to meet the high demand for PLA, including anionic [47–50], cationic [51–54], organocatalytic [55–57], and coordination-insertion [58] methods. The coordination-insertion process is now commonly regarded as the most efficient method for the well-controlled synthesis of PLA with regard to composition, molecular weight, and microstructure [10, 16, 17, 58–65]. The coordination-insertion mechanism of lactide polymerization involves the coordination of the monomer to the metal center, followed by a nucleophilic attack of the alkoxide on the acryl carbon atom and the insertion of lactide into the metal-alkoxide species with retention of configuration [66]. A new metal-alkoxide species is formed, which is capable of further insertion reactions. A vast multitude of well-defined Lewis acid catalysts following a coordination-insertion mechanism have been developed for this reaction, mainly based on tin [67], zinc [68–71], aluminum [72–74], and rare earth metals [75–79]. Moreover, organocatalysts [20, 55–57] have been successfully used in ROP and even some heterogeneous catalysts such as tin-substituted mesoporous silicas [80] have been proposed. It is very important to control the stereochemistry of PLA because it plays a very important role in PLAs' physico-chemical and mechanical properties, and biodegradability. Remarkable progress has been made on the synthesis of various stereo-controlled polymer architectures by ROP of enantiomerically pure

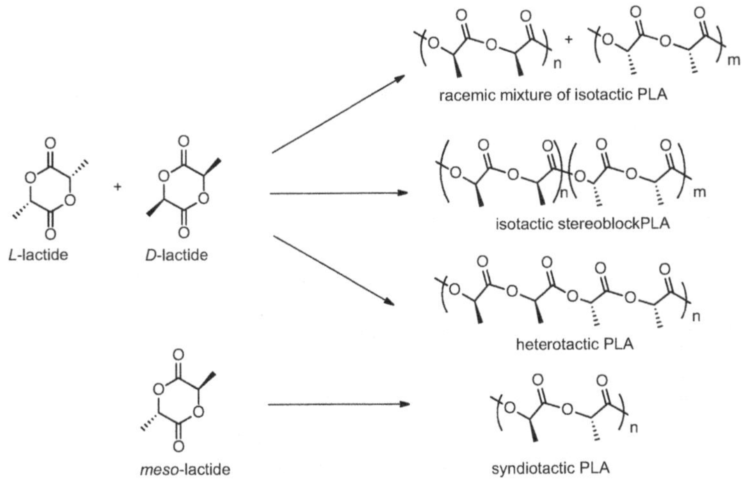

Scheme 2 Lactide stereochemistry and typical PLA microstructures

monomer, racemic mixtures, or meso lactide. So far, highly isotactic and heterotactic PLA materials have been formed from *rac*-LA, while highly syndiotactic PLA is prepared from *meso*-LA (Scheme 2) ([66, 81] and reference therein). Described below are the selected examples that allow the synthesis of highly stereo-controlled polymers.

2.2.1 Formation of Isotactic/Stereoblock PLA Materials from *rac*-LA

Some of the most significant advances in stereocontrolled polymerization of lactide have been demonstrated by using aluminum-based catalysts. In 1994, Spassky and coworkers reported the first stereocontrolled example of Schiff base (SALEN type) aluminum complexes (*R*, *R*)-**1a** [82], which selectively promote ROP of (*R*, *R*)-LA from *rac*-LA to form isotactic stereoblock PLA and leave (*S*, *S*)-LA largely unreacted (at 70°C, $k_{RR}/k_{SS} = 20{:}1$). This material exhibited a melting-transition temperature (T_m) of 187°C, higher than that of the enantiopure isotactic (*S*)-PLA or (*R*)-PLA (between 170 and 180°C). Since then, more aluminum Schiff base systems have been developed and exhibit high stereoselectivity in the ROP of LA. For example, Coates et al. found that complex ***rac*-1b** [83, 84] mediates ROP of *rac*-LA to give a stereoblock PLA material with a T_m of 179°C, while complex (*R*, *R*)-**2** [73, 74] reported by Feijen et al. has a moderate polymerization activity but a strong preference for the polymerization of (*S*, *S*)-LA ($k_{SS}/k_{RR} = 14$) from *rac*-LA.

In 2004, Chen and co-workers [85] reported that complex **3a** exhibits high isoselectivity in the ROP of *rac*-LA, generating a stereoblock PLA with a P_m value of 0.90. Thermal analysis revealed that this stereoblock PLA has a T_m of 201°C. Replacing the ethyl group with the isopropyl group, complex **3b** [86] showed the same isoselectivity and polymerization rate constant as that of **3a**. It

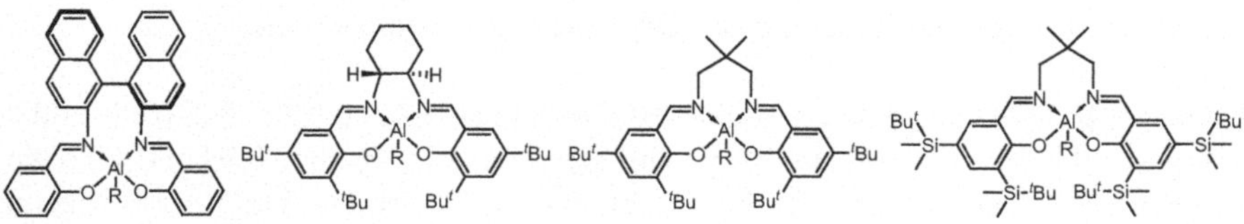

1a, R= OMe
1b, R= O*i*Pr

2, R= O*i*Pr

3a, R= Et
3b, R= O*i*Pr

4 R= OCH2Ph

Fig. 1 Structure of complexes **1–4**

is noted that complex **3b** can maintain its high isoselectivity in ROP of *rac*-LA, even at high temperature [87]. With increasing polymerization temperature, T_m and isotacticity of the corresponding PLA materials decreased somewhat (at 130°C, $T_m = 169°C$, $P_m = 0.84$; at 150°C, $T_m = 158°C$, $P_m = 0.82$; at 180°C, $T_m = 155°C$, $P_m = 0.80$). In 2007, Nomura et al. reported the synthesis of Schiff base aluminum complex **4** [88] with flexible but bulky ^{t}BuMe$_2$Si substituents, which exhibited the highest isoselectivity in the ROP of *rac*-LA to form isotactic stereoblock PLA materials with a P_m value of 0.98 and a T_m of 210°C. More recently, highly active yttrium phosphasalen initiators were reported for the stereocontrolled ROP of *rac*-lactide [89]. Changing the phosphasalen structure enables access to isoselectivities ($P_m = 0.84$) or hetero-selectivities ($P_r = 0.87$) (Fig. 1).

Since the first report of the use of 4-dimethylaminopyridine (DMAP) for the ROP of lactide in 2001 [90], the field of organocatalysts for ROP has received much more attention since it provides an alternative to the metal-based catalysts stereo-controlled ROP of lactides. In 2007 Wade et al. found that a dimeric phosphazene base stereoselectively produced isotactic PLA from *rac*-LA ($P_m = 0.95$ at −75°C) [91]. Recently, Chen reported that β-isocupreidine (ICD) was applied in stereoselective ROP of *rac*-lactide to form an isotactic-enriched PLA ($P_m = 0.74$) at room temperature [92].

2.2.2 Formation of Sydiotactic PLA Materials from *meso*-LA

In 1999, Coates and Ovitt reported the first example of highly syndiotactic PLA (with syndiotacticity up to 96%) formation from the ROP of *meso*-LA catalyzed by chiral aluminum isoperoxide complex (*R*, *R*)-**1b** [93]. Due to the high degree of stereoregularity, this PLA has a high T_m of 152°C. When catalyzed by achiral metal complexes through a chain-end-control mechanism, the ROP of *meso*-LA did not produce highly stereo-controlled PLAs. For example, β-diketiminate zinc complex **5a** [69] afforded syndiotactic PLA with a P_r value of 0.76 and yttrium amido complex **6** [94, 95] yielded a moderate syndiotactic PLA material with a P_r value of 0.76.

2.2.3 Formation of Heterotactic PLA Materials from *rac*-LA

Coates et al. has developed a new class of *β*-diiminate zinc complexes **5** [69] for the synthesis of highly heterotactic PLA (with heterotacticity P_r up to 0.94 at 0°C) from *rac*-LA by alternately incorporating (*R*, *R*)- and (*S*, *S*)-lactides. The substituents on the *β*-diiminate ligand affected both the degree of stereoselectivity and the rate of polymerization; at 20°C, changing the ligand substituent from isopropyl (**5a**) to ethyl group (**5b**) or to *n*-propyl (**5c**) groups resulted in the heterotacticity decreasing from 0.90 to 0.79 or from 0.90 to 0.76, respectively.

Achiral Salan-type Schiff base aminophenoxide aluminum complexes **7** [96] reported by Gibson et al. also exhibited a high level of heteroselectivity in the polymerization of *rac*-LA. The tacticity of these PLA materials is significantly influenced by the substituents at the *ortho* and *para* positions of the phenol groups in the complexes. In the presence of benzyl alcohol as an initiator, isotactic PLAs were obtained when phenoxide groups are unsubstituted (for complexes **7a** and **7b**, P_m is up to 0.79), whereas highly heterotactic PLAs were produced when phenoxide units contain substituents in the *ortho* and *para* position (for complexes **7c** and **7d**, P_r is up to 0.96) (Fig. 2). The tacticity is also significantly influenced by the substituents R^1 attached to the amino nitrogen donors. Switching from benzylamine derivative **7d** to their methylamine analogue **7c** decreased the heterotacticity from 0.96 to 0.88.

Exciting advances were made when rare earth metal complexes were applied in *rac*-LA polymerizations. For example, lanthanoid complexes **8** [75, 76] reported by Okuda et al. showed excellent heterotactic-control in *rac*-LA polymerization and the heteroselectivity improved as the size of the bisphenolato ligand at the *ortho* position increased. Most notably, incorporation of one more carbon atom into the bridge resulted in the P_r value increasing from 0.78 to 0.95. Carpentier et al. discovered that yttrium amido complexes **6** [94, 95, 97] also exhibit enhanced heteroselectivity in the ROP of *rac*-LA by introducing bulky substituents at the *ortho* and *para* positions of the phenol group and changing the donor group on the pendant chain from methoxy ether to a dimethyl substituted amine group (P_r up to 0.90 at 20°C). In 2007, Cui et al. [98] also reported a series of THF-solvated achiral lanthanide alkyl complexes **9** that displayed modest activity but high stereoselectivity in the ROP of *rac*-LA to give highly heterotactic PLA materials with P_r values ranging from 0.95 to 0.99, the highest value reported to date. Unlike the other sophisticated stereoselective catalyst systems, Hillmyer et al. developed a simple but highly stereoselective robust system [99] with a combination of indium trichloride, benzyl alcohol, and triethylamine to yield highly heterotactic PLA from *rac*-LA under a variety of reaction conditions ($0.86 < P_r < 0.94$ at 25°C, $P_r = 0.97$ at 0°C) (Fig. 3).

Some organocatalysts were also shown to be highly heteroselective in the ROP of lactides. For example, NHC **10a** produced isotactic enriched PLA with a P_m value of 0.83 from ROP of *rac*-LA at −70°C. With the more bulky phenyl substituents on the imidazol ligand, NHC **10b** yielded a heterotactic PLA material from *rac*-LA with a higher P_r value of 0.90 at −70°C (Fig. 3) [100].

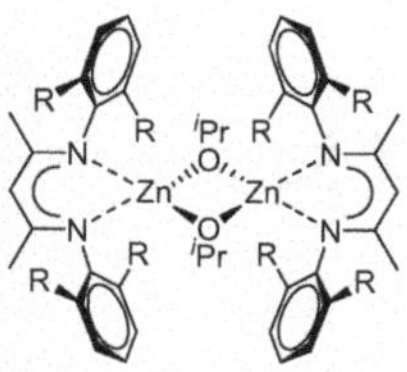

5a, R = ^{i}Pr
5b, R = Et
5c, R = ^{n}Pr

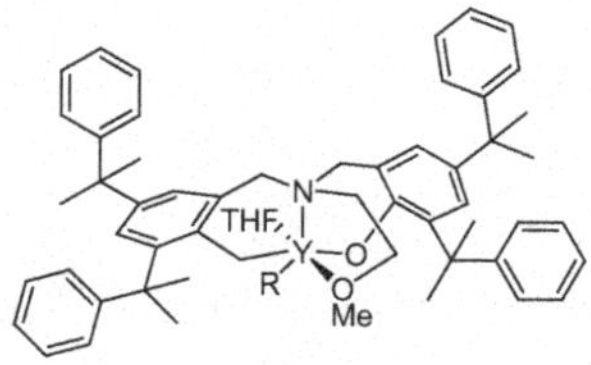

6, R = $N(SiHMe_2)_2$

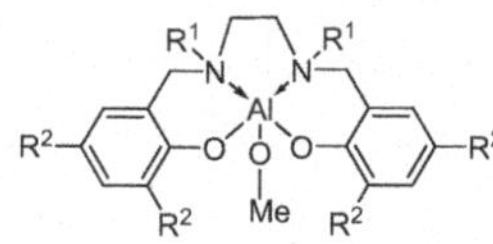

7a, R^1 = Me, R^2 = H
7b, R^1 = CH_2Ph, R^2 = H
7c, R^1 = Me, R^2 = Cl
7d, R^1 = CH_2Ph, R_2 = Cl

Fig. 2 Structure of complexes **5–7**

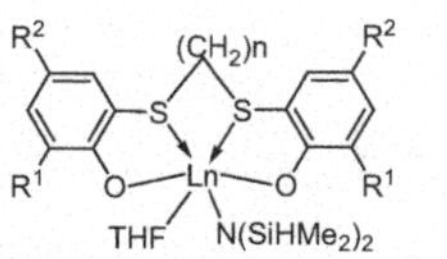

8a, n=2, R^1 = ^{t}Bu, R^2 = Me, Ln = Sc
8b, n=3, R^1 = ^{t}Bu, R^2 = Me, Ln = Lu
8c, n=2, R^1 = ^{t}Bu, R^2 = Me, Ln = Y
8d, n=3, R^1 = R^2 = ^{t}Bu, Ln = Sc
8e, n=3, R^1 = R^2 = ^{t}Bu, Ln = Lu
8f, n=3, R^1 = R^2 = ^{t}Bu, Ln = Y

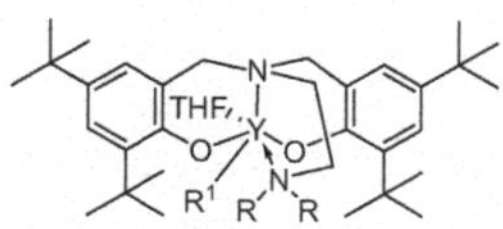

9a, R = Me, R^1 = CH_2SiMe_3
9b, R = Et, R^1 = CH_2SiMe_3

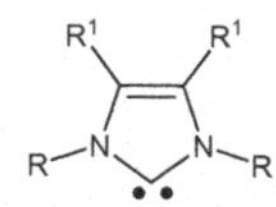

10a, R = 2,4,6-$Me_3C_6H_2$, R^1 = H
10b, R = 2,4,6-$Me_3C_6H_2$, R^1 = Ph

Fig. 3 Structure of complexes **8–10**

2.3 *Furfural and HMF Platform*

As furan derivatives, both furfural and 5-hydroxymethylfurfural (HMF) are readily prepared from renewable biomass. Furfural can be easily obtained from a variety of biomass containing pentoses, mainly including corn cobs, oats and rice hulls, sugar cane bagasses, cotton seeds, olive husks and stones, and wood chips. Furfuryl was first produced in the early nineteenth century and right now the annual production is 300,000 tons [101]. On the other hand, HMF is another major promising furan derivative due to its rich chemistry and potential availability from hexose carbohydrates or from their precursors such as fructose, glucose, sucrose, cellulose, and inulin [14].

As two non-petroleum chemicals readily accessible from renewable resources, both furfural and HMF are suitable starting materials for the preparation of versatile fine chemicals [14, 102–106] and can also serve as renewable monomers for preparation of sustainable polymer products [107]. Schemes 3, 4, and 5 depict the structures of the selected furan-based monomers [107–113]. As a typical precursor, furfural can be converted to a vast array of furan-based monomers bearing a moiety which can normally be polymerized by chain-growth polymerization mechanisms [108–113]. As shown in Scheme 3, these monomers are all readily polymerizable by chain-growth reactions. However, depending on their specific structure, the nature of the polymerization mechanism is different, ranging from free radical, cationic, anionic, to stereospecific initiation [108–113]. On the other hand, furfuryl

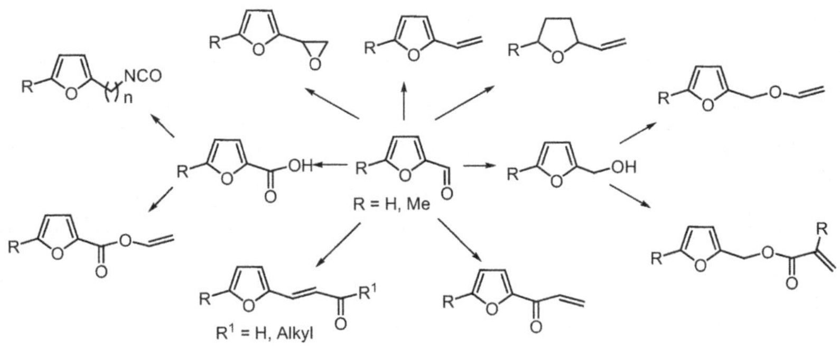

Scheme 3 Selected monomers derived from furfural

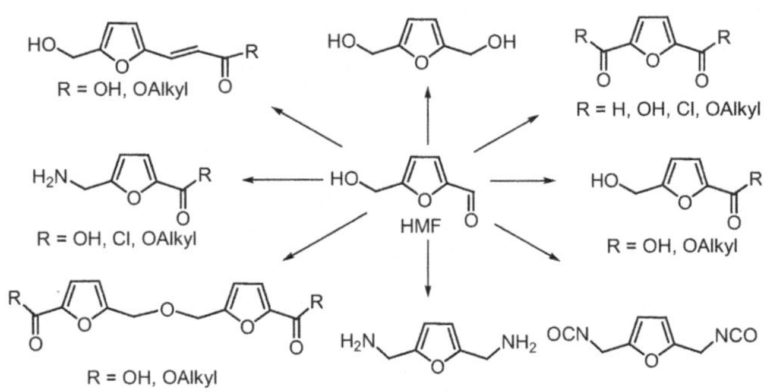

Scheme 4 Selected monomers derived from HMF

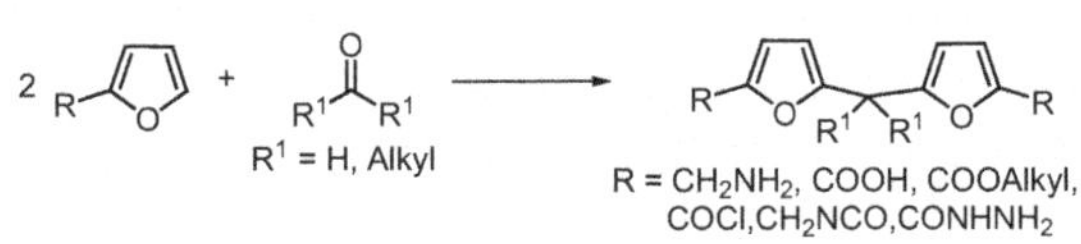

Scheme 5 Di-furan monomers derived from 2-substituted furans

alcohol (FA), which is in fact today the most important commercially available furan compound, derived from furfural (Scheme 3), is not polymerized by chain-growth reactions but by self-condensation [108–113]. Turning into materials, FA has attracted considerable attention in recent investigations aimed at synthesis and characterization of carbonaceous and other materials [114–129].

HMF is a suitable precursor for the synthesis of bifunctional furan monomers as summarized in Scheme 4. All these monomers can be used to prepare polycondensates by step-growth reactions with the other corresponding bifunctional monomers derived either from petrochemical precursors or from renewable resources. The polycondensates obtained, such as polyesters, polyamides, and polyurethanes, etc. have been characterized [107–113]. Scheme 5 shows another approach for the synthesis of bifunctional monomers through acid-catalyzed condensation of the corresponding mono-functional furan derivatives with an aldehyde

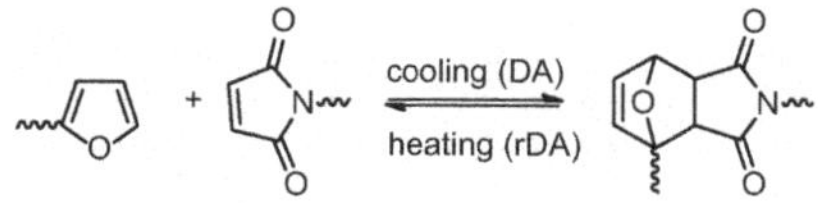

Scheme 6 Diels–Alder equilibrium in furan/ maleimide system

or a ketone group. Those monomers can also be polymerized by step-growth reactions and the corresponding polymeric materials have been explored [108–113].

The transformation of furfural or HMF to selected fine chemicals and monomers, as well as the characterization and polymerization of the corresponding renewable monomers, have been well documented by several comprehensive reviews [107–113, 130, 131]. Among these, Gandini et al. recently wrote reviews on this topic almost every other year. Therefore the synthesis of the furan-based monomers and the corresponding polymers are not covered in this review. With the presence of the reactive furfuryl group, furan-based polymers can form thermally amendable cross-linked polymers through the Diels–Alder (DA) reaction using a suitable dienophile. Here we will concentrate our attention on bio-based furan polymers with self-healing abilities prepared by DA reactions, a prosperous research area which has been exploited recently.

The DA reaction is one of the most important reactions in organic chemistry. The DA reaction between a diene and a dienophile forms covalent bonds which could be cleaved upon heating due to the thermoreversibility of the DA reaction (Scheme 6). Therefore, the DA reaction could be utilized for the preparation of self-healing polymers with well-defined architectures and properties. The most appealing part is that the healing process could be initiated simply by increasing the temperature without addition of chemical or healing reagents and theoretically the repetition number of such a healing process could be infinite. One of the most frequently used systems involves the furan/maleimide pair [132–135].

In 1969, Craven reported the first example of the DA reaction with the furan/ maleimide system [136]. He synthesized polymers which consisted of chains of the saturated condensation polymer backbone bearing the furan group reacted with maleimides. Since then several patents and papers have been published, all concerning the fabrication of a thermally reversible polymer network bearing DA-reactive furan and maleimide units, either as pendant groups (for reversible cross-linking) [137–147], or as part of the polymer backbone (for reversible polymerization) [148–154].

Moreover, the DA chemistry was widely applied in polymer science as well as materials science [138–143, 155]. Wudl et al. utilized a tetra-furan (4F) and a tris-maleimide (3M) to produce a clear solid DA-step-growth polymer (3M4F) [156]. The occurrence of the rDA reaction at ca. 120°C and the healed polymer exhibited ca. 57% of the original polymer strength. Subsequently, an improved system of 2MEP4F constructed from the DA reaction between 4F and 2MEP (1,8-bis(maleimido)-1-ethylpropane) was developed, which exhibited crack-healing with as much as 83% recovery of the polymer's original strength [157].

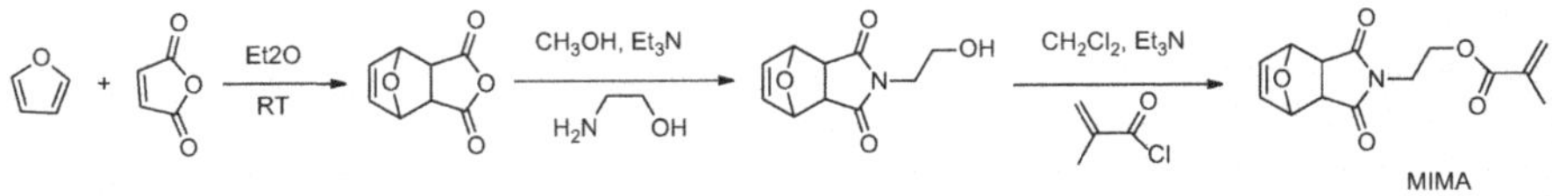

Scheme 7 Synthesis of the maleimide methacrylate (MIMA)

Singha et al. reported DA cross-linked products [158] using furan-modified polymethacrylate (PFMA) as the polymeric precursor, which was prepared through atom transfer radical polymerization (ATRP) and free-radical polymerization (FRP). Furthermore, the self-healing behavior of a triblock copolymer (PFMA-*co*-MMA) prepared by ATRP was demonstrated by means of scanning electron microscopy (SEM). With the modification, an almost fully recovered surface from knife-cut samples has been observed [159]. Chen et al. also reported the DA polymer product of PFMA-BM possessing thermal reversibility, whereas the homopolymer was prepared from anionic polymerization [160].

Furthermore, Schubert et al. synthesized new monomer maleimide methacrylate (MIMA) (Scheme 7), which can be copolymerized with MMA and FMA to form linear self-healing polymers with a methacrylate-based backbone containing the two corresponding functionalities, both furan and maleimide units, for the DA reaction in the side chain. No additional crosslinker is required to obtain a self-healing polymeric material, which represented the first one-component self-healing material and the healing process could be repeated multiple times [161].

Most recently, Yoshie and co-workers reported the first example of room-temperature-healable network polymers [162], prepared by the DA reaction between a bio-based furan polymer, poly(2,5-furandimethylene succinate, PFS, and a bismaleimide. The mechanical properties of these network polymers can be controlled over a wide range by adjusting the bismaleimide content and self-healing efficiency was as high as 74%, whereas in most of the previous reports, higher efficiencies were achieved at higher temperatures. In a subsequent publication [163], using a similar approach, a series of bio-based self-healing polymers were prepared to investigate the effectiveness of molecular structure of the bismaleimide on the DA reaction and the mechanical and healing properties of the resulting polymers (Scheme 8).

The DA reaction has also been used in thermoreversible non-linear polymerization and dendrimer chemistry [164] as well, for assembly [165, 166] or periphery modification [167–169], representing a facile example of covalent structural modification. McElhanon et al. reported the first thermally labile-reassembling DA dendrimer, which exhibited roughly 40% dissociation of the DA links after 1 h at 110°C and full restoration of the original structure upon cooling to 65°C over a couple of days [170]. Adopting this approach, McElhanon et al. also prepared the first through fourth generation dentrimers (Scheme 9), which were found to undergo thermal degradation at 95°C and thermal reassembly at 60°C [171]. Since then, a number of related polymer systems have been reported [172–178].

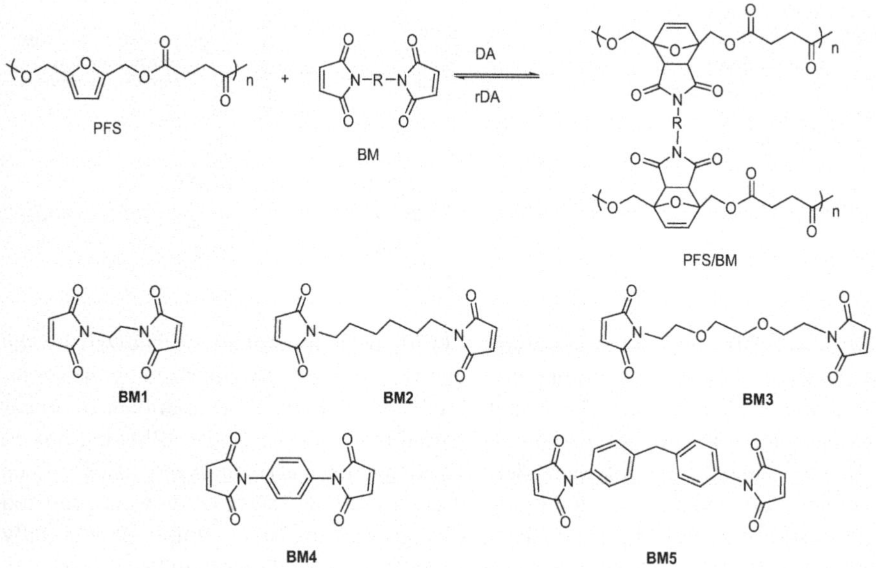

Scheme 8 Diels–Alder reaction of PFS with different bismaleimide

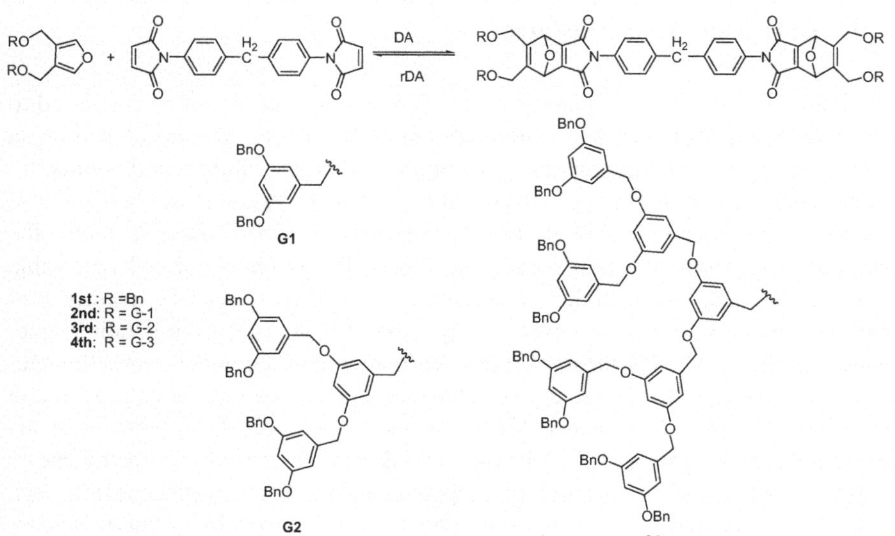

Scheme 9 First through fourth generation dendrimers

Most recently, the use of two click-chemistry mechanism was introduced to build up thermally reversible dendronized step-polymers. In 2010, McElhanon et al. reported the preparation and characterization of first through third generation

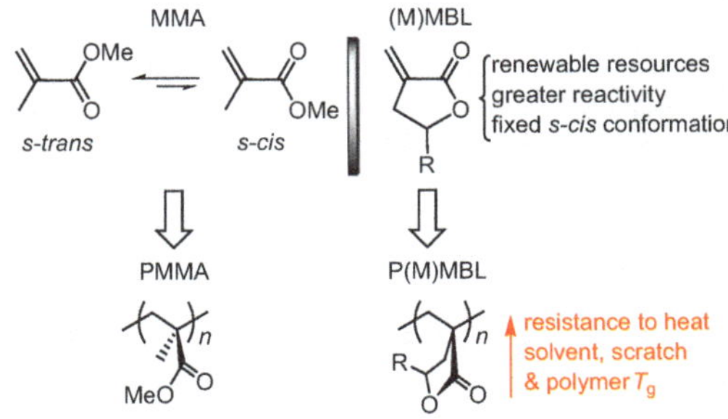

Scheme 10 Renewable butyrolactone-based vinylidene monomers (M) MBL (R = H, Me) and polymers P(M)MBL vs MMA and PMMA

linear AA–BB dendroid step-polymers [179] from monomers derivatized by the azide-alkyne Huisgen 1,3-dipolar cycloaddition and then polymerized by the furan-maleimide DA reaction. These materials represent the first examples of linear covalent dendroinized polymers with thermal reversibility. In the following report, AB dendronized step-polymers derivatized by the Cu(I)-catalyzed azide-alkyne cycloaddition and polymerized by the furan-maleimide DA reaction were prepared and characterized [180]. These materials represent the first example of thermally reversible dendronized polymers originating from a single monomeric species.

2.4 *Levulinic Acid Platform*

Levulinic acid has been of interest for many years because it can be converted to valuable chemicals [181]. The biorefining renewable process was developed on an industrial scale for the production of levulinic acid from cellulose and hemicelluloses present in agricultural or forest residues [182, 183]. A kinetic study of glucose conversion to levulinic acid in H_2SO_4 solutions was performed to study the influence of various reaction parameters [184]. The biomass-derived renewable monomer γ-methyl-α-methylene-γ-butyrolactone ($_{\gamma}$MMBL) can be readily prepared via a two-step process developed by DuPont [181, 185] from levulinic acid, which is produced at 450 tons/year [186]. Its homologous compound α-methylene-γ-butyrolactone (MBL) is also a renewable monomer, so-called Tulipalin A, found in tulips [187, 188]. Chemically, MBL can also be produced from biomass sugar-based itaconic anhydride [189]. The β-methyl derivative, β-methyl-α-methylene-γ-butyrolactone ($_{\beta}$MMBL), is currently prepared by a multi-step synthesis [190, 191], but can be potentially prepared from condensation of 3-methyl-γ-butyrolactone, available from hydrogenation of the biomass-derived itaconic acid [192], with formaldehyde [193]. In this context, renewable butyrolactone-based vinylidene monomers, such as MBL, $_{\gamma}$MMBL, and $_{\beta}$MMBL, are of particular interest in exploring the prospects of substituting the petroleum-based methacrylate monomers for specialty chemicals and polymers production [194, 195]. Structurally, MBL can be described as the cyclic analogue of MMA (methyl methacrylate) (Scheme 10); however, it exhibits greater reactivity in free radical polymerization

[196] than typical methacrylate monomers such as MMA due to the presence of both the nearly planar five-membered lactone ring, which provides a high degree of resonance stabilization for the active radical species, and the higher energy exocyclic C=C double bond, as a result of the ring strain and the fixed *s-cis* conformation [197]. The cyclic ring in MBL also imparts significant enhancements to the materials properties of the resulting PMBL (Scheme 10), as compared to PMMA, thanks to the conformational rigidity of the polymer chain through incorporation of the butyrolactone moiety. Thus, the T_g of PMBL produced by the radical polymerization is 195°C [198], which is about 90°C higher than that of atactic PMMA. Additionally, PMBL has improved optical properties as well as resistance to solvents, heat, and scratching [199–201]. Some of these materials property enhancements have also been observed for $_{\gamma}$MMBL [193, 202, 203].

Various types of polymerization processes have been adopted to polymerize MBL to polymers with low to high molecular weight, proceeding through radical polymerization [196–198, 204–207], anionic polymerization [198], group-transfer polymerization [208], and coordination polymerization with metallocene complexes [202]; MBL has been copolymerized with various comonomers [197] such as MMA [209], styrene [206, 210], methoxystyrene [211], and vinyl thiophenes [212]. The polymerization of $_{\gamma}$MMBL has not been studied as extensively as MBL; nevertheless, it has also been polymerized by free-radical emulsion polymerization [213, 214] as well as by radical, anionic, and group-transfer polymerization methods [215]. Free radical copolymerizations of $_{\gamma}$MMBL with MMA and styrene have been investigated in detail through photoinitiation using the pulsed laser polymerization/size exclusion chromatography technique [216, 217]. Recently reported literature for the polymerization of $_{\gamma}$MMBL included: coordination polymerization with metallocene complexes [202, 203], half-sandwich indenyl rare-earth metal dialkyls [218], *ansa*-rare-earth metal catalysts [193], and group 4 non-metallocene benzyl complexes [219], group-transfer polymerization with bifunctional silicon propagators [220] and dinuclear silylium-enolate bifunctional catalysts [221], anionic polymerization with potassium salts ($KH/Al(C_6F_5)_3$) [222], and zwitterionic polymerization with *N*-heterocyclic carbenes [223, 224] and alane-based classical and frustrated Lewis pairs [225, 226]. Several patents were filed based on MBL/$_{\gamma}$MMBL/$_{\beta}$MMBL polymerization, copolymerization, and their applications [201, 227–229], for example, as plastic optical fiber materials, due to their excellent transparency and heat resistance [230]. $_{\beta}$MMBL was initially polymerized radically to atactic materials [190] but more recently through metal-catalyzed coordination polymerization to highly stereoregular (isotactic) materials [191, 193, 203]. Stereoselective polymerization of $_{\beta}$MMBL into stereo-defect-free P$_{\beta}$MMBL by single-site chiral metallocene catalysts [203] that are known to promote stereospecific coordination polymerization of polar vinyl monomers [231] has also been reported.

Based on the above overview, there is abundant research on the polymerization of MBL-containing compounds including several comprehensive reviews [81, 232]. Here we will focus on the recent advances in this field during the last 5 years.

2.4.1 Polymerizations of MBL, $_{\gamma}$MMBL, and $_{\beta}$MMBL by Rare Earth Metal Complexes

The first metal-catalyzed coordination polymerization of MBL and $_{\gamma}$MMBL was reported in 2010 [202] using neutral divalent lanthanocene(II) $Cp^*{}_2Sm(THF)_2$ (**11**, $Cp^* = \eta^5$-C_5Me_5) [233] and trivalent non-lanthanocene(III) complexes $Ln[N(SiMe_3)_2]_3$ (Ln = La, Nd, Sm, Er) [234] (Fig. 4). Samarocene(II) complex **11** catalyzes rapid, efficient, and controlled coordination polymerization of MBL and $_{\gamma}$MMBL in DMF at RT, as demonstrated by its high turnover frequency (TOF) of up to 3,000 h^{-1}, typically near quantitative initiator efficiency, and its ability to control the polymer MW with the monomer-to-catalyst ratio or monomer conversion. The resulting atactic PMBL and P$_{\gamma}$MMBL exhibit high T_g of 194 and 227°C, respectively, and P$_{\gamma}$MMBL also shows greatly enhanced thermal properties. More remarkably, the T_g and onset decomposition temperatures of the P$_{\gamma}$MMBL are ~120 and 40°C higher than that of the atactic PMMA with comparable MW. Thanks to the living/controlled characteristics of this polymerization, defined random and block copolymers of MBL with MMA and MMBL can be readily synthesized through statistical and sequential block copolymerization procedures.

Kinetic studies revealed that the polymerization by $Cp^*{}_2Sm(THF)_2$ is zero-order in [$_{\gamma}$MMBL] and second-order in [catalyst], as a result of two samarium centers working in tandem to produce one polymer chain. This result, coupled with the polymerization initiator efficiency result, which also pointed to the bimetallic nature of the propagation, conforms to the proposed MMA polymerization mechanism by the same divalent catalyst involving a redox-then-radical-coupling initiation process, with the true active species being the two trivalent samarocene centers attached to the single growing polymer chain.

The MBL polymerization by non-lanthanocene(III) silylamides, $Ln[N(SiMe_3)_2]_3$ (Ln = La, Nd, Sm, Er), is much slower (>130 times) than the polymerization by $Cp^*{}_2Sm(THF)_2$. The polymerization by these lanthanide silylamides is also ill-controlled and can involve more than one silylamide ligand in chain initiation.

Subsequently, the characteristics of MBL, $_{\gamma}$MMBL, and $_{\beta}$MMBL polymerizations were investigated by discrete half-sandwich rare-earth metal (REM, which includes lanthanides and group 3 metals) dialkyl catalysts [191, 218] incorporating the disilylated indenyl ligand, $(1,3\text{-}(SiMe_3)_2C_9H_5)RE(CH_2SiMe_3)_2(THF)$ [RE = Sc, **12** [235]; Y, **13** [236]; Dy, **14** [236]; Lu, **15** [236]] (Fig. 4). All four half-sandwich REM dialkyl catalysts investigated are extremely active for polymerization of $_{\gamma}$MMBL in DMF. Specifically, these catalysts can achieve a quantitative monomer conversion in DMF in <1 min with a catalyst loading of 0.20 mol%, giving a high TOF > 30,000 h^{-1} for this catalyst system, which is at least ten times higher than the sandwich REM catalyst $Cp^*{}_2Sm(THF)_2$ [202]. The polymerization in DCM is slower, but all catalysts can achieve a quantitative monomer conversion. The activity trend of Dy ≥ Y > Lu > Sc is the same for both MBL and $_{\gamma}$MMBL polymerizations: the largest Dy and Y metals are most active, whereas the smallest

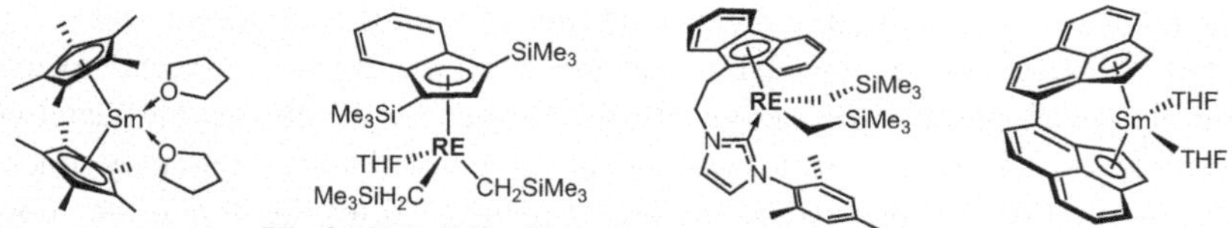

11 RE = Sc, **12**; Y, **13**; Dy, **14**; Lu, **15** RE = Y, **16**; Lu, **17** **18**

Fig. 4 Divalent lanthanide(II) and trivalent rare-earth (RE) metal complexes employed for initial investigation of coordination polymerization of MBL, γMMBL, and $_{\beta}$MMBL

Sc catalyst is least active, with the Lu catalyst lying somewhere in the middle. The P_{γ}MMBLs produced are syndio-rich atactic materials with high T_g values ranging from 217 (by Sc) to 222°C (by Lu).

Kinetic experiments using the Dy catalyst have revealed a first order dependence on [$_{\gamma}$MMBL] but a second-order dependence on [Dy], indicating a bimolecular propagation involving two Dy metal centers in the rate-limiting C–C bond-forming step. Each metal center can carry more than one polymer chain and grow only one at a time, which originated from initiation with both alkyl groups on the metal and the first-order dependence on the monomer. The polymer has a structural formula of Me_3SiCH_2-($_{\gamma}$MMBL)$_n$-H and has been characterized by MALDI-TOF mass spectrometry, whereas the more-than-one-chain-per-metal scenario has been evidenced by the results of NMR studies and by typically greater or much greater than 100% catalyst efficiencies.

Half-sandwich RE dialkyl complexes **12–15** also promote rapid polymerization of $_{\beta}$MMBL at ambient temperature [191]. The complex of Dy, the largest ion of this Ln series, exhibits the highest activity, thus achieving nearly quantitative polymer yield (97%) within 1 min of reaction that employs a low loading of catalyst (0.25 mol%), corresponding to a high TOF of 390 min^{-1}. More significantly, this highly active coordination polymerization system also affords the highly stereoregular polymer $_{\beta}$MMBL having an isotacticity of 91.0% *mm*, in contrast to the atactic polymer produced by radical polymerization initiated by AIBN. Other half-sandwich RE catalysts of the current series are also highly active and produce polymers with a similarly high isotacticity. The resulting isotactic P_{β}MMBL is thermally robust, with a high T_g of 280°C, and is resistant to all common organic solvents at ambient or elevated temperatures.

Intriguingly, simple homoleptic hydrocarbyl RE complexes, RE $(CH_2SiMe_3)_3(THF)_2$ (RE = Sc, Y, Dy, Lu), also produce highly isotactic polymer P_{β}MMBL. However, their polymerization activity is much lower than that of the corresponding half-sandwich dialkyl complexes, with the exception of the Lu complex, which maintains its high activity for both types of complexes.

Computational studies [191] of both half-sandwich and simple hydrocarbyl yttrium complexes have led to a stereocontrol mechanism that well explains the observed high stereoselectivity of $_{\beta}$MMBL polymerization by the current catalysts. Concisely, in the proposed monometallic propagation mechanism, formation of an isotactic polymer originates chiefly from interactions between the methyl groups on the chiral β-C atom of the five-membered ring of both the monomer and the last inserted $_{\beta}$MMBL unit of the chain, and the auxiliary ligand on metal exerts a negligible contribution to the stereocontrol exhibited by the current half-sandwich RE complexes. This mechanism is in good agreement with the current experimental results and is further supported by the above-mentioned results in that the coordination polymerization of MBL and $_{\gamma}$MMBL, catalyzed by the same half-sandwich RE dialkyl complexes under the same reaction conditions, yields essentially atactic polymers.

Bridged *ansa*-REM complexes have also been employed for the coordination polymerizations of MBL, $_{\gamma}$MMBL, and $_{\beta}$MMBL [193], including ethylene-bridged, NHC-functionalized half-sandwich C_2H_4(Flu-NHC)RE(CH_2SiMe_3)$_2$ (RE = Y, **16**; Lu, **17**) [237] and C_2-symmetric *ansa*-samarocene **18** [238]. *Ansa*-REM complexes **16**, **17** exhibited exceptional activity for the polymerization of racemic $_{\gamma}$MMBL at room temperature in DMF, achieving 100% monomer conversion in <1 min with a high TOF of >24,000 h^{-1}. This TOF value represents a rate enhancement, by a factor of 8, 22, or 2,400, over the polymerizations by unbridged samarocene **11** [202], bridged *ansa*-samarocene **18**, or the corresponding REM trialkyls without the *ansa*-Flu-NHC ligation [202], respectively. Kinetic experiments have revealed the first-order dependence on both [monomer] and [catalyst], thus establishing unimolecular propagation for this coordination polymerization, which is in sharp contrast to the bimolecular and bimetallic propagation mechanism for $Cp^*{}_2Sm(THF)_2$ and [η5-(1,3-($SiMe_3$)$_2C_9H_5$)]-RE(CH_2SiMe_3)$_2$(THF) systems, respectively. More significantly, catalyst **16** is also highly active for the polymerization of racemic $_{\beta}$MMBL at room temperature. The resulting P$_{\beta}$MMBL is highly stereoregular (91% *mm*). This material is thermally robust and resistant to common organic solvents at ambient or elevated temperature, and exhibits an extremely high T_g of 290°C.

2.4.2 Polymerizations of MBL, $_{\gamma}$MMBL, and $_{\beta}$MMBL by Group 4 Metallocene Catalysts

The polymerizations of MBL, $_{\gamma}$MMBL, and $_{\beta}$MMBL by cationic group 4 metallocene and half-metallocene catalysts incorporating C_2 and C_s symmetric ligands [203, 210] were also investigated. Figure 5 depicts the structures of such catalysts, including the C_{2v}-ligated two-component catalyst system **19** consisting of the neutral zirconocene $Cp_2ZrMe[OC(OR) = CMe_2]$ [239, 240] as initiator and the cationic zirconocene $Cp_2ZrMe^+MeB(C_6F_5)_3^-$ [241] as catalyst, C_s-ligated titanium complex {(CGC)Ti(THF)[OC(O^{i}Pr) = CMe_2]}$^+$ $MeB(C_6F_5)_3^-$ [**20**, CGC = $Me_2Si(\eta^5$-(Me_4C_5)(^{t}BuN)] [242], C_s-ligated zirconocene catalyst {[(*p*-Et_3SiPh)$_2$C(Cp)(2,7-^{t}Bu$_2$-Flu)]Zr[OC(O^{i}Pr) = CMeCH$_2$C(Me$_2$)C(O^{i}Pr) = O]}$^+$[B(C_6F_5)$_4$]$^-$ (**21**) [243], as well as C_2-ligated zirconocenium catalysts, *rac*-(EBI)Zr$^+$(THF)[OC(O^{i}Pr) = CMe_2][MeB(C_6F_5)$_3$]$^-$ [**22**, EBI = $C_2H_4(\eta^5$-indenyl)$_2$] [244, 245], and *rac*-(EBDMI)Zr$^+$(THF)[OC(O^{i}Pr) = CMe_2][MeB(C_6F_5)$_3$]$^-$ [**23**, EBDMI = $C_2H_4(\eta^5$-4,7-dimethylindenyl)$_2$] [203]. Coordination-addition polymerization of

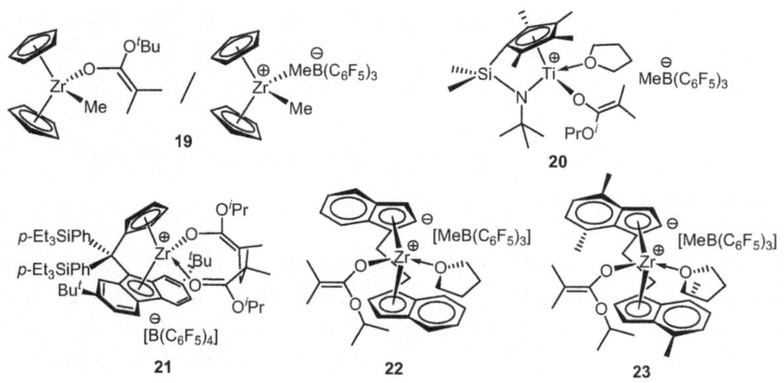

Fig. 5 Cationic group 4 metallocene catalysts employed for investigation of coordination polymerization of MBL, γMMBL, and $_{\beta}$MMBL

acrylic monomers by cationic group 4 metallocenium catalysts is typically carried out in hydrocarbons such as toluene and polar non-coordinating solvents such as CH_2Cl_2, whereas polar coordinating solvents such as THF and DMF usually shut down the polymerization [231]. Owing to the insolubility of P(M)MBL in toluene or CH_2Cl_2, polymerization of (M)MBL by group 4 catalysts in such solvents proceeds in a heterogeneous fashion, thereby negatively impacting the catalyst activity and control over the polymerization. Compared to the REM catalysts the MBL and MMBL polymerizations by group 4 metallocene catalysts show low monomer conversion or need longer times to get quantitative yields. In addition, unlike the precision polymerization of MMA by the various types of group 4 metallocene complexes, MBL and $_{\gamma}$MMBL polymerizations are neither controlled nor stereospecific.

However, most remarkably, polymerization of $_{\beta}$MMBL in CH_2Cl_2 produced a highly isotactic polymer ($mm = 95.2\%$) by **23** or a perfectly isotactic polymer ($mmmm > 99\%$) by **22** [203]. It is striking that the same chiral catalysts MBL and $_{\gamma}$MMBL into stereo-random polymers, but polymerized $_{\beta}$MMBL into highly isotactic or perfectly isotactic polymers. Computational studies on the competitive monomer addition transition-state geometries have revealed that steric interactions involving the monomer, the chain, and the catalyst ligand are responsible for achieving or lacking the observed stereocontrol [203]. Calculations indicate that for the *R*,*R*-ligated EBI catalyst the *R* chain clearly favors addition of another *R* $_{\beta}$MMBL molecule on the *re*-face of the chain, while the *S* chain clearly favors addition of another *S* $_{\beta}$MMBL molecule on the *si*-face of the chain. As the calculation yielded a high *re* over *si* and *R* over *S* selectivity for an *R* chain, and a high *si* over *re* and *S* over *R* selectivity for an *S* chain, the resulting P$_{\beta}$MMBL should display a high regularity in the sequence of the configurations of the chiral β-C atom of the five-membered ring and should be highly isotactic. In fact, for both chains the calculated ΔE_{stereo} were about 4 kcal/mol, which corresponds to >99% of *mmmm* pentads at 25°C, in agreement with the experimental results. In contrast, in the case of MBL or $_{\gamma}$MMBL addition such steric interactions become negligible, thus rendering the polymerization non-stereoselective by these catalysts.

2.4.3 Living Polymerization of MBL and $_{\gamma}$MMBL by Ambiphilic Silicon Propagators

A highly active bifunctional propagating species for efficient living/controlled (meth)acrylate polymerization has been developed by instantaneous oxidative activation of trialkylsilyl methyl dimethylketene acetal (^{R}SKA, $Me_2C = C(OMe)OSiR_3$) initiators [246, 247], which are commonly employed in the conventional GTP [208, 248, 249], but both chain initiation and propagation are fundamentally different from those steps of GTP. This ambiphilic silicon catalyst system has consequently been applied to the MBL and $_{\gamma}$MMBL polymerization [220]. Specifically, the initiation is uniquely "monomer-less," which involves vinylogous hydride abstraction of ^{R}SKA by Ph_3C^+, leading to the R_3Si^+-activated MMA (i.e., activation of the initiator simultaneously generates the silylium catalyst and the activated monomer); subsequent Michael addition of ^{R}SKA to the silylated MMA generates the bifunctional active propagating species **A** (Scheme 11). The chain propagation consists of a fast step of recapturing the silylium catalyst from the ester group of the growing chain by the incoming monomer, followed by a rate-determining step (*r.d.s.*) of C–C bond coupling via intermolecular Michael addition of the polymeric SKA to the silylated monomer. Investigations into the effects of SKA and activator structures found that the $Me_2C = C(OMe)OSi^{i}Bu_3/[Ph_3C][CB(C_6F_5)_4]$ combination is the most active and best controlled system for MBL and $_{\gamma}$MMBL polymerizations. The polymerization of MBL in CH_2Cl_2 is heterogeneous and achieves typically low yields of polymers with bimodal MWDs. Thanks to the solubility of P$_{\gamma}$MMBL in CH_2Cl_2, the living/controlled polymerization of $_{\gamma}$MMBL is homogeneous with the quantitative yield obtained in 10 min even with a low catalyst loading of 0.05 mol%. The controlled low to high ($M_n = 5.43 \times 10^5$ kg/mol) MW and narrow MWDs (1.01–1.06) were obtained depending on the [$_{\gamma}$MMBL]/[^{iBu}SKA] ratio. The copolymerization approach of MBL and $_{\gamma}$MMBL not only confirmed the living nature of this system but also solved the insolubility and bimodality issue of PMBL, which successfully leads to the well-defined MBL-containing copolymers.

In order to overcome some limitations of the bimolecular, activated monomer propagation mechanism (Scheme 11), such as limitations on polymerizations under highly dilute initiator or catalyst conditions and on the stereochemical control of polymerization, two types of di-SKA compounds (**24**, **25**) having different linkages have been synthesized (Scheme 12) [221]. Their activation chemistry for the generation of the corresponding dinuclear silylium-enolate active species **26** and **27** (Scheme 12) and investigation of their behavior in the polymerization of $_{\gamma}$MMBL have been examined as well. The kinetics study identified that this

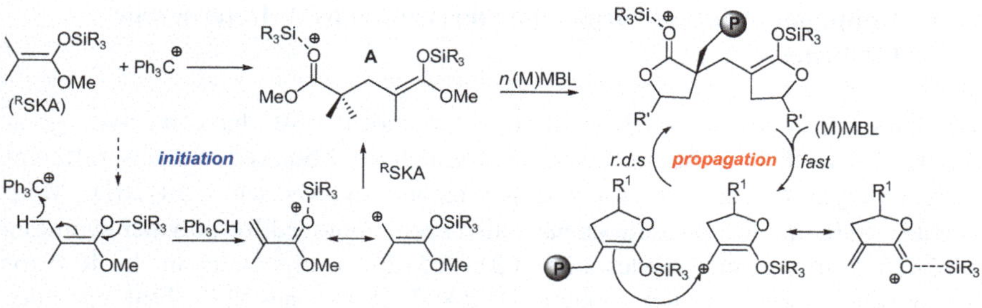

Scheme 11 Initiation and propagation involved in living/controlled (M)MBL polymerization catalyzed by R_3Si^+ (R^1 = H or Me)

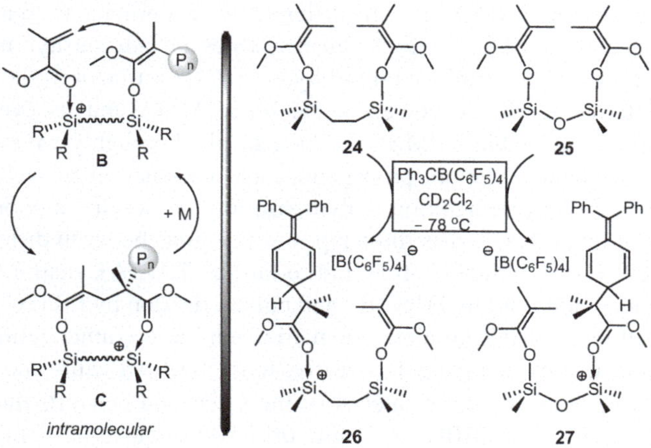

Scheme 12 Proposed intramolecular Michael addition pathway for dinuclear silylium-enolate active species **B** and activation of di-SKA by $[Ph_3C][B(C_6F_5)_4]$ for generation of dinuclear silylium-enolate active species

polymerization system is consistent with an intramolecular Michael addition propagation mechanism depicted in Scheme 12. Specifically, this unimolecular process involves propagating intermediate C, formed by an intramolecular delivery of the polymeric enolate nucleophile to the monomer activated by the silylium ion electrophile in the same silylium-enolate active species B. The first-order dependence on [M] also implies that the release of the silylium catalyst from its coordination to the penultimate ester group of the growing polymer chain (i.e., intermediate C) to the incoming monomer is a rate-determining step. Both the ethyl-and oxo-bridged dinuclear species are much more active for the polymerization of MMA than the mononuclear SKA-based active species. The oxo-bridged silylium-enolate species is considerably more active and controlled than the ethyl-bridged one. The activity difference between these two dinuclear systems is even greater for the polymerization of $_{\gamma}$MMBL with the TOF of **27** being about 6.7 times higher than that of **26**.

2.4.4 Conjugate-Addition Organopolymerization by *N*-Heterocyclic Carbenes

N-Heterocyclic carbenes (NHCs), used as organocatalysts, have received great interest due to their unique reactivity and selectivity observed in many different types of organic reactions (for selected recent reviews see [250–261]. More recently, NHC-mediated reactions have also been employed for polymer synthesis ([262–264]; for selected reviews see [20, 265–267]), especially in the ROP of heterocyclic monomers, such as lactides [268–272], lactones [273–276], epoxides [277–279], cyclic carbonates [280], cyclic siloxanes [281, 282], and *N*-carboxyl-anhydrides [283, 284]. NHC-mediated step-growth polymerization has also been reported [285–287].

Using three common NHCs of vastly different nucleophilicity as organocatalysts – $I^{t}Bu$ (1,3-di-*tert*-butylimidazolin-2-ylidene), IMes (1,3-di-mesitylimidazolin-2-ylidene), and TPT (1,3,4-triphenyl-4,5-dihydro-1*H*-1,2,4-triazol-5-ylidene) – and two representative acrylic monomers – the linear MMA and its cyclic analog, biomass-derived renewable $_{\gamma}$MMBL – Chen et al. thoroughly investigated the mechanisms of chain initiation, propagation, and termination of NHC-mediated organocatalytic conjugate-addition polymerization of acrylic monomers [223, 224]. It is noted that there exists three types of reactions between different NHCs and MMA (Scheme 13): dimerization (tail-to-tail) by TPT (A), enamine formation (single-monomer addition) by IMes (B), and polymerization by $I^{t}Bu$ (C). However, for MMBL all three NHCs promote polymerization but no dimerization, with the polymerization activity being highly sensitive to the NHC structure and the solvent polarity. $I^{t}Bu$ is the most active catalyst of the series and converts quantitatively 1,000–3,000 equiv. of MMBL in 1 min or 10,000 equiv. in 5 min at room temperature to MMBL-based bioplastics with a narrow range of molecular weights of $M_n = 70$–85 kg/mol, regardless of the [MMBL]/[$I^{t}Bu$] ratio employed as long as it is larger than 800. The $I^{t}Bu$-catalyzed MMBL polymerization reaches an exceptionally high TOF up to 122 s^{-1} and a high initiator efficiency value up to 1,600%. The production of relative high molecular weight linear polymers and the catalytic nature of this NHC-mediated conjugate-addition polymerization are attributed to the unique chain-termination mechanisms. Computational studies have provided mechanistic insights into reactivity and selectivity between two competing pathways for each NHC-monomer zwitterionic adduct, namely enamine formation/dimerization through proton transfer vs polymerization through conjugate addition.

2.4.5 Zwitterionic Polymerization by Classical and Frustrated Lewis Pairs

The seminal works [288–294] of Stephan and Erker uncovered the concept of "Frustrated Lewis Pairs" (FLPs) to describe sterically encumbered borane Lewis acid (most commonly $B(C_6F_5)_3$) and base (e.g., $^{t}Bu_3P$) pairs that are sterically

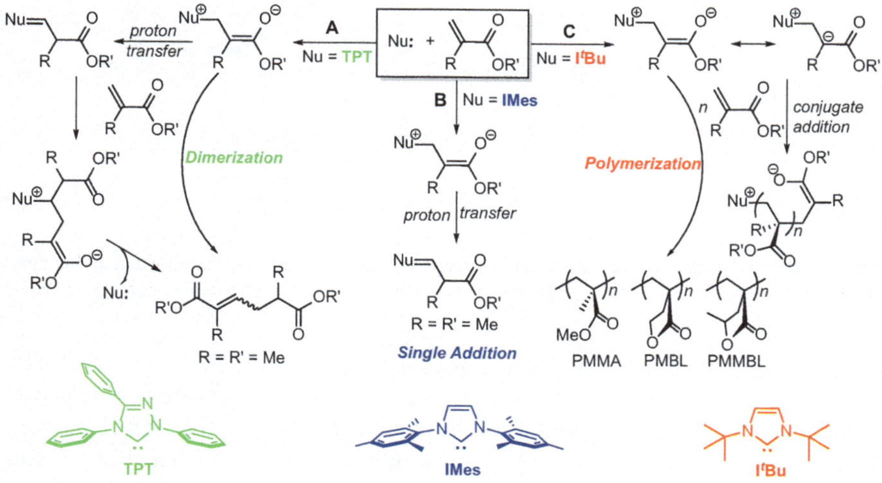

Scheme 13 Outline of three distinctive reaction pathways involved in the reaction of NHCs and acrylic monomers [224]

precluded from forming classical donor-acceptor adducts. Instead, the unquenched, opposite reactivity of FLPs can carry out unusual reactions or reactions that were previously known to be possible only by transition metal complexes [295]. The first polymerization of polar vinyl monomers such as MBL and $_{\gamma}$MMBL directly using FLPs was reported in 2010 [225]. Highly active and effective LPP systems for polymerization of MBL and $_{\gamma}$MMBL have been achieved with classical and frustrated LPs based on the strong Lewis acid $Al(C_6F_5)_3$. Subsequently, a full account of combined experimental and theoretical study on Lewis Pairs polymerization (LPP), including experimental investigations into LA, LB, and monomer scopes and computational study of active species formation and polymerization mechanism were presented [226, 296]. As showed in Scheme 14, the zwitterionic phosphonium or imidazolium enolaluminate species are the active propagating species because they rapidly polymerize the subsequently added monomer. The structures of such active species derived from the stoichiometric reaction of the FLP with MMA (**28**) and MBL (**29**) have been isolated, characterized by NMR, and confirmed by X-ray diffraction analysis [226]. Both experimental and computational results indicated that the bimetallic mechanism involves addition of a LA activated monomer to the zwitterions.

2.5 *Sorbitol Platform*

Sorbitol is a fine chemical with a wide range of applications, such as cosmetics, food, medicine, and others [297–299]. It is derived from hydrogenation of glucose, and industrial production is ca. 700,000 tons per year [300]. Most of the industrial

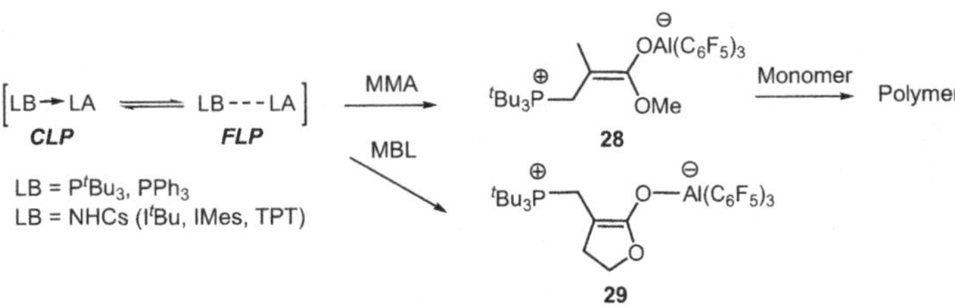

Scheme 14 Generalized scheme for the zwitterionic polymerization of conjugated polar vinyl monomers by classical or frustrated Lewis pairs (CLP or FLP) through zwitterionic active specie

processes rely on batch-type hydrogenation with Raney nickel catalysts [300]. A disadvantage of the use of skeletal nickel in the hydrogenation of glucose-containing feedstocks to sorbitol is the fact that some of the nickel leaches. To generate non-leaching catalyst systems, ruthenium on carbon catalyst systems were developed and used for sorbital production in >99% yield [301, 302]. Very recently, sorbitol was reported to be synthesized directly from cellulose via C–C and C–O bond cleavage in so-called hydrogenolysis reactions over supported metal catalysts [303].

Isosorbide, 1,4:3,6-dianhydro-D-glucitol, also attracted increasing research interest due to its potential industrial applications such as the preparation of isosorbide nitrates used in cardiac or vascular disease and the preparation of alkyl derivatives used as solvents in pharmaceutical or cosmetic compositions. Several reports described conversion of sorbitol to isosorbide through double dehydration from sorbitol [304]. For example, the dehydration of sorbitol in the presence of acidic zeolites afforded isosorbide in 50% yield [305] (Scheme 15). The sulfated copper oxide was employed to convert sorbitol to isosorbide in 67% yield at 200°C [306]. More recently, a two-step sequential process was developed, in that cellulose was first depolymerized with Ru/$NbOPO_4$-pH2 catalyst by hydrolysis, followed by hydrogenation, and then the resultant sorbitol and sorbitan were directly converted into isosorbide in 56.7% yield in the presence of the same solid acid catalyst $NbOPO_4$-pH2 [307]. Moreover, this system avoids using liquid acid and exhibits excellent cycling stability, factors that are important for industrial isosorbide production.

In addition, isosorbide and other 1,4:3,6-dianhydrohexitols (isomannide derived from D-mannose, isoidide derived from L-fructose) are also attractive to serve as monomers for polymer production due to their rigidity, chirality, and non-toxicity (Fig. 6). Such features may introduce special properties into the polymers formed, such as enhanced T_g and/or special optical properties. Their innocuous nature also opens the possibility of applications in packaging or medical devices. As a bifunctional monomer, isosorbide can be polymerized with other bifunctional monomers via condensation polymerization. A recent review described various isosorbide-based polymers synthesized, including polyesters, polyamides, poly(ester amide)s, poly(ester imide)s, polycarbonates, polyurethanes, and so on [308], and the present

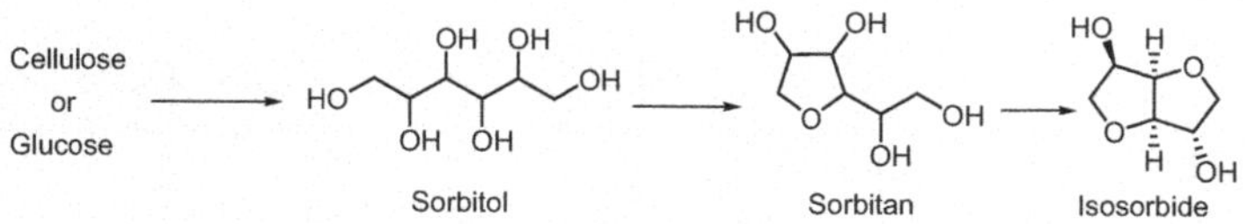

Scheme 15 Conversion of cellulose or glucose to isosorbide

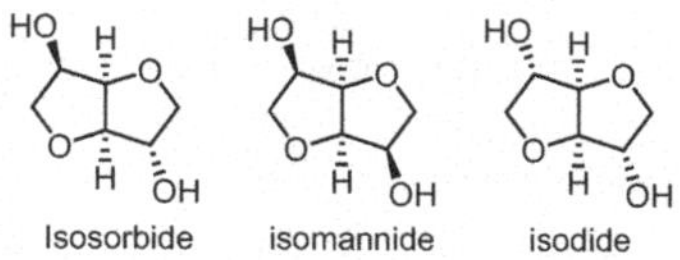

Fig. 6 Three diastereomers of 1,4:3,6-dianhydrohexitols: isosorbide, isomannide, and isoidide

review focuses on recent literature related to the use of isosorbide as a (co)monomer for the synthesis of novel polymers with unique properties.

Oligo(isosorbide adipate) (OSA), oligo(isosorbide suberate) (OSS), and isosorbide dihexanoate (SDH) were synthesized by reaction of isosorbide with the corresponding diacids and evaluated as renewable resource alternatives to traditional phthalate plasticizers. The blends plasticized with SDH had properties almost identical to PVC/diisooctyl phthalate (DIOP) blends. Compared to the PVC/DIOP or PVC/SDH blends, the blends containing OSA and OSS plasticizers based on dicarboxylic acids had somewhat lower strain but higher stress at break and better thermal stability. All the synthesized isosorbide plasticizers have been developed as an alternative to replace phthalates [309].

Through a simple catalyst-free, one-shot polymerization, biocompatible and biodegradable polyurethanes (Scheme 16) were prepared with a fixed aliphatic diisocyanate level and varying ratios of isosorbide and PCL diol (M_w = 2,000 g/mol) [310]. In order to avoid the potential toxicity, catalysts were not used. The mechanical properties, degradation rate, and cytocompatibility were measured, which in general correlated with the isosorbide to PCL diol ratio. It turned out that these biodegradable polyurethanes can serve as promising materials for cardiovascular, trachea, and bladder applications [311].

Besides aliphatic diisocyanates, an aromatic isocyanate was also used to prepare polyurethanes. A new renewable copolymer (Scheme 17) was synthesized from reactive bis-hydroxylated poly(3-hydroxybutyrate-*co*-3-hydroxyvalerate) oligomers (PHBHV-diol), isosorbide, and 1,4-phenylene diisocyanate. The molar number-average molecular weights (M_n) of most of the copolyesters were about 10,000 g/mol with polydispersities (PDI) in the range 1.2–1.9. The incorporation of isosorbide units into the PHBHV backbone increased the T_g from 5°C to 34°C [312].

However, isocyanate reactants are very toxic and could entail adverse health effects such as asthma, dermatitis, conjunctivitis, and acute poisoning [313]. Hence, the synthesis of isocyanate-free polyurethanes is of great interest. Accordingly, an isocyanate-free method and new synthesis procedure has been developed to prepare

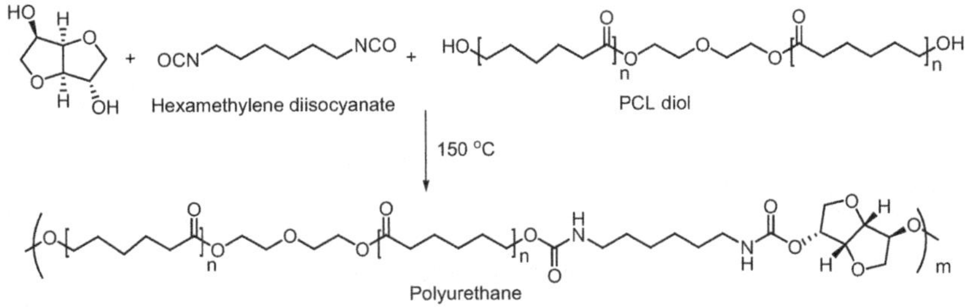

Scheme 16 Polyurethane series synthesis

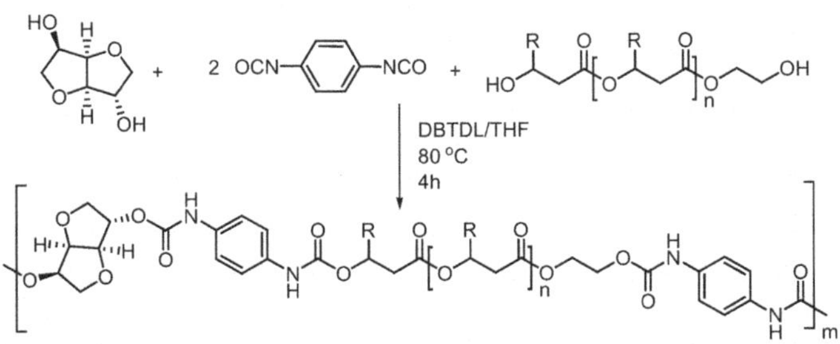

Scheme 17 Structure of poly(HBHV-*co*-isosorbide)

isosorbide-based polyurethanes [314]. In this new procedure, isosorbide was functionalized with glycidyl ether groups to produce functionalized oligoisosorbides, which were carbonated using mild conditions with a total conversion. Then polyhydroxyurethanes (PHUs) (Scheme 18) were synthesized by a cyclocarbonate-aliphatic amine step growth polyaddition with different diamines. This reaction was found to be very effective as the reaction of cyclocarbonate groups is completed within 12 h. Linear and branched PHUs were obtained with T_g values in the range −8 to 59°C, and low T_g PHUs are suitable for coatings application. An acceptable thermal stability (T_d between 234 and 255°C) was determined via TGA for all PHUs. This work demonstrates the easy synthesis of new biobased PHUs with ether bonds and with isosorbide as the hard segment. Interestingly, the carbon dioxide and secondary amines were released by the degradation of PHUs. Non-isocyanates were detected during thermal degradation. Choices of other aliphatic biobased diamines, shorter ones like C_5 or longer ones like C_{36}, will allow for preparation of harder or softer PHUs.

A combination of ROP and polycondensation in a "one-pot procedure" using the same catalysts, such as $SnCl_2$, $ZnCl_2$, or Zn-lactate was proven feasible to produce the copolymers of isosorbide, lactide, and isophthalic acid (Scheme 19). High weight average molecular weights (M_ws, in the range 60–95 kg/mol) were obtained and T_g values were up to 160 or 199°C for copolyesters or homopolyesters from isosorbide, respectively. Variation of the lactide content allows for systematic

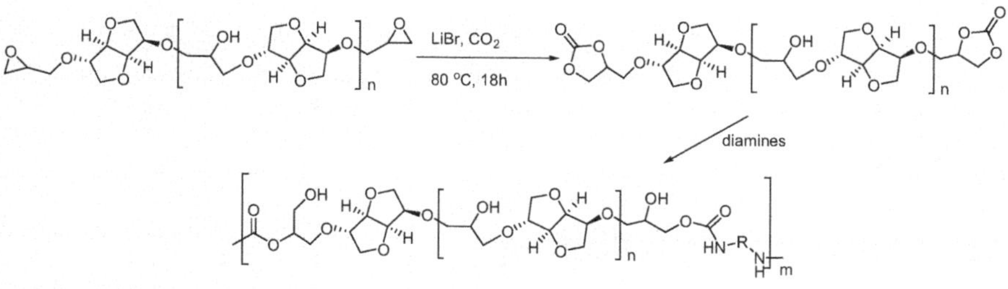

Scheme 18 Structure of polyhydroxyurethanes (PHUs)

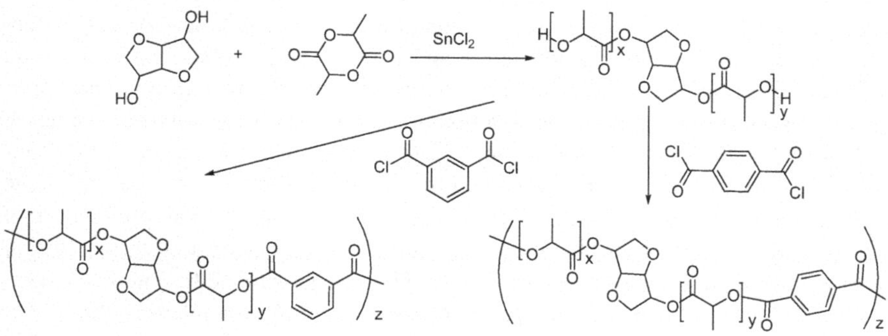

Scheme 19 Synthesis of copolyesters from isosorbide, L-lactide, and isophthaloyl chloride [315] or terephthaloyl chloride, respectively

variation of the T_g over a broad temperature range (from 87 to 199°C). Differential thermoanalysis demonstrated that the thermostability decreases with higher fractions of lactide, but processing from the melt seems to be feasible up to temperatures of 260°C without risking degradation [315]. The $SnCl_2$-catalyzed ROPs of isosorbide and L-lactide combined with polycondensation of terephthaloyl chloride in a "one-pot procedure" were also reported by the same authors (Scheme 19) [316]. M_w values in the range 80–130 kg/mol were obtained. On average, these values were higher than those achieved under identical conditions with isophthaloyl chloride [315]. This difference may be ascribed to a lower cyclization tendency of the repeat units based on terephthalic acid. The T_gs were higher than those reported for analogous polyesters of isophthalic acid, but the difference disappeared when the isosorbide/L-LA ratio decreased to 2/8 [315]. The successful incorporation of phenyl phosphate resulted in copolyesters of lower inflammability. Several authors reported the syntheses of terephthalic acid from renewable resources [317–319]. Hence, this type of copolyester could potentially be synthesized by 100% biomass derived monomers.

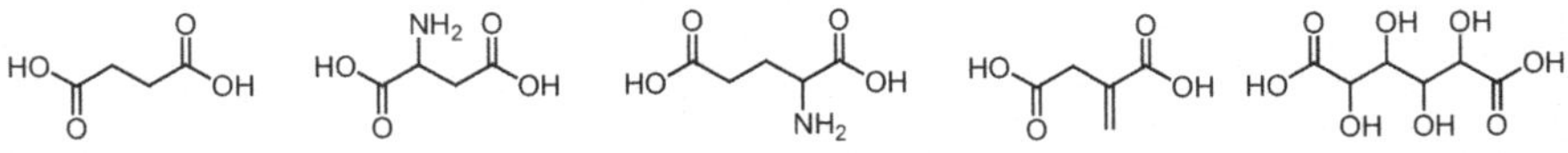

Fig. 7 Structures of dicarboxylic acids

2.6 Dicarboxylic Acids Platform

In 2004, the US Department of Energy (DOE) identified 12 building block chemicals that can be produced from carbohydrates or sugars via biological or chemical conversions [320]. Most of them are bifunctional dicarboxylic acids. The structures of five dicarboxylic acids containing different numbers of carbon are depicted in Fig. 7, including succinic acid, aspartic acid, glutamic acid, itaconic acid, and glucaric acid. The following section presents the preparation and application of such dicarboxylic acids as bifunctional monomers for polymer synthesis.

Nowadays, the bulk of succinic acid (SA) is produced at a rate of 25,000 tons per year from maleic anhydride obtained by oxidation of the C_4 fraction (*n*-butane or butadiene) of crude oil [321]. However, the biotechnological production of SA by fermentation of carbohydrates has attracted more recent attention (http://www.dsm.com/en_US/downloads/media/12e_09_dsm_and_roquette_commercialize_bio_based_succinic_acid.pdf, http://www.bio-amber.com/press_releases.php, [322–326]). The economic and environmental analysis of a biorefinery producing SA indicated that bio-SA is a promising intermediate provided that its production cost could be lowered further [322].

SA is of great interest as a platform chemical for polymer production [327]. Besides serving as monomer, SA could also be converted to other monomers in the following four classes: (1) acyclic O-containing: 1,4-butanediol, SA esters; (2) acyclic O,N-containing: 1,4-butanediamine, succinamide, succinonitrile; (3) cyclic O-containing: tetrahydrofuran (THF), SA anhydride, dihydrofuran-2(3*H*)-one (*c*-butyrolactone); (4) cyclic O,N-containing: pyrrolidin-2-one (2-pyrrolidone) and derivatives, pyrrolidine-2,5-dione (succinimide). Of these classes, the bifunctional acyclic compounds such as 1,4-butandiol and 1,4-butandiamine in combination with SA or other dicarboxylic acids are of interest for the production of polyesters and polyamides. Methods for the production of polyamides by polycondensation of aliphatic diamines and dicarboxylic acids or the polyaddition of lactams were reviewed by Kabasci et al. [327]. The basic structure of the polyamides is either on the basis of 1,4-butanediamine [323–325, 328] or on the basis of SA [326, 329–331]. Up to now, only one polyamide based on both SA and 1,4-butanediamine is manufactured on the laboratory scale. Both of monomers are produced from biomass by fermentation as mentioned above. Mitsubishi Chemical Corporation and Showa Denko have developed the polymers GS Pla®, a poly (butylene succinate), and Bionolle®, a polyester based on ethylene glycol and 1,4-butanediol together with SA or adipic acid [327]. Both polymers are commercially available now. More recently, the polycondensation of SA and 1,4-butanediol

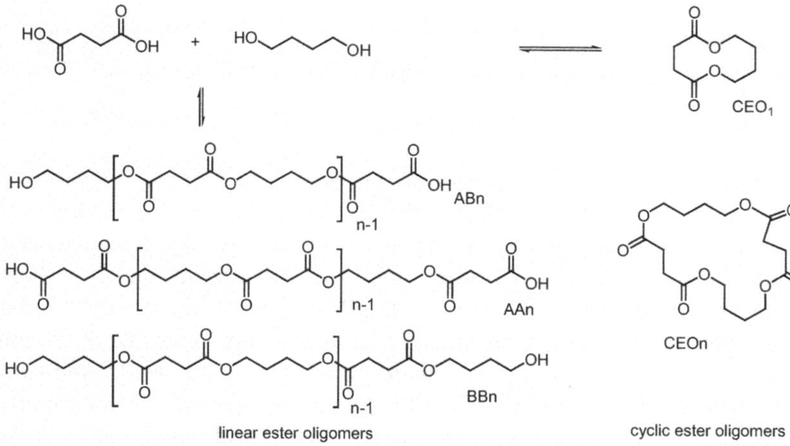

Scheme 20 Products from succinic acid (A) and 1,4-butanediol (B) polycondensation

was investigated. Both linear and cyclic ester oligomers were obtained in the presence of immobilized *Candida antarctica* lipase B (Scheme 20) [332].

Co- and terpolyesters based on SA and isosorbide in combination with other renewable monomers such as 2,3-butanediol, 1,3-propanediol, and citric acid have been synthesized and characterized. Those copolymers are good candidates for coating applications due to avoiding usage of aromatic monomers, which are more susceptible to photodegradation [333–336] causing yellowing of the coating over time. Copolyesters derived from isosorbide and 1,4-cyclohexane dicarboxylic acid and SA were also synthesized. However, it is difficult to obtain the high molar mass copolyesters by an inexpensive approach [337]. Copolymerizations of isosorbide, SA, and isophthalic acid in the presence of $SnCl_2$ or $ZnCl_2$ led to the corresponding copolymers with M_n in the range of 7,000–15,000 g/mol with PDI values in the range 3–9. The T_g increased with the content of isophthalic acid from 75 to 180°C and the thermostabilities also followed this trend [338].

Glutamic acid is a five-carbon α amino acid and occurs naturally in many foods. As another platform chemical, glutamic acid is also a fermentation product of glucose [320]. By reduction and/or hydrogenation, glutamic acid can be converted to several bifunctional compounds such as 1,5-pentanedicarboxylic acid (glutaric acid), 1,5-pentanediol, and 2-amino-1,5-pentandiol. These compounds can potentially serve as monomers for the production of novel polyamides and polyesters. Since the amino group exists at the α position, glutamic acid can undergo a self-condensation polymerization to poly(glutamic acid) (PGA), which is naturally formed by bacterial fermentation [339]. γ-PGA (the form where the peptide bonds are between the amino group of GA and the carboxyl group at the end of the GA side chain) is a major constituent of the Japanese food natto. This polymer is soluble in water, biodegradable, and edible. γ-PGA has a large number of potential applications including food, medicine, and water treatment. It is widely used as a drug delivery system for cancer treatment [340].

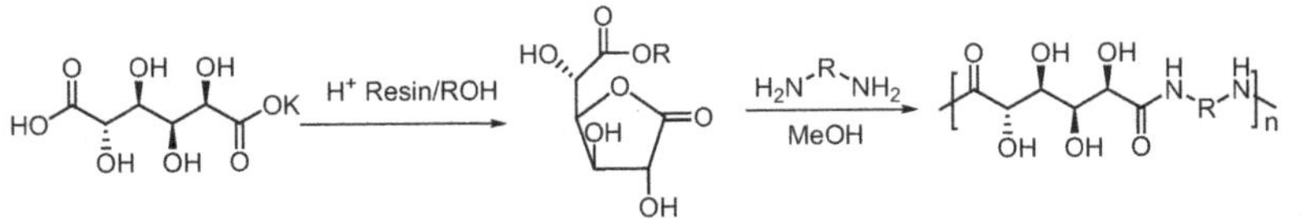

Scheme 21 Synthesis of polyhydroxypolyamides from glucaric acid

Aspartic acid is a four-carbon α amino acid. There are two forms or enantiomers of aspartic acid. Of these two forms, only one, "L-aspartic acid," is directly incorporated into proteins. The biological roles of its counterpart, "D-aspartic acid," are more limited. Where enzymatic synthesis will produce one or the other, most chemical syntheses will produce both forms, "DL-aspartic acid," known as a racemic mixture. The preferred method for producing L-aspartic acid currently is the enzymatic route, reacting ammonia with fumaric acid, catalyzed by a lyase enzyme [320]. As a homologous compound of glutamic acid, aspartic acid can be polymerized with itself to produce poly(aspartic acid) (PASA). PASA and its derivatives are environmentally friendly and biodegradable alternatives to traditional polyanionic materials, in particular as potential replacements for polyacrylic acid [341]. PASA also has an ability to inhibit deposition of calcium carbonate, calcium sulfate, barium sulfate, and calcium phosphate salts, and can be used as an antiscaling agent in cooling water systems, water desalination processes, and wastewater treatment operations [342]. In addition, due to its ability to chelate metal ions, it provides corrosion inhibition [343]. It could act as a super-swelling material in diaper/feminine-hygiene products and food packaging [344]. There is a broad interest in this material from the biomedical and material research communities.

Itaconic acid is a C_5 unsaturated dicarboxylic acid with one carboxyl group conjugated to the methylene group [320]. It can be regarded as α-substituted acrylic, methacrylic acid, or methylene SA. Hence, it can be readily incorporated into polymers and may serve as a substitute for petrochemical-based acrylic or methacrylic acid. It is produced by fermentation of carbohydrates such as glucose by fungi (Aspergillus terreus) with a current market volume of about 15,000 tons per year. Itaconic acid is primarily used as a co-monomer at 1–5% in the production of resins, such as styrene-butadiene-acrylonitrile and acrylate latexes with applications in manufacture of synthetic fibers, in coatings, adhesives, thickeners, and binders [345].

Glucaric acid is a C_6 dicarboxylic acid with four hydroxyl groups. It can be produced from glucose by selective oxidation of both the hydroxyl function of the terminal carbon and the aldehyde function of the first carbon to carboxylic acid functions by nitric acid [320]. Glucaric acid was used as a monomer for the production of new hydroxylated nylons (polyhydroxypolyamides) by condensation reaction with a variety of diamines (Scheme 21) [346, 347]. Glucaric acid (and its esters) is also a potential starting material for new types of hyperbranched polyesters addressing markets of similar sizes to nylons with a similar value structure. Finally, glucaric acid could also address the very large detergent surfactant market, as it should exhibit useful chelating properties for cations [320].

3 Conclusions

The utilization of cellulose as the raw material for production of monomers and polymers is reviewed and discussed. As the most abundant nonfood biomass resource on Earth, cellulose can be catalytically depolymerized to glucose, while glucose is a versatile starting material for a large variety of platform chemicals including ethanol, lactic acid, HMF, levulinic acid, sorbitol, succinic acid, aspartic acid, glutamic acid, itaconic acid, glucaric acid, and so on. These platforms can be used as monomers directly or further converted to polymerizable monomers for polymer synthesis.

Ethanol, categorized as a supercommodity, can be converted to olefins such as ethylene for the production of the commodity plastic polyethylene, which provides a direct interface between the biorefinery and the conversion infrastructure of the petrochemical industry. The Braskem company has established the product line for biopolyethylene production.

Lactic acid is currently produced by fermentation of carbohydrates and is one of the high potential and versatile biomass-derived platform chemicals, leading to various useful polymer products. PLA is produced by ROP of lactide (derived from lactic acid) and exhibits mechanical properties similar to poly(ethylene terephthalate) and polypropylene. Representative examples discussed herein included the synthesis of highly stereo-controlled PLAs, such as isotactic, heterotactic, and syndiotactic PLA materials, rendered by different catalyst/initiator systems.

Furfural and HMF are readily prepared from various catalytic biomass conversion processes. Both furfural and HMF can be readily converted to a large variety of monomers for polymerizations by chain-growth and/or condensation mechanisms. As the transformation of furfural and HMF to fine chemicals or monomers for polymers has been well documented by several comprehensive reviews [107–113, 130, 131], this chapter has mainly focused on the bio-based furan polymers with self-healing ability through thermally reversible Diels–Alder reactions, which is a recently exploited prosperous research area. In addition, the furan-based DA reaction has also been used in the thermoreversible nonlinear polymerization and dendrimer chemistry.

The biomass-derived renewable monomer $_{\gamma}$MMBL can be readily prepared from levulinic acid, another platform from cellulose or glucose, by a two-step process developed by DuPont. Polymerization of $_{\gamma}$MMBL and its homologous MBL and $_{\beta}$MMBL can be achieved by various the polymerization methods, including coordination polymerization by various metal catalysts, living polymerization by the ambiphilic silicon catalyst system, zwitterionic polymerization by classical and frustrated Lewis pairs, and conjugate-addition organopolymerization by NHCs. These biomass-derived methylene butyrolactones offer a sustainable alternative to their cyclic analogue MMA; significantly, these sustainable methylene butyrolactone polymers exhibit superior materials properties to the PMMA materials such as much higher T_g (195, 225, and 290°C for PMBL, P$_{\gamma}$MMBL, and P$_{\beta}$MMBL, respectively) and enhanced resistance to solvents, heat, and scratching.

Isosorbide, a double dehydration product derived from sorbitol, is an attractive monomer due to its rigidity, chirality, and non-toxicity. Various isosorbide-based polymers have been synthesized by condensation polymerization, including polyesters, polyamides, poly(ester amides), polycarbonates, and polyurethanes. Incorporation of isosorbide into polymers' backbones will introduce special properties, such as enhanced glass transition temperatures and/or optical properties.

Last, but not least, the application and polymerization of five biomass dicarboxylic acids including succinic acid, aspartic acid, glutamic acid, itaconic acid, and glucaric acid have been described. Carrying two carboxylic groups, they can be readily transformed to other bifunctional compounds, such as diimines and dialcohols. Therefore, these compounds can potentially work as monomers for the production of polyamides and polyesters. In addition, there exist other functional groups in such dicarboxylic acids which will induce unique reactions. For example, itaconic acid has one carboxyl group conjugated to the methylene group; thus, it can serve as a substitute for petrochemical-based acrylic or methacrylic acid for conjugate addition polymerization. Glutamic acid with the amino group at the α position can undergo self-condensation polymerization to poly(glutamic acid) which is naturally formed by bacterial fermentation. In addition, the multiple hydroxyl groups give glucaric acid (and its esters) the ability to form new types of hyperbranched polyesters.

Overall, the recent developments highlighted in this chapter have documented increasing academic and industrial efforts in the utilization of biomass-derived renewable monomers for the production of synthetic polymers that offer sustainable alternatives to the current petroleum-based polymers. Furthermore, some of the sustainable polymers also exhibit enhanced or unique materials properties over the polymers derived from the depleting resources. Such efforts will continue in the future, with an emphasis being placed on making biomass-derived polymers not only renewable but also technically and economically practicable and competitive.

Acknowledgments This work was supported by the 1,000 Talents Young program and National Natural Science Foundation of China (NSFC) program (grant #21374040 to Y. Zhang) and by the US Department of Energy-Office of Basic Energy Sciences, grant DE-FG02-10ER16193 (to E. Y.-X. Chen).

References

1. IEA (2007) World energy outlook world energy outlook. International Energy Agency, Paris
2. Rass-Hansen J, Falsig H, Jorgensen B, Christensen CH (2007) J Chem Technol Biotechnol 82:329–333
3. Olah GA, Goeppert A, Prakash GKS (2006) Beyond oil and gas: the methanol economy. Wiley-VCH, Weinheim. ISBN 3-527-31275-7
4. Weisz PB (2004) Phys Today 57:47–52
5. IPCC (2007) Climate change 2007: the physical science basis. In: Solomon S, Qin D, Manning M, Chen Z, Marquis M, Averyt KB et al (eds) Contribution of working group

1 to the fourth assessment report of the intergovernmental panel on climate change. Cambridge University Press, Cambridge/New York
6. Morschbacker A (2009) Polym Rev 49:79–84
7. Dusselier M, Van Wouwe P, Dewaele A, Makshina E, Sels BF (2013) Energy Environ Sci 6:1415–1442
8. Jin F, Enomoto H (2011) Energy Environ Sci 4:382–397
9. Corma A, Iborra S, Velty A (2007) Chem Rev 107:2411–2502
10. Drumright RE, Gruber PR, Henton DE (2000) Adv Mater 12:1841–1846
11. Vink ETH, Rábago KR, Glassner DA, Gruber PR (2003) Polym Degrad Stab 80:403–419
12. Datta R, Henry M (2006) J Chem Technol Biotechnol 81:1119–1129
13. Fan Y, Zhou C, Zhu X (2009) Catal Rev Sci Eng 51:293–324
14. Bozell JJ, Petersen GR (2010) Green Chem 12:539–554
15. MäKi-Arvela P, Simakova IL, Salmi T, Murzin DY (2014) Chem Rev 114(3):1909–1971
16. Auras R, Lim LT, Selke SEM, Tsuji H (2010) Poly(lactic acid) – synthesis structures, properties, processing, and applications. Wiley, Hoboken
17. Inkinen S, Hakkarainen M, Albertsson AC, Södergard A (2011) Biomacromolecules 12:523–532
18. Takasu A, Narukawa Y, Hirabayashi T (2006) J Polym Sci Pol Chem 44:5247–5253
19. Kim KW, Woo SI (2002) Macromol Chem Phys 203:2245–2250
20. Kamber NE, Jeong W, Waymouth RM, Pratt RC, Lohmeijer BGG, Hedrick JL (2007) Chem Rev 107:5813–5840
21. Hayashi Y, Sasaki Y (2005) Chem Commun 2716–2718
22. Janssen KPF, Paul JS, Sels BF, Jacobs PA (2007) Stud Surf Sci Catal 170B:1222–1227
23. Roman-Leshkov Y, Moliner M, Labinger JA, Davis ME (2010) Angew Chem Int Ed 49:8954–8957
24. Pescarmona P, Janssen KPF, Delaet C, Stroobants C, Houthoofd K, Philippaerts A, Jonghe CD, Paul JS, Jacobsand PA, Sels BF (2010) Green Chem 12:1083–1089
25. Taarning E, Saravanamurugan S, Holm MS, Xiong JM, West RM, Christensen CH (2009) ChemSusChem 2:625–627
26. West RM, Holm MS, Saravanamurugan S, Xiong JM, Beversdorf Z, Taarning E, Christensen CH (2010) J Catal 269:122–130
27. Rasrendra CB, Fachri BA, Makertihartha GBN, Adisasmito S, Heeres HJ (2011) ChemSusChem 4:768–777
28. Wang J, Masui Y, Onaka M (2011) Appl Catal B 107:135–139
29. de Clippel F, Dusselier M, Rompaey RV, Vanelderen P, Dijkmans J, Makshina E, Giebeler L, Oswald S, Baron GV, Denayer JFM, Pescarmona PP, Jacobs PA, Sels BF (2012) J Am Chem Soc 134:10089–10101
30. Holm MS, Saravanamurugan S, Taarning E (2010) Science 328:602–605
31. Rasrendra CB, Makertihartha IGBN, Adisasmito S, Heeres HJ (2010) Top Catal 53:1241–1247
32. Onda A, Ochi T, Kajiyoshi K, Yanagisawa K (2008) Catal Commun 9:1050–1053
33. Onda A, Ochi T, Kajiyoshi K, Yanagisawa K (2008) Appl Catal A 343:49–54
34. Zeng W, Cheng DG, Chen F, Zhan X (2009) Catal Lett 133:221–226
35. Epane G, Laguerre JC, Wadouachi A, Marek D (2010) Green Chem 12:502–506
36. Holm MS, Pagan-Torres YJ, Saravanamurugan S, Riisager A, Dumesic JA, Taarning E (2012) Green Chem 14:702–706
37. Roman-Leshkov Y, Davis ME (2011) ACS Catal 1:1566–1580
38. Lobo RF (2010) ChemSusChem 3:1237–1240
39. Yan X, Jin F, Tohji K, Moriya T, Enomoto H (2007) J Mater Sci 42:9995–9999
40. Liu Z, Li W, Pan C, Chen P, Lou H, Zheng X (2011) Catal Commun 15:82–87
41. Yan XY, Jin FM, Tohji K, Kishita A, Enomoto H (2010) AIChE J 56:2727–2733
42. Barros dos Santors J, Lins da Silva F, Altino FMRS, Moneira da Silva T, Meneghetti MR, Meneghetti SMP (2013) Catal Sci Technol 3:673–678

43. Wang FW, Huo ZB, Wang YQ, Jin FM (2011) Res Chem Intermediat 37:487–492
44. Zhang S, Jin F, Hu J, Huo Z (2011) Bioresour Technol 102:1998–2003
45. Sanchez C, Eguees I, Garcia A, Llano-Ponte R, Labidi J (2012) Chem Eng J 181:655–660
46. Wang FF, Liu CL, Dong WS (2013) Green Chem 15:2091–2095
47. Stere C, Iovu M, Boborodea A, Vasilescu DS, Fazakas-Anca IS (1998) Polym Adv Technol 9:322–325
48. Bhaw-Luximon A, Jhurry D, Spassky N, Pensec S, Belleney J (2001) Polymer 42:9651–9656
49. McGuinness DS, Marshall EL, Gibson VC, Steed JW (2003) J Polym Sci Pol Chem 41:3798–3803
50. Csihony S, Beaudette TT, Sentman AC, Nyce GW, Waymouth RM, Hedrick JL (2004) Adv Synth Catal 346:1081–1086
51. Barskaya IG, Lyudvig YB, Shifrina RR, Izyumnikov AL (1983) Vysokomol Soedin Ser A 25:1283–1288
52. Emig N, Nguyen H, Krautscheid H, Reau R, Cazaux JB, Bertrand G (1998) Organometallics 17:3599–3608
53. Bourissou D, Martin-Vaca B, Dumitrescu A, Graullier M, Lacombe F (2005) Macromolecules 38:9993–9998
54. Basko M, Kubisa P (2006) J Polym Sci Pol Chem 44:7071–7081
55. Pratt RC, Lohmeijer BGG, Long DA, Waymouth RM, Hedrick JL (2006) J Am Chem Soc 128:4556–4557
56. Chuma A, Horn HW, Swope WC, Pratt RC, Zhang L, Lohmeijer BGG, Wade CG, Waymouth RM, Hedrick JL, Rice JE (2008) J Am Chem Soc 130:6749–6754
57. Becker JM, Tempelaar S, Stanford MJ, Pounder RJ, Covington JA, Dove AP (2010) Chem Eur J 16:6099–6105
58. Dechy-Cabaret O, Martin-Vaca B, Bourissou D (2004) Chem Rev 104:6147–6176
59. Platel RH, Hodgson LM, Williams CK (2008) Polym Rev 48:11–63
60. Wheaton CA, Hayes PG, Ireland BJ (2009) Dalton Trans 4832–4846
61. Wu J, Yu TL, Chen CT, Lin CC (2006) Coord Chem Rev 250:602–626
62. Mehta R, Kumar V, Bhunia H, Upadhyay SN (2005) J Macromol Sci C Polym Rev 45:325–349
63. Kricheldorf HR (2001) Chemosphere 43:49–54
64. Dijkstra PJ, Dum H, Feijen J (2011) Polym Chem 2:520–527
65. Stanford MJ, Dove AP (2010) Chem Soc Rev 39:486–494
66. Thomas CM (2010) Chem Soc Rev 39:165–173
67. Dove AP, Gibson VC, Marshall EL, Rzepa HS, White AJP, Williams DJ (2006) J Am Chem Soc 128:9834–9843
68. Williams CK, Breyfogle LE, Choi SK, Nam W, Young VG, Hillmyer MA, Tolman WB (2003) J Am Chem Soc 125:11350–11359
69. Chamberlain BM, Cheng M, Moore DR, Ovitt TM, Lobkovsky EB, Coates GW (2001) J Am Chem Soc 123:3229–3238
70. Ejfler J, Szafert S, Mierzwicki K, Jerzykiewicz LB, Sobota P (2008) Dalton Trans 6556–6562
71. Wu JC, Huang BH, Hsueh ML, Lai SL, Lin CC (2005) Polymer 46:9784–9792
72. Jones MD, Davidson MG, Keir CG, Hughes LM, Mahon MF, Apperley DC (2009) Eur J Inorg Chem 635–642
73. Zhong ZY, Dijkstra PJ, Feijen J (2002) Angew Chem Int Ed 41:4510–4513
74. Zhong ZY, Dijkstra PJ, Feijen J (2003) J Am Chem Soc 125:11291–11298
75. Ma H, Spaniol TP, Okuda J (2008) Inorg Chem 47:3328–3339
76. Ma H, Spaniol TP, Okuda J (2006) Angew Chem Int Ed 45:7818–7821
77. Peckermann I, Kapelski A, Spaniol TP, Okuda J (2009) Inorg Chem 48:5526–5534
78. Lian B, Ma H, Spaniol TP, Okuda J (2009) Dalton Trans 9033–9042
79. Amgoune A, Thomas CM, Ilinca S, Roisnel T, Carpentier JF (2006) Angew Chem Int Ed 45:2782–2784
80. Abdel-Fattah TM, Pinnavaia TJ (1996) Chem Commun 665–666

81. Tschan MJL, Brulé E, Haquette P, Thomas CM (2012) Polym Chem 3:836–851
82. Spassky N, Wisniewski M, Pluta C, LeBorgne A (1996) Macromol Chem Phys 197:2627–2637
83. Ovitt TM, Coates GW (2000) J Polym Sci Pol Chem 38:4686–4692
84. Ovitt TM, Coates GW (2002) J Am Chem Soc 124:1316–1326
85. Tang Z, Chen X, Pang X, Yang Y, Zhang X, Jing X (2004) Biomacromolecules 5:965–970
86. Tang Z, Chen X, Yang Y, Pang X, Sun J, Zhang X, Jing X (2004) J Polym Sci Pol Chem 42:5974–5982
87. Ishii R, Nomura N, Kondo T (2004) Polym J 36:261–264
88. Nomura N, Ishii R, Yamamoto Y, Kondo T (2007) Chem Eur J 13:4433–4451
89. Bakewell C, Cao T, Long N, Le Goff XF, Auffrant A, Wiliams CK (2012) J Am Chem Soc 134:20577–20580
90. Nederberg F, Connor EF, Moller M, Glauser T, Hedrick JL (2001) AIChE J 40:2712–2715
91. Zhang L, Nederberg F, Messman JM, Pratt RC, Hedrick JL, Wade CG (2007) J Am Chem Soc 129:12610–12611
92. Miyake GM, Chen EY-X (2011) Macromolecules 44:4116–4124
93. Ovitt TM, Coates GW (1999) J Am Chem Soc 121:4072–4073
94. Amgoune A, Thomas CM, Roisnel T, Carpentier JF (2006) Chem Eur J 12:169–179
95. Amgoune A, Thomas CM, Carpentier JF (2007) Macromol Rapid Commun 28:693–697
96. Hormnirun P, Marshall EL, Gibson VC, White AJP, Williams DJ (2004) J Am Chem Soc 126:2688–2689
97. Cai CX, Amgoune A, Lehmann CW, Carpentier JF (2004) Chem Commun 330–331
98. Liu X, Shang X, Tang T, Hu N, Pei F, Cui D, Chen X, Jing X (2007) Organometallics 26:2747–2757
99. Pietrangelo A, Hillmyer MA, Tolman WB (2009) Chem Commun 2736–2737
100. Dove AP, Li H, Pratt RC, Lohmeijer BGG, Culkin DA, Waymouth RM, Hedrich JL (2006) Chem Commun 2881–2883
101. Win DT (2005) Au J T 8:185–190
102. Thananatthanachon T, Rauchfuss TB (2010) ChemSusChem 3:1139–1141
103. Thananatthanachon T, Rauchfuss TB (2010) Angew Chem Int Ed 49:6616–6618
104. Werpy T, Petersen G (2004) In: U.S. Department of Energy (DOE) report: DOE/GO-102004-1992 (ed) Top value added chemicals from biomass, vol I
105. Jacoby M (2009) C&EN July 6:26–28
106. Rosatella AA, Simeonov SP, Frade RFM, Afonso CAM (2011) Green Chem 13:754–793
107. Gandini A (2010) Polym Chem 1:245–251
108. Gandini A (1977) Adv Polym Sci 25:47–96
109. Gandini A (1990) ACS Sym Ser 433:195–208
110. Gandini A, Belgacem MN (1997) Prog Polym Sci 22:1203–1379
111. Moreau C, Belgacem MN, Gandini A (2004) Top Catal 27:11–30
112. Gandini A (2008) Macromolecules 41:9491–9504
113. Gandini A, Belgacem MN (2008) Furan derivatives and furan chemistry at the service of macromolecular materials. Elsevier, Amsterdam, pp 115–152
114. Kawashima D, Aihara T, Kobayashi Y, Kyotani T, Tomita A (2000) Chem Mater 12:3397–3401
115. Yao J, Wang H, Liu J, Chan KY, Zhang L, Xu N (2005) Carbon 43:1709–1715
116. Zarbin AJG, Bertholdo R, Oliveira MAFC (2002) Carbon 40:2413–2422
117. Wang H, Yao J (2006) Ind Eng Chem Res 45:6393–6404
118. Yi B, Rajagopalan R, Foley HC, Kim UJ, Liu X, Ecklund PC (2006) J Am Chem Soc 128:11307–11313
119. Hirasaki T, Meguro T, Wakihara T, Tatami J, Komeya K (2007) J Mater Sci 42:7604–7606
120. Cesano F, Scarano D, Bertarione S, Bonino F, Damin A, Bordiga S, Prestipino C, Lamberti C, Zecchina A (2008) J Photochem Photobiol A 196:143–153

121. Bertarione S, Bonino F, Cesano F, Jain S, Zanetti M, Scarano D, Zecchina A (2009) J Phys Chem B 113:10571–10574
122. Tondi G, Pizzi A, Pasch H, Celzard A, Rode K (2008) Eur Polym J 44:2938–2943
123. Tondi G, Pizzi A, Pasch H, Celzard A (2008) Polym Degrad Stab 93:968–975
124. Pizzi A, Tondi G, Pasch H, Celzard A (2008) J Appl Polym Sci 110:1451–1456
125. Yao J, Wang H (2007) Ind Eng Chem Res 46:6264–6268
126. Zhai Y, Tu B, Zhao D (2009) J Mater Chem 19:131–140
127. Grund S, Kempe P, Baumann G, Seifert A, Spange S (2007) Angew Chem Int Ed 46:628–632
128. Spange S, Grund S (2009) Adv Mater 21:2111–2116
129. Pranger L, Tannenbaum R (2008) Macromolecules 41:8682–8687
130. Tong X, Ma Y, Li Y (2010) Appl Catal Gen 385:1–13
131. Gallezot P (2012) Chem Soc Rev 41:1538–1558
132. Gandini A (2013) Prog Polym Sci 38:1–29
133. Liu Y, Chuo T (2013) Polym Chem 4:2194–2205
134. Syrett JA, Becer CR, Haddleton DM (2010) Polym Chem 1:978–987
135. Bergman SD, Wudl F (2008) J Mater Chem 18:41–62
136. Craven JM (1969) US Pat., 3,435,003
137. Stevens M, Jenkins A (1979) J Polym Sci 17:3675–3685
138. Kuster BFM (1990) Starch Stärke 42:314–321
139. Chujo Y, Sada K, Saegusa T (1990) Macromolecules 23:2636–2641
140. Imai Y, Itoh H, Naka K, Chujo Y (2000) Macromolecules 33:4343–4346
141. Canary SA, Stevens MP (1992) J Polym Sci Polym Chem 30:1755–1760
142. Laita H, Boufi S, Gandini A (1997) Eur Polym J 33:1203–1211
143. Gheneim R, Perez-Berumen C, Gandini A (2002) Macromolecules 35:7246–7253
144. Goiti E, Huglin MB, Rego JM (2001) Polymer 42:10187–10193
145. Goiti E, Huglin MB, Rego JM (2003) Macromol Rapid Commun 24:692–696
146. Goiti E, Huglin MB, Rego JM (2004) Eur Polym J 40:219–226
147. Goiti E, Huglin MB, Rego JM (2004) Eur Polym J 40:1451–1460
148. Tesoro GC, Sastri VR (1986) Ind Eng Chem Prod Res Dev 25:444–448
149. He X, Sastri VR, Tesoro GC (1989) Makromol Chem Rapid Commun 9:191–194
150. O'Dell R (1990). Ph.D. Thesis, Lancaster University, UK
151. Kuramoto N, Hayashi K, Nagai K (1994) J Polym Sci Polym Chem 30:2501–2504
152. Diakoumakos CD, Mikroyannidis JA (1992) J Polym Sci Polym Chem 30:2559–2563
153. Diakoumakos CD, Mikroyannidis JA (1994) Eur Polym J 30:465–472
154. Goussé C, Gandini A (1999) Polym Int 48:723–731
155. Gandini A, Coelho D, Silvestre AJ (2008) Eur Polym J 44:4029–4036
156. Chen X, Dam MA, Ono K, Mal A, Shen H, Nutt SR, Sheran K, Wudl F (2002) Science 295:1698–1702
157. Chen X, Wudl F, Mal AK, Shen H, Nutt SR (2003) Macromolecules 36:1802–1807
158. Kavitha AA, Singha NK (2007) J Polym Sci Pol Chem 45:4441–4449
159. Kavitha AA, Singha NK (2009) ACS Appl Mater Interfaces 1:1427–1436
160. He J, Zhang Y, Chen EYX (2013) J Polym Sci Pol Chem 51:2793–2803
161. Kötteritzsch J, Stumpf S, Hoeppener S, Vitz J, Hager MD, Schubert US (2013) Macromol Chem Phys 214:1636–1649
162. Zeng C, Seino H, Ren J, Hatanaka K, Yoshie N (2013) Macromolecules 46:1794–1802
163. Zeng C, Seino H, Ren J, Hatanaka K, Yoshie N (2013) Polymer 54:5351–5357
164. Franc G, Kakkar AK (2009) Chem Eur J 15:5630–5639
165. Bauer RE, Grimsdale AC, Müllen K (2005) Top Curr Chem 245:253–286
166. Bernhardt S, Baumgarten M, Müllen K (2006) Eur J Org Chem 11:2523–2529
167. Kim C, Kim H, Park K (2003) J Organomet Chem 667:96–102
168. Kim C, Kim H, Park K (2004) J Polym Sci Part A 42:2155–2161
169. Kim C, Lim KI, Song CG (2005) J Organomet Chem 690:3278–3285
170. McElhanon JR, Wheeler DR (2001) Org Lett 3:2681–2683

171. Szalai ML, McGrath DV, Wheeler DR, Zifer T, McElhanon JR (2007) Macromolecules 40:818–823
172. Sullivan PA, Olbricht BC, Akelaitis AJP, Mistry AA, Liao Y, Dalton LR (2007) J Mater Chem 17:2899–2903
173. Kose MM, Yesilbag G, Sanyal A (2008) Org Lett 10:2353–2356
174. Vieyres A, Lam T, Gillet R, Franc G, Castonguay A, Kakkar A (2010) Chem Commun 46:1875–1877
175. Costanzo PJ, Demaree JD, Beyer FL (2006) Langmuir 22:10251–10257
176. Costanzo PJ, Beyer FL (2007) Chem Mater 19:6168–6173
177. Aumsuzan N, Urban MW (2009) Polymer 50:33–36
178. Shi M, Wosnick JH, Ho K, Keating A, Shoihet MS (2007) Angew Chem Int Ed 46:6126–6131
179. Polaske NW, McGrath DV, McElhanon JR (2010) Macromolecules 43:1270–1276
180. Polaske NW, McGrath DV, McElhanon JR (2011) Macromolecules 44:3203–3210
181. Manzer LE (2006) Feedstocks for the future. ACS Sym Ser 921:40–51
182. Fitzpatrick SW (2006) Feedstocks for the future. ACS Sym Ser 921:271–287
183. Hayes DJ, Fitzpatick S, Hayes MHB, Ross JRH (2006) In: Kamm B, Gruber P, Kamm M (eds) Biorefineries-industrial processes and products. Wiley-VCH, Weinheim
184. Girisuta B, Janssen LPBM, Heeres HJ (2006) Chem Eng Res Des 84:339–349
185. Manzer LE (2004) Appl Catal A Gen 272:249–256
186. Sauer M, Porro D, Mattanovich D, Branduardi P (2008) Trends Biotechnol 26:100–108
187. Rossum MWPCV, Alberda M, Plas LHWVD (1998) Phytochemistry 49:723–729
188. Hoffman HMR, Rabe J (1985) Angew Chem Int Ed 24:94–110
189. Yokota K, Hirabayashi T (1992) (Iwata Kagaku Kogyo KK, Japan) J.P. Pat. 04049288 A
190. Pittman CU, Lee JH (2003) J Polym Sci Pol Chem 41:1759–1777
191. Hu Y, Wang X, Chen Y, Caporaso L, Cavallo L, Chen EYX (2013) Organometallics 32:1459–1465
192. Geilen FMA, Engendahl B, Harwardt A, Marquardt W, Klankermayer J, Leitner W (2010) Angew Chem Int Ed 49:5510–5514
193. Hu Y, Miyake GM, Wang B, Cui D, Chen EYX (2012) Chem Eur J 18:3345–3354
194. Agarwal S, Jin Q, Maji S (2012) ACS Sym Ser 1105:197–212
195. Mullin R (2004) Sustainable specialties. Chem Eng News 82(45):29–37
196. Akkapeddi MK (1979) Polymer 20:1215–1216
197. Stansbury JW, Antonucci JM (1992) Dent Mater 8:270–273
198. Akkapeddi MK (1979) Macromolecules 12:546–551
199. Kimura Y, Nakamura S (2009) JP,046,560 A
200. Pickett JE, Ye Q (2007) U.S. Patent 2007/0,122,625
201. Brandenburg CJ (2004) (to E. I. du Pont de Nemours and Company) WO 2004/069,926 A1
202. Miyake GM, Newton SE, Mariott WR, Chen EYX (2010) Dalton Trans 39:6710–6718
203. Chen X, Caporaso L, Cavallo L, Chen EYX (2012) J Am Chem Soc 134:7278–7281
204. Mosnáček J, Yoon JA, Juhari A, Koynov K, Matyjaszewski K (2009) Polymer 50:2087–2094
205. Mosnáček J, Matyjaszewski K (2008) Macromolecules 41:5509–5511
206. Ueda M, Takahashi M, Imai Y, Pittman CU Jr (1982) J Polym Sci Polym Chem Ed 20:2819–2828
207. Gridnev AA, Ittel SD (2000) (to E. I. du Pont de Nemours and Company) WO 00/35,960 A2
208. Sogah DY, Hertler WR, Webster OW, Cohen GM (1987) Macromolecules 20:1473–1488
209. Van den Brink M, Smulders W, van Herk AM, German AL (1999) J Polym Sci Pol Chem 37:3804–3816
210. Koinuma H, Sato K, Hirai H (1982) Makromol Chem Rapid Commun 3:311–315
211. Lee C, Hall HK Jr (1989) Macromolecules 22:21–25
212. Trumbo DL (1991) Polym Bull 26:271–275
213. Brandenburg CJ (2005) (to E. I. du Pont de Nemours and Company) U.S. Pat. 6,841,627 B2
214. Qi G, Nolan M, Schork FJ, Jones CW (2008) J Polym Sci Pol Chem 46:5929–5944

215. Suenaga J, Sutherlin DM, Stille JK (1984) Macromolecules 17:2913–2916
216. Cockburn RA, McKenna TFL, Hutchinson RA (2010) Macromol Chem Phys 211:501–509
217. Cockburn RA, Siegmann R, Payne KA, Beuermann S, McKenna TFL, Hutchinson RA (2011) Biomacromolecules 12:2319–2326
218. Hu Y, Xu X, Zhang Y, Chen Y, Chen EYX (2010) Macromolecules 43:9328–9336
219. Gowda RR, Chen EYX (2013) Dalton Trans 42:9263–9273
220. Miyake GM, Zhang Y, Chen EYX (2010) Macromolecules 43:4902–4908
221. Zhang Y, Gustafson LO, Chen EYX (2011) J Am Chem Soc 133:13674–13684
222. Hu Y, Gustafson LO, Zhu H, Chen EYX (2011) J Polym Sci Pol Chem 49:2008–2017
223. Zhang Y, Chen EYX (2012) Angew Chem Int Ed 51:2465–2469
224. Zhang Y, Schmitt M, Falivene L, Caporaso L, Cavallo LI, Chen EYX (2013) J Am Chem Soc 135:19725–19742
225. Zhang Y, Miyake GM, Chen EYX (2010) Angew Chem Int Ed 49:10158–10162
226. Zhang Y, Miyake GM, John MG, Falivene L, Caporaso L, Cavallo L, Chen EYX (2012) Dalton Trans 41:9119–9134
227. Brandenburg CJ (2003) (to E. I. du Pont de Nemours and Company) WO 03/048220 A2
228. Brandenburg CJ, Manzer LE, Subramanian P (2005) (to E. I. du Pont de Nemours and Company) WO 2005/028,529 A2
229. Ohrbom WH (2008) (to BASF Corporation) U.S. Pat. 2008/0,293,901 A1
230. Sakashita K (2007) (to Mitsubishi Rayon Co., Ltd) E.P. Pat. 1,834,968 A1
231. Chen EYX (2009) Chem Rev 109:5157–5214
232. Gowda RR, Chen EYX (2013) Encycl Polym Sci Technol, in press
233. Evans WJ, Grate JW, Choi HW, Bloom I, Hunter WE, Atwood JL (1985) J Am Chem Soc 107:941–946
234. Schuetz SA, Day VW, Sommer RD, Rheingold AL, Belot JA (2001) Inorg Chem 40:5292–5295
235. Xu X, Chen Y, Sun J (2009) Chem Eur J 15:846–850
236. Xu X, Chen Y, Feng J, Zou G, Sun J (2010) Organometallics 29:549–553
237. Wang B, Cui D, Lv K (2008) Macromolecules 41:1983–1988
238. Fedushikin IL, Dechert S, Schumann H (2001) Angew Chem Int Ed 40:561–563
239. Li Y, Ward EG, Reddy SS, Collins S (1997) Macromolecules 30:1875–1883
240. Ning Y, Zhu H, Chen EYX (2007) J Organomet Chem 692:4535–4544
241. Yang X, Stern CL, Marks TJ (1994) J Am Chem Soc 116:10015–10031
242. Rodriguez-Delgado A, Mariott WR, Chen EYX (2004) Macromolecules 37:3092–3100
243. Zhang Y, Ning Y, Caporaso L, Cavallo L, Chen EYX (2010) J Am Chem Soc 132:2695–2709
244. Rodriguez-Delgado A, Chen EYX (2005) Macromolecules 38:2587–2594
245. Bolig AD, Chen EYX (2004) J Am Chem Soc 126:4897–4906
246. Zhang Y, Chen EYX (2008) Macromolecules 41:36–42
247. Zhang Y, Chen EYX (2008) Macromolecules 41:6353–6360
248. Webster OW (2004) Adv Polym Sci 167:1–34
249. Webster OW, Hertler WR, Sogah DY, Farnham WB, RajanBabu TV (1983) J Am Chem Soc 105:5706–5708
250. Ryan SJ, Candish L, Lupton DW (2013) Chem Soc Rev 42:4906–4917
251. Grossmann A, Enders D (2012) Angew Chem Int Ed 51:314–325
252. Bugaut X, Glorius F (2012) Chem Soc Rev 41:3511–3522
253. Biju AT, Kuhl N, Glorius F (2011) Acc Chem Res 44:1182–1195
254. Dröge T, Glorius F (2010) Angew Chem Int Ed 50:6940–6952
255. Chiang PC, Bode JW (2010) In: RSC catalysis series. Royal Society of Chemistry, Cambridge, pp 399–435
256. Moore JL, Rovis T (2009) Top Curr Chem 291:77–144
257. Hahn FE, Jahnke MC (2008) Angew Chem Int Ed 47:3122–3172
258. Nair V, Vellalath S, Babu BP (2008) Chem Soc Rev 37:2691–2698
259. Marion N, Díez-González S, Nolan SP (2007) Angew Chem Int Ed 46:2988–3000

260. Enders D, Niemeier O, Henseler A (2007) Chem Rev 107:5606–5655
261. Bourissou D, Guerret O, Gabbaï FP, Bertrand G (2000) Chem Rev 100:39–91
262. Nyce GW, Glauser T, Connor EF, Möck A, Waymouth RM, Hedrick JL (2003) J Am Chem Soc 125:3046–3056
263. Connor EF, Nyce GW, Myers M, Möck A, Hedrick JL (2002) J Am Chem Soc 124:914–915
264. Nederberg F, Connor EF, Möller M, Glauser T, Hedrick JL (2001) Angew Chem Int Ed 40:2712–2715
265. Brown HA, Waymouth RM (2013) Acc Chem Res 46:2585–2596
266. Fèvre M, Pinaud J, Gnanou Y, Vignolle J, Taton D (2013) Chem Soc Rev 42:2142–2172
267. Kiesewetter MK, Shin EJ, Hedrick JL, Waymouth RM (2010) Macromolecules 43:2093–2107
268. Jeong W, Shin EJ, Culkin DA, Hedrick JL, Waymouth RM (2009) J Am Chem Soc 131:4884–4891
269. Culkin DA, Jeong W, Csihony S, Gomez ED, Balsara NP, Hedrick JL, Waymouth RM (2007) Angew Chem Int Ed 46:2627–2630
270. Dove AP, Li H, Pratt RC, Lohmeijer BGG, Culkin DA, Waymouth RM, Hedrick JL (2006) Chem Commun 2881–2883
271. Coulembier O, Dove AP, Pratt RC, Sentman AC, Culkin DA, Mespouille L, Dubois P, Waymouth RM, Hedrick JL (2005) Angew Chem Int Ed 44:4964–4968
272. Csihony S, Culkin DA, Sentman AC, Dove AP, Waymouth RM, Hedrick JL (2005) J Am Chem Soc 127:9079–9084
273. Shin EJ, Brown HA, Gonzalez S, Jeong W, Hedrick JL, Waymouth RM (2011) Angew Chem Int Ed 50:6388–6391
274. Sen TK, Sau SC, Mukherjee A, Modak A, Mandal SK, Koley D (2011) Chem Commun 47:11972–11974
275. Kamber NE, Jeong W, Gonzalez S, Hedrick JL, Waymouth RM (2009) Macromolecules 42:1634–1639
276. Jeong W, Hedrick JL, Waymouth RM (2007) J Am Chem Soc 129:8414–8415
277. Raynaud J, Absalon C, Gnanou Y, Taton D (2010) Macromolecules 43:2814–2823
278. Raynaud J, Ottou WN, Gnanou Y, Taton D (2010) Chem Commun 46:3203–3205
279. Raynaud J, Absalon C, Gnanou Y, Taton D (2009) J Am Chem Soc 131:3201–3209
280. Nederberg F, Lohmeijer BGG, Leibfarth F, Pratt RC, Choi J, Dove AP, Waymouth RM, Hedrick JL (2007) Biomacromolecules 8:153–160
281. Rodriguez M, Marrot S, Kato T, Stérin S, Fleury E, Baceiredo A (2007) J Organomet Chem 692:705–708
282. Lohmeijer BGG, Dubois G, Leibfarth F, Pratt RC, Nederberg F, Nelson A, Waymouth RM, Wade C, Hedrick JL (2006) Org Lett 8:4683–4686
283. Guo L, Lahasky SH, Ghale K, Zhang D (2012) J Am Chem Soc 134:9163–9171
284. Guo L, Zhang D (2009) J Am Chem Soc 131:18072–18074
285. Coutelier O, El Ezzi M, Destarac M, Bonnette F, Kato T, Baceiredo A, Sivasankarapillai G, Gnanou Y, Taton D (2012) Polym Chem 3:605–608
286. Pinaud J, Vijayakrishna K, Taton D, Gnanou Y (2009) Macromolecules 42:4932–4936
287. Nyce GW, Lamboy JA, Connor EF, Waymouth RM, Hedrick JL (2002) Org Lett 4:3587–3590
288. McCahill JSJ, Welch GC, Stephan DW (2007) Angew Chem Int Ed 46:4968–4971
289. Chase PA, Welch GC, Jurca T, Stephan DW (2007) Angew Chem Int Ed 46:8050–8053
290. Welch GC, Stephan DW (2007) J Am Chem Soc 129:1880–1881
291. Spies P, Erker G, Kehr G, Bergander K, Frohlich R, Grimme S, Stephan DW (2007) Chem Commun 5072–5074
292. Welch GC, Juan RRS, Masuda JD, Stephan DW (2006) Science 314:1124–1126
293. Erker G, Stephan DW (2013) Frustrated Lewis pairs I—uncovering and understanding. Topics in current chemistry 332

294. Erker G, Stephan DW (2013) Frustrated Lewis pairs II-expanding the scope. Top Curr Chem 334
295. Stephan DW, Erker G (2010) Angew Chem Int Ed 49:46–76
296. Chen EYX (2013) Top Curr Chem 334:239–260
297. Vogel R (2000) Ullmann's encyclopedia of industrial chemistry. Wiley, on line
298. Hill K, Rhode O (1999) Fett/Lipid 101:25–33
299. Corma A, Hamid SBA, Iborra S, Velty A (2008) ChemSuSChem 1:85–90
300. Gallezot P, Cérino P, Blanc B, Flèche G, Fuertes P (1994) J Catal 146:93–102
301. van Gorpa K, Boermana E, Cavenaghib CV, Berben PH (1999) Catal Today 52:349–361
302. Gallezot P, Nicolaus N, Flèche G, Fuertes P, Perrard A (1998) J Catal 180:51–55
303. Van de Vyver S, Geboers J, Jacobs PA, Sels BF (2011) ChemCatChem 3:82–94
304. Fléche G, Huchette M (1986) Starch/Staerke 38:26–30
305. Kurszewska M, Skorupowa E, Madaj J, Konitz A, Wojnowski W, Wisniewski A (2002) Carbohydr Res 337:1261–1268
306. Xia J, Yu D, Hu Y, Zou B, Sun P, Li H, Huang H (2011) Catal Commun 12:544–547
307. Xi J, Zhang Y, Ding D, Xia Q, Wang J, Liu X, Lu G, Wang Y (2014) Appl Catal Gen 469:108–115
308. Fenouillot F, Rousseau A, Colomines G, Saint-Loup R, Pascault JP (2010) Prog Polym Sci 35:578–622
309. Yin B, Hakkarainen M (2011) J Appl Polym Sci 119:2400–2407
310. Rivera-Armenta JL, Heinze TH, Mendoza-Martínez AM (2004) Eur Polym J 40:2803–2812
311. Park HS, Gong MS, Knowles JC (2013) J Mater Sci Mater Med 24:281–294
312. Lorenzini C, Renard E, Bensemhoun J, Babinot J, Versace DL, Langlois V (2013) React Funct Polym 73:1656–1661
313. Karol MH, Kramarik JA (1996) Toxicol Lett 89:139–146
314. Besse V, Auvergne R, Carlotti S, Boutevin G, Otazaghine B, Caillol S, Pascault JP, Boutevin B (2013) React Funct Polym 73:588–594
315. Kricheldorf H, Weidner SM (2013) Eur Poly J 49:2293–2302
316. Kricheldorf H, Weidner SM (2013) Macromol Chem Phys 214:726–733
317. Williams CL, Chang CC, Do P, Nikbin N, Caratzoulas S, Vlachos DG, Lobo RF, Fan W, Dauenhauer PJ (2012) ACS Catal 2:935–939
318. Shiramizu M, Toste MFD (2012) Chem Eur J 17:12452–12457
319. Colonna M, Berti C, Fiorini M, Binassi F, Mazzacurati M, Vanini M, Kuranam S (2012) Presented at the ACS meeting "Polycondensation 2012", San Francisco, September 2012
320. Werpy T, Petersen G (2004) Top value added chemicals from biomass, vol 1. U.S. Department of Energy, Oak Ridge
321. Fumagalli C (1997) Kirk–Othmer encyclopedia of chemical technology, vol 22, 4th edn. Wiley, New York, p 1074
322. Luo L, Voet EVD, Huppes G (2010) Bioresour Technol 101:5023–5032
323. Pipper G, Koch EM (1990) US Patent 5,030,709
324. Gaymans RJ, Venkatraman VS, Schuijer J (1984) J Polym Sci Polym Chem Ed 22:1373–1382
325. Carothers WH (1938). US Patent 2,130,948
326. Katsarava RD, Kharadze DP, Aualkhuili LM (1986) Makromol Chem 187:2053–2062
327. Bechthold I, Bretz K, Kabasci S, Kopitzky R, Springer A (2008) Chem Eng Technol 31:647–654
328. Coffman DD, Berchet GJ, Peterson WR, Spanagel EW (1947) J Polym Sci 2:306–313
329. Katsarava RD, Katsarava RD, Kharadze DP, Bendiashvili TM, Urman YG, Slonim IY, Alekseeva SG, Cefelin P, Janout V (1988) Acta Polym 39:523–533
330. Katsarava RD (2003) Macromol Sym 199:419–429
331. Wang Q, Shao Z, Yu T (1996) Polym Bull 36:659–665
332. Habeych DI, Eggink G, Boeriu CG (2011) Biocatal Biotransfor 29:299–310
333. Decker C, Moussa K, Bendaikha T (1991) J Polym Sci Pol Chem 29:739–747

334. Allen NS, Edge M, Mahammadian M, Jones K (1993) Polym Degrad Stab 41:191–196
335. Luda MP, Tauriello R, Camino G (2000) Eur Coat J 74:76–79
336. Hult A, Raanby B (1984) Polym Degrad Stab 8:75–87
337. Garaleh M, Yashiro T, Kricheldorf HR, Simon P, Chatti S (2010) Macromol Chem Phys 211:1206–1214
338. Chatti S, Weidner SM, Fildier A, Kricheldorf HR (2013) J Polym Sci Pol Chem 51:2464–2471
339. Shih IL, Van YT (2001) Bioresour Technol 79:207–225
340. Li C, Yu DF, Newman A, Cabra FL, Stephens C, Hunter NR, Milas L, Wallace S (1998) Cancer Res 58:2404–2409
341. Gross RA, Kalra B (2002) Science 297:803–807
342. Hasson D, Shemer H, Sher A (2011) Ind Eng Chem Res 50:7601–7607
343. Low KC, Wheeler AP, Koskan LP (1996) Adv Chem ACS 248:99–111
344. Zahuriaan-Mehr MJ, Pourjavadi A, Salimi H, Kurdtabar M (2009) Polym Adv Technol 20:655–671
345. Willke KD (2001) Appl Microbiol Biotechnol 56:289–295
346. Kiely DE, Chen L, Lin TH (1994) Polymers from agricultural coproducts. ACS Sym Ser 575:149–158
347. Kiely DE, Chen L, Lin TH (1994) J Am Chem Soc 116:571–578

Top Curr Chem (2014) 353: 229–256
DOI: 10.1007/128_2014_535

Published online: 3 April 2014

Approaches to the Selective Catalytic Conversion of Lignin: A Grand Challenge for Biorefinery Development

Joseph J. Bozell

Abstract Lignin comprises 15–25% of terrestrial biomass and is the second most abundant source of renewable carbon after cellulose. However, its structural heterogeneity frustrates efforts for its selective conversion into biobased chemicals. Catalyst design for lignin transformation offers an opportunity to improve selectivity, and, hence, improve lignin's utility as a raw material in chemical production. Catalytic deconstruction and conversion of lignin has been examined using a variety of thermochemical treatments, analogous to those used in the petrochemical industry. However, the complex nature of these products limits their utility. More recently, greater focus has been given to an understanding of lignin's molecular level structure, and designing catalysts that can be targeted to key individual structural units within the biopolymer. This review gives a sense of the field by providing a representative description of recent developments in some of the primary technologies employed for lignin conversion and approaches that promise to improve selectivity.

Keywords Biobased chemicals · Biomass · Biorefinery · Catalysis · Heterogeneous · Homogeneous · Lignin · Lignin models · Selectivity · Thermochemistry

Contents

J.J. Bozell (✉)
Center for Renewable Carbon, Center for the Catalytic Conversion of Biomass (C3Bio), University of Tennessee, Knoxville TN 37996, USA
e-mail: jbozell@utk.edu

Abbreviations

Ac	Acetyl
BTX	Benzene, toluene, xylene
COD	Cyclooctadiene
dipic	Pyridine-2,6-dicarboxylic acid
DMBQ	Dimethoxybenzoquinone
DMSO	Dimethylsulfoxide
G	Guaiacyl
GC/MS	Gas chromatography/mass spectrometry
HDO	Hydrodeoxygenation
HSQC	Heteronuclear single quantum coherence
MMBQ	Monomethoxybenzoquinone
NMR	Nuclear magnetic resonance
ppm	Parts per million
S	Syringyl
TEMPO	Tetramethylpiperidinoxyl
wt%	Weight percent
xantphos	4,5-Bis(diphenylphosphino)-9,9-dimethylxanthene

1 Introduction

Lignin is a source of great frustration for the biorefinery.

Within biomass, it exists as a barrier, barring the way of the pulp and paper manufacturer or ethanol producer to the more valuable polysaccharides contained within lignocellulosic feedstocks. To the chemist, it presents a seemingly intractable array of multiple functional groups exhibiting similar reactivity, spread over a wide range of fragments bearing a similarly wide range of molecular weights. As a raw material, its structure and reactivity varies as a function of its source and the process used for its isolation. Large, dependable supplies of lignin could potentially be available from kraft pulping, but because of the pulp industry's reliance on the lignin extracted from wood as fuel for the operation of chemical recovery boilers, those with the greatest access to lignin have historically exhibited the least interest in developing selective methodology for its transformation into higher value chemicals. Moreover, the emerging biorefining industry, with its continued focus

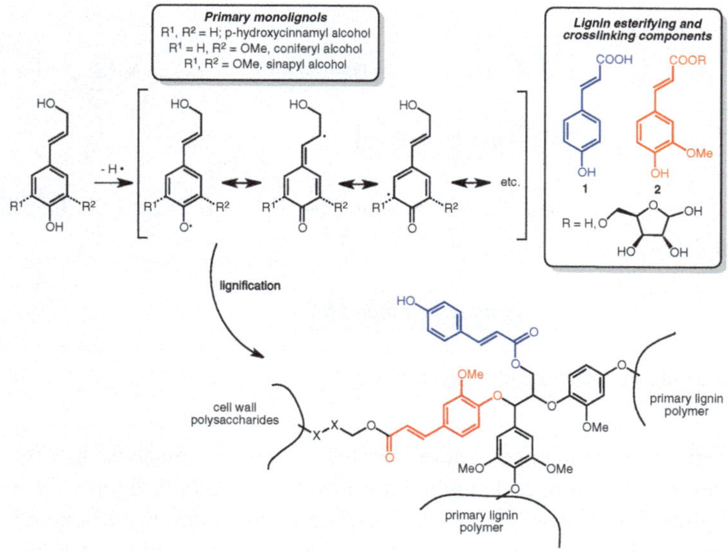

Fig. 1 Overview of lignin biosynthesis

on converting carbohydrates to biofuels, has followed this lead and defaults to burning lignin for process energy.

Nonetheless, lignin remains an exceptionally attractive biobased raw material whose potential is lost when it is used simply as boiler fuel. After cellulose, lignin is the second most abundant source of renewable carbon in Nature, comprising 15–25 wt% of lignocellulosic feedstocks. Projections show it to be an inexpensive raw material that, if upgraded to high value chemicals, could provide a significant economic boost to lignocellulosic biorefineries currently focused on ethanol production [1]. Technology such as organosolv fractionation is realizing production of lignin as a separate process stream that need not be integrated with the overall steam and energy balance of the operation [2]. Moreover, lignin is cheap. Multiple evaluations project an internal transfer cost for biorefinery lignin of \$0.03–0.06/lb [3–5].

Lignin's primary disadvantage is a high level of structural heterogeneity that arises from two sources. First, biosynthesis introduces heterogeneity as lignin is manufactured in the plant cell wall (Fig. 1) [6]. Lignin biosynthesis uses the three primary monolignols, *p*-hydroxycinnamyl alcohol, coniferyl alcohol, and sinapyl alcohol, leading to the well-recognized *p*-hydroxyphenyl, guaiacyl, and syringyl units, respectively, in the resulting lignin biopolymer. During biosynthesis, these monolignols are converted into highly delocalized phenoxy radicals that undergo radical–radical coupling and conversion to the lignin polymer. Softwoods are primarily constructed of guaiacyl units, while hardwoods contain significant amounts of both guaiacyl and syringyl units. Herbaceous feedstocks (grasses) incorporate *p*-hydroxyphenyl functionality and offer additional complexity, as their structure includes extensive crosslinking between lignin and hemicelluloses

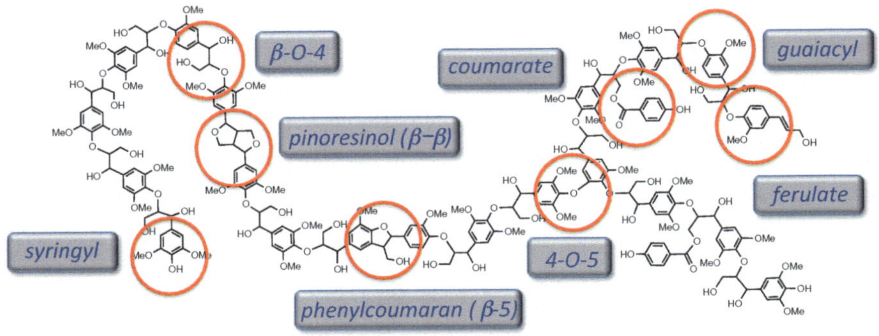

Fig. 2 Suggested structure of native poplar lignin

through ferulate ester linkages. Moreover, the ferulates themselves undergo dimerization and trimerization, and can be introduced into the bulk lignin. The result of these coupling reactions is the production of lignin as a complex aromatic biopolymer [7]. Figure 2 provides a representation of lignin from poplar, a hardwood, and shows many of the well-recognized substructural units that result from radical coupling between monolignol building blocks.

The second source of lignin's structural heterogeneity results when it is isolated from the lignocellulosic matrix. Any process used to isolate lignin inevitably induces structural changes in the native material through loss of some substructural units and introduction of new interunit linkages. For example, switchgrass organosolv lignin showed a marked change in the concentration of β-O-4 units as the severity of the fractionation increased [8]. Quantitative ^{13}C analysis of the NMR spectral region between 88 and 77 ppm identifies lignin's β-O-4 units, which can be 50% or more of the interunit linkages in native lignin. After organosolv fractionation, NMR spectra revealed a dramatic reduction of these linkages to almost zero at the highest fractionation severities. Other candidate processes for producing a lignin process stream within the biorefinery result in similar changes. Dilute acid pretreatment was reported to reduce β-O-4 units by 36% [9], and steam explosion can nearly eliminate them under proper conditions [10]. In parallel, lignin's structure realizes an increase in the number of free phenolic –OH groups as the β-O-4 units are cleaved. The use of lignin as a chemical feedstock, therefore, depends on identifying transformational processes able to accommodate multiple functional groups introduced during biosynthesis as well as those that are present after lignin isolation.

A further complicating factor is that the biorefining industry has not settled on an optimum pretreatment/fractionation process, leading to investigation of both a wide array of lignins as starting materials and multiple conversion processes. All lignins are not created equal, and the functional group profile present in a given starting material can vary markedly as a function of biomass source, pretreatment or fractionation methodology, and isolation technique. This functional group profile may be modified using methods such as sequential extraction [11–13] or selective precipitation [14, 15] to prepare fractions of differing molecular weight and hence, different reactivity. However, despite the wide diversity in lignins available as chemical feedstocks, there is a surprising tendency to use illustrations similar to that in Fig. 2 as descriptors of lignin's structure, especially in the more recent literature examining new catalytic methodology for lignin conversion. It is important to remember that these figures are only approximations (and indeed, possibly poor approximations), as the actual structure depends strongly on the source and handling employed for the lignin.

Achieving selectivity in transforming a heterogeneous starting material into a single, low molecular weight structure thus becomes the grand challenge facing lignin utility in the biorefinery. Catalysis and intelligent catalyst design are playing a critical role in developing fundamental knowledge regarding the navigation of the functional diversity of different biorefinery lignins needed for the emergence of selective lignin conversion processes. This literature reflects this growing interest in lignin, as evidenced by a number of excellent reviews on various processes used for catalytic lignin valorization [16–24].

Nonetheless, *truly selective lignin transformations remain elusive*. Decades of work have been carried out on the nonselective deconstruction of lignin in the pulp and paper industry [25–27], its thermochemical transformation and upgrading to liquid feedstocks positioned as fuels, or conversion to blends of materials enriched in higher value chemicals. However, a greater focus is now being given to lignin's molecular level structure and how catalysts may be designed for attack on and transformation of specific key structural units within lignin. Much of this work is still being carried out on lignin model compounds in order to understand and better control these processes on transformation of both specialized lignin samples and actual biorefinery lignin.

Catalysis designed for lignin transformation has grown rapidly over the last 5 years, and this review attempts to give a sense of the field by providing a representative, but not exhaustive, review of recent developments in some of the primary technologies employed for lignin conversion. The review will focus primarily on catalytic transformations of lignin as an isolated material, and thus transformations that start with whole biomass are not covered, except for a few pertinent examples. The review will begin with a brief initial examination of catalyzed, but nonselective, thermochemical transformations to illustrate some of the existing landscape for lignin conversion. More recent work will then be presented, describing efforts based on understanding the molecular level structure of lignin, and targeting specific interunit linkages to design selective processes for the catalytic conversion of both lignin and lignin models. Although not covered here, a potentially dramatic contribution to the biorefinery's ability to convert lignin selectively may result from ongoing work focused on modifying lignin's biosynthesis. The ability to use biotechnology and genetic engineering to define the structure of a raw material *before* it reaches a refining process is unprecedented in the petrochemical industry, and a critical advantage that renewable carbon sources have over nonrenewable. Such work could have a significant impact on

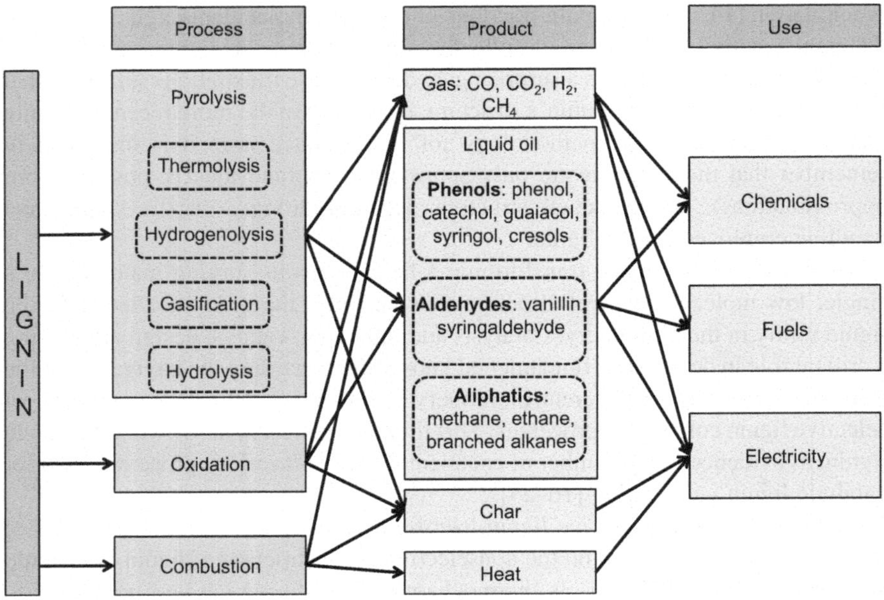

Fig. 3 The matrix of thermochemical lignin transformation processes (from Pandey [31])

selective lignin transformations by reducing lignin's heterogeneity and improving its utility as a chemical feedstock [28–30].

2 Catalytic Thermochemical Transformations of Lignin

The complexity of lignin's structure suggests comparison with nonrenewable feedstocks bearing similar complexity, such as coal or crude oil. Accordingly, a large body of work exists for the thermochemical treatment and conversion of lignin analogous to distillation or cracking processes carried out in the petrochemical industry [31]. A wide range of general experimental processes has been employed, and includes fast pyrolysis (both thermal and in the presence of catalysts), hydrothermal treatment (hydrolysis), hydrodeoxygenation (which can be a combination of thermal lignin deconstruction and catalytic post-treatments), oxidation, gasification, and hydrogenolysis (Fig. 3). However, as will be shown in the following examples, this broad spectrum of processing technologies (which normally involve high temperatures or harsh conditions), not surprisingly, produces a large number of monomeric and oligomeric compounds in different product classes: gases, aromatics, aliphatics, and, in many cases, a large amount of char. Although these processes lead to lignin deconstruction, a reduction in lignin's molecular weight, and enrichment of the resulting condensable fractions in certain product classes, true selectivity has yet to be achieved.

2.1 Pyrolytic Transformations of Lignin

Pyrolysis is a process of rapidly heating a feedstock under anaerobic conditions to induce depolymerization, and results in the production of a pyrolysis oil, normally accompanied by gas and char fractions. The absence of air assures that the lower molecular weight fractions do not undergo combustion. The complex nature of lignin's structure is reflected in its more complex pyrolytic behavior compared to cellulose or hemicellulose. While biomass hemicellulose and cellulose pyrolyze at 220–315 and 315–400°C, respectively, lignin's thermal conversion occurs over a range of 150–900°C [32]. Moreover, an extensive study found that at bench scale, isolated, purified lignin does not pyrolyze well in conventional reactor systems designed for potential industrial use [33]. The mechanism of pyrolysis is complex, and is dominated by multiple, parallel radical and rearrangement pathways [34]. Thermogravimetric analysis of an oligomeric β-O-4 lignin model suggests three separate stages for lignin pyrolysis, with the β-O-4 linkage being cleaved between 250 and 350°C, formation of solid products at about 350°C, and eventual conversion to polyaromatics between 350 and 550°C [35].

The portfolio of products obtained from pyrolysis can be modified when pyrolysis vapors are passed over a heterogeneous catalyst. Catalytic pyrolysis of Alcell (organosolv) lignin (a generally cleaner and lower molecular weight feedstock than other lignin sources) in the presence of HZSM-5 zeolite gave a maximum 43 wt% yield of liquid at 550°C. Of that liquid component, 78.1% was BTX. An interesting feature of this process is that the solubility of organosolv lignin in organic solvents allowed feeding of the material in solution [36]. The use of a zeolite with a larger pore size (H-USY) allowed conversion of alkaline (kraft) lignin to a liquid in 75 wt% yield with little char formation at a catalyst/lignin ratio of 4. Of that liquid, approximately 40% was BTX [37]. Simple aromatic hydrocarbons are also observed from pyrolysis of kraft lignin over $Mo_2N/\gamma\text{-}Al_2O_3$. Pyrolysis at 850°C gave a 14% yield of condensable liquid fraction that was 70% benzene and 15% toluene. Pyrolysis at 700°C gave a slightly lower selectivity to benzene and toluene, but the overall yield of aromatic hydrocarbons was higher (17.5%). A two-stage mechanism was suggested, beginning with pyrolytic depolymerization of the lignin followed by catalytic cracking and deoxygenation to afford aromatic hydrocarbons [38]. Quantum chemical calculations have been used to correlate the selectivity of zeolite catalysts for the production of various phenols and aromatics from the fast pyrolysis of kraft lignin. The results suggest that a strict correlation between known pore size of the zeolite and molecular dimensions of the substrate is not accurate as the zeolite's pore size can increase at elevated temperatures. Modeling allowed estimation of the effective pore size of the zeolite and provided a tool for evaluating the shape selectivity of a given catalyst [39].

Catalytic upgrading of pyrolysis oils or in situ catalytic upgrading of pyrolysis vapors can be carried out to provide mixtures of potential value as chemical feedstocks. Wheat straw lignin was converted to ethylbenzene in a two-step sequence, first via pyrolytic depolymerization over a composite Re-Y/HZSM-5 (25) catalyst and subsequent conversion to ethylbenzene via treatment with EtOH in the presence of an HZSM-5(25) catalyst [40]. The first stage gave approximately 15 wt% yield of organic liquid, which contained 90% benzene. The selectivity to ethylbenzene in the subsequent step was optimized at about 72%. Production of coke in the initial pyrolytic step was high, approaching 25 wt%. Related work has examined lignin as a feedstock in catalytic pyrolysis for the production of light olefins. Treatment of a commercially available lignin with HZSM-5 impregnated with 6 wt% La at 600°C afforded about 9% light olefins and 6 wt% aromatics, spread over nine different products [41]. Pyrolysis of kraft lignin over anatase TiO_2 at 550°C led to a 7.5 wt% yield of simple phenols, cresols, and xylenols via catalytic defunctionalization of the more complex structures present in the pyrolysis vapor [42].

2.2 Lignin Hydrodeoxygenation

Hydrodeoxygenation (HDO) of lignin is carried out catalytically as a means to upgrade lignin primarily for fuel use, as lignin's phenolic groups introduce corrosivity and immiscibility with fuel supplies. The general process treats the substrate with a catalyst and H_2 (or alternatively, hydrogen donors such as formic acid or tetralin) at high pressures and temperatures, analogously to HDO processes originally targeted for deoxygenation of petroleum or coal based liquids. Considerable investigation regarding optimization and mechanism of HDO processes has been carried out on low molecular weight materials representing intermediate structures formed during initial lignin depolymerization (phenols, cresols, etc.). These model studies are valuable as they address compounds or compound classes found in high proportion after lignin pyrolysis. Excellent overviews of this work can be found in reviews by Laskar et al. [23] and Furimsky [43].

Early examples of HDO on isolated lignin illustrate how these processes are often carried out, and the types of results obtained. The Noguchi process, developed in the 1950s, found that lignin could be liquefied at 250–400°C using a co-catalyst system such as sulfided iron/copper/zinc at H_2 pressures between 15 and 46 MPa, using phenol as the solvent [17]. Phenol yields as high as 21% were reported, partially as a result of solvent consumption, but the process never proved economically viable [44]. The Hydrocarbon Research Institute Lignol Process converted kraft lignin into a phenol-rich mixture using earth-abundant metal oxides as catalysts at temperatures between 650 and 850°C and 3.4–17.2 MPa H_2. The product mixture was dominated by phenols, which were 37.5 wt% of the starting lignin charged, and included *p*-ethylphenol (33%), *p*-propylphenol (20%), and *m*-cresol (12%) [45]. The reaction of kraft pine lignin with hydrogen (7–10 MPa) using a water soluble ammonium heptamolybdate catalyst for 60 min at 430°C afforded a 61 wt% yield of low molecular weight oil (based on the starting lignin charge). Depending on the pyrolysis conditions, up to 74% of the oil had a measured molecular weight range below 200, 22% of the material had a molecular weight between 200 and 500, and a small amount ranged as high as 2,000. Analysis of the oil showed that low molecular weight materials included phenols (8.7%), cyclohexanes (5.0%), benzenes (3.8%), naphthalenes (4.0%), and phenanthrenes (1.2%) [46]. Expansion of this methodology to organosolv and hardwood kraft lignin using commercial mixed metal oxide catalysts based on NiMo or Cr_2O_3 on different supports gave similar results, with the oil yields (49–71 wt% of the starting lignin) decreasing in the order organosolv>softwood kraft>hardwood kraft [47]. Between 13% and 39% of this oil was identifiable by GC, and included monocyclic and polycyclic aromatics, as well as a phenol fraction. As with pyrolysis, these processes are nonselective, and convert lignin into a mixture of low molecular weight aromatics or more highly reduced derivatives potentially valuable as mixed hydrocarbon feedstocks.

HDO can be carried out directly on lignin, or can be part of a two-stage process, for example, pyrolytic treatment of lignin to generate pyrolysis oil followed by upgrading via HDO. Pine organosolv lignin was pyrolyzed and the resulting oil was subjected to a two-stage HDO using Ru/C and 14 MPa H_2 at 300°C and 250°C, respectively, for the two stages of hydrogenation. The overall carbon content for each stage was 33% and 35% of the carbon content in the original pyrolysis oil and afforded a mixture of materials suggested as a renewable gasoline [48]. Two-step depolymerization/HDO has been carried out on organosolv and kraft lignin. Liquid phase reforming of lignin in alkaline solution over a Pt/γ-Al_2O_3 catalyst at 5.8 MPa Ar and 225°C afforded an extractable lignin oil composed of oxygenated monomeric species in about 12 wt% yield. In a subsequent step, the extracted oil was treated under HDO conditions, using either a CoMo/Al_2O_3 or Mo_2C/CNF catalyst in dodecane at 300°C and 5.5 MPa H_2. As expected, significant deoxygenation occurred, leading to a mixture of arenes and oxygenated arenes. Under optimal conditions, up to 9% aromatic products, 24% of which were oxygen free, could be prepared from organosolv lignin [49]. Related work on the catalyst revealed that it was unstable under the reaction conditions, forming crystalline regions and reduced surface area. In parallel, the supported Pt underwent sintering, reducing the catalyst's activity. Interestingly, in the presence of lignin models or high concentrations of lignin, substrate binding to the catalyst surface occurred and reduced these catalyst deactivation pathways [50]. Two-stage pyrolysis/HDO was also carried out on herbaceous and deciduous organosolv lignins. A large bench scale (up to 100 g lignin fed per hour) continuous pyrolysis fluidized bed reactor gave 13–21% condensable organic oils containing 7–9% low molecular weight phenolics. Subsequent HDO over a Ru/C catalyst and 10 MPa H_2 at 350°C in dodecane reduced most of the aromatic content and gave small amounts of cycloalkanes [51]. Lignin isolated from corn stover residue after conversion of the sugars was used as a feedstock for catalytic hydrogenolysis. Treatment of the lignin with 2 MPa H_2 in the presence of a Ru/C catalyst at 275°C afforded 73 wt% liquid products, which contained 4-ethylphenol and 4-ethylguaiacol at levels of 3% and 1.4%, respectively [52].

A more recent approach to HDO substitutes H_2 gas with formic acid as the hydrogen source and converts lignin into a mixture of hydrocarbons and phenols in a single step. Treatment of lignin from enzymatic and weak acid hydrolysis of biomass with formic acid in either EtOH or *i*-PrOH solvent at 380°C for 24 h affords a high yield of oil from these lignins, but, again, as a complex mixture of materials [53]. A subsequent mechanistic study supports a radical formation/recombination process leading to the final product mixture. Although the mechanistic investigation used structurally-defined lignin model compounds as substrates, the final HDO product mixtures were highly complex [54]. Treatment of Alcell hardwood and wheat straw organosolv lignin under supercritical conditions in the presence of formic acid afforded up to 45% of an aromatic enriched liquid containing up to 12 wt% identifiable phenols [55]. However, as is often seen in these transformations, a large amount of char (40–50 wt%) was also formed from nonselective condensation of reactive intermediates formed under the reaction conditions.

By changing the reaction medium to supercritical EtOH, a 92 wt% yield of liquid was obtained at 98% conversion of organosolv lignin in a one-step process using 5% Ru/γ-Al_2O_3 as catalyst and 2 MPa H_2 at 300°C [56]. A blank reaction under the same conditions afforded a 63% yield of oil at 70% conversion, but the presence of the catalyst significantly decreased the O/C ratio, showing that HDO was taking place. Again, GC/MS analysis of the resulting liquid showed a complex mixture of materials, with propylsyringol being observed as 23% of the total GC peak area.

2.3 *Lignin Hydrotreating (Hydrolysis)*

Lignin can be depolymerized and liquefied hydrolytically through the use of water at high temperatures and pressures. A recent review on hydrothermal treatment of lignin has appeared and describes a number of different approaches and methodologies [57]. As with other thermochemical treatments, catalytic hydrothermal treatment is inherently nonselective, despite its wide use in lignin conversion and upgrading. As an example, subjecting pure vanillin (a derivative of lignin conversion) to hydrothermal treatment at 200°C in water under O_2 in the presence of Co_3O_4 or Mn_3O_4 nanoparticles afforded a large number of water soluble monomeric acids and dimeric coupling products arising from multiple deconstruction pathways [58]. The primary mechanistic transformations were suggested to include simple oxidation of vanillin to the corresponding vanillic acid, formation of intermediate radical species, and Fenton-like formation of hydroxylated intermediates.

Direct liquefaction of organosolv lignin was examined in hot compressed water, both in the absence and presence of catalysts. Treatment of organosolv lignin at 250°C for 1 h afforded a 97% conversion of the lignin, approximately 53% of which was an oil containing phenolics and neutrals. GC/MS analysis determined that approximately 74% of this oil was phenolic material as part of a mixture containing at least 20 components [59]. The process was also examined in the presence of $Ba(OH)_2$ and $RbCO_3$ catalysts, which reduced the amount of oil production by 15% and 23% respectively.

To address some of the separation issues associated with nonselective conversion processes, alkaline lignin treated under similar conditions (300°C in water) was subjected to a more extensive post-purification sequence employing a designed series of pH shifts and solvent extractions to classify the families of compounds obtained from deconstruction. A 28% yield of lignin oil was obtained, and afforded more than 100 compounds (identified by GC/MS) spread over 6 fractions [60]. A mechanism for lignin cleavage was presented and postulated the preferential cleavage of side chain ether linkages. Kraft and organosolv lignin were also hydrotreated at 374°C for 10 min and afforded 58–72% oil from the kraft lignin and 79% oil from the organosolv, again demonstrating the strong dependence of oil formation on the initial source of the lignin and its preparation. GC/MS analysis indicated as many as 25 identifiable products in the reaction mixture [61].

A related two-step sequence treated organosolv lignin propionate in a 4:1 BuOH/H_2O mixture to depolymerize the lignin at 200°C over a silica/alumina catalyst. This first step afforded greater than 90% liquid product, but only about 15% of the mixture was identifiable by GC. Catalytic lignin depolymerization was suggested to occur in the aqueous phase, and the lower molecular weight components were then extracted into the BuOH phase. In comparison, similar treatment of kraft lignin also gave about 90% liquid product of which 18% was identifiable by GC/MS. The BuOH soluble fraction (which contained the great majority of soluble material) was then cracked over a $ZrO_2/Al_2O_3/FeO_x$ catalyst at 200°C for the preparation of a mixture containing primarily phenol and cresol (~70% of the mixture). Under optimized conditions, the amount of phenolic material reached about 6.6% for organosolv lignin and 8.6% for kraft lignin [62].

Hydrothermal treatment has also been carried out in the presence of oxidizing agents. Alkali (kraft) lignin was treated with 0.1% HOOH in hot compressed water at temperatures between 150 and 200°C, leading to the formation of mixed organic acids (formic, acetic, and succinic) in yields as high as 45 wt% (based on the starting lignin), 19% CO_2, and a high molecular weight lignin residue. In contrast, organosolv lignin underwent a more effective depolymerization giving an oligomeric mixture with a molecular weight of about 300 and the production of 20 wt% organic acids [63].

Kraft lignin was subjected to alkaline (NaOH) hydrolysis in the presence of phenol (to cap and capture intermediate radical species from lignin depolymerization) and ethanol. The process reduced the M_n and M_w from 10,000 and 60,000 to 440–480 and 900–1,200, respectively, and converted all the lignin with negligible formation of char. Under optimized conditions (300°C) a combined phenol yield of 35% was observed, which was dominated by the phenol added at the beginning of the treatment [64]. Alkaline hydrothermal treatment of lignin from corn stover catalyzed by disodium tetraborate decahydrate gave a liquid product composed mainly of phenols. The highest product yields were for 2,6-dimethoxyphenol (8 wt%), guaiacol (8%), and phenol (7%) [65].

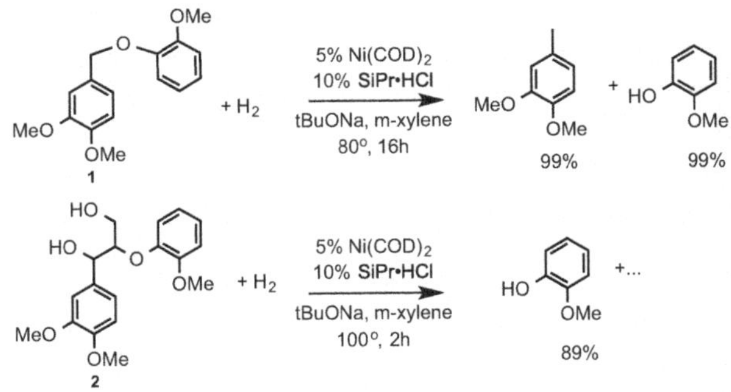

Fig. 4 Ni catalyzed cleavage of lignin model β-aryl ethers

3 Selective Catalytic Transformations of Lignin Models

An understanding of lignin's gross chemical structure as an aromatic biopolymer and an interest in demonstrating it as a viable industrial source of mixed phenols, aromatics, and hydrocarbons have driven research into the thermochemical treatment of lignin. Much less attention has been paid to lignin's structure at the molecular level and developing catalysts that target abundant functional groups common to all lignins, regardless of source. However, recently, the organic synthesis and catalysis community seems to have discovered the β-O-4 linkage. As a result, catalytic methodology for cleaving aryl ethers is now prevalent in the literature, as is work focusing on transformation of lignin-like model compounds. While this work's impact on selective transformation of whole lignin is still small, the approaches being examined with lignin models are tackling the bigger questions of mechanism, structure, and selectivity at the molecular level that will lead to catalytic systems able to convert lignin more effectively.

3.1 *Reductive β-Aryl Ether Cleavage*

Hartwig has reported reductive cleavage of aryl ether bonds similar to those in lignin. By using homogeneous $Ni(COD)_2$ (20%) in the presence of an N-heterocyclic carbene ligand (10–40%) and t-BuONa in xylene at 120°C, diphenyl ethers were cleaved in excellent yield to afford a phenol and arene as cleavage products with no formation of the corresponding arene reduction product. When the process was applied to α-O-4 lignin model **1**, nearly quantitative yields of the corresponding cleavage products were obtained. Similar reaction with β-O-4 model **2** afforded a high yield of guaiacol, but the cleaved side chain unit reacted nonselectively to give multiple products (Fig. 4) [66].

Hartwig's group also reported the hydrogenolysis of simple benzyl aryl and diarylethers representative of 4-O-5 units in lignin using a ligandless Ni $(CH_2TMS)_2$(TMEDA) precatalyst. Reaction of the substrate with 1 atm of H_2 and as little as 0.25 mol% of the catalyst at 120°C in the presence of 2.5 equiv. t-BuONa

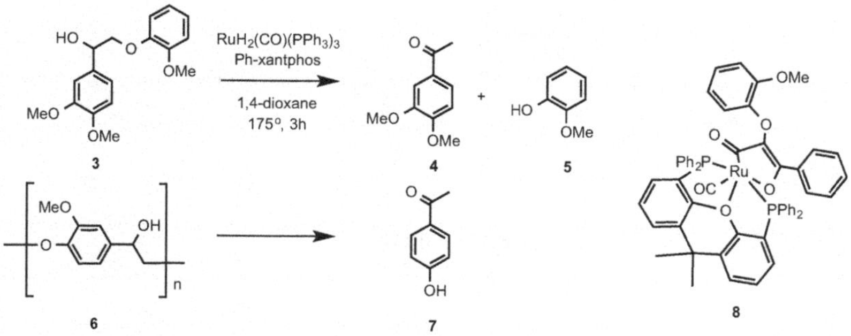

Fig. 5 Redox-neutral cleavage of model β-aryl ethers

affords excellent yields of the corresponding cleavage products. Preliminary mechanistic investigation suggests that the reaction is heterogeneous, proceeding through the formation of Ni nanoparticles during the reaction. The *t*-BuONa is thought to stabilize the nanoparticles [67].

An elegant redox neutral aryl ether cleavage catalyzed by Ru recognizes that certain β-aryl ether cleavages can be represented by the formal dehydrogenation of a C-OH bond at the α-position of the side chain and use of that H_2 for subsequent cleavage of the β-ether bond (Fig. 5). Treatment of **3** with 5% $RuH_2CO(PPh_3)_3$ and Ph-xantphos as a large bite angle external ligand afforded production of cleavage products **4** and **5**. When applied to synthetic lignin model polymer **6**, cleavage to acetophenone **7** occurred in 99% isolated yield. The proposed mechanism for the process is thought to start with dehydrogenation of the benzylic alcohol to the corresponding carbonyl group [68]. In contrast, models bearing a β-CH_2OH substituent (e. g., analogous to **2**) undergoes cleavage in yields <20% as a result of catalyst deactivation via formation of complex **8** [69]. DFT evaluation of the process provided good agreement between calculated and measured kinetics, offering support for the proposed mechanism [70].

A unique heterogeneous Pd/C/Zn catalyst also induces β-aryl ether cleavage in a number of lignin models. Treatment of **9** with 5 wt% catalyst and 2 MPa H_2 in MeOH at 150°C produces propylguaiacol **10** and guaiacol **11** in 85% yield. A small amount of alcohol **12** is also formed as a byproduct. Hydrogenolysis of a polymeric lignin model under these conditions afforded **10** and **11** as major products in 56% and 44% yield, respectively (Fig. 6) [71].

3.2 *Oxidative β-Aryl Ether Cleavage*

Highly simplified β-O-4 lignin models are also cleaved under oxidative conditions using V(dipic) catalysts. Upon reaction of **13** with catalyst **14** in air at 100 ºC in DMSO for 1 week, 95% conversion of **13** was observed, and afforded benzoic acid,

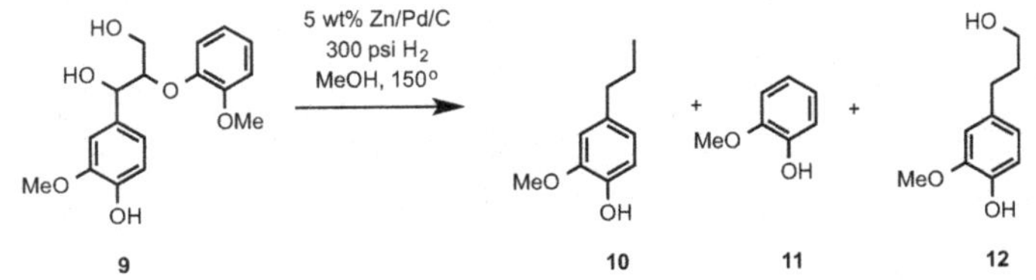

Fig. 6 β-Aryl ether hydrogenolysis over a Zn/Pd/C catalyst

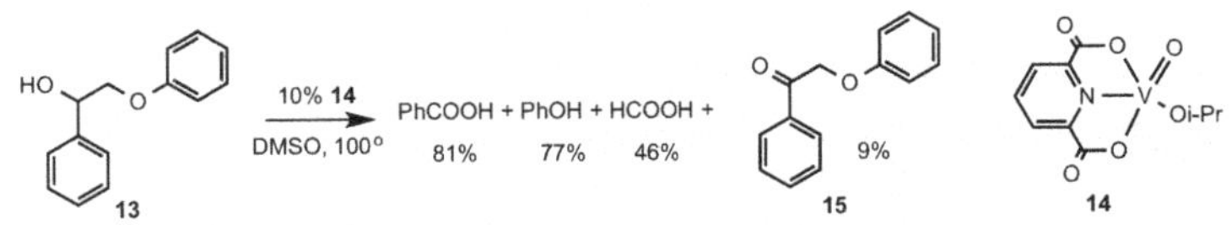

Fig. 7 Oxidative β-aryl ether cleavage with V(dipic) catalysts

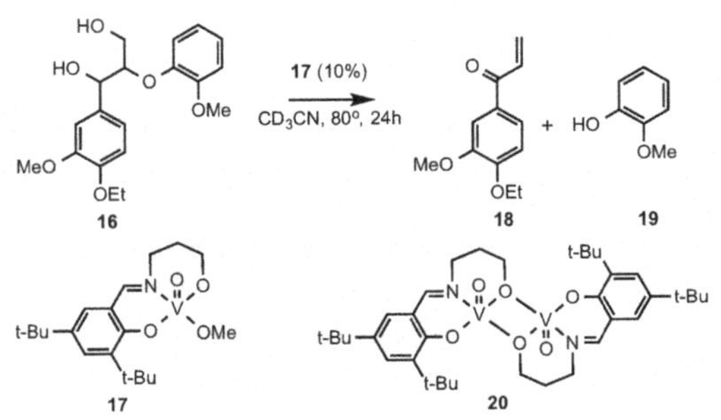

Fig. 8 Vanadium catalyzed non-oxidative cleavage of β-aryl ethers

phenol and formic acid in 81%, 77%, and 46% yield, respectively. A small amount of the benzylic oxidation product **15** was also observed (Fig. 7).

Vanadium catalysts bearing a salen-like ligand also cleave the β-O-4 linkage in lignin model dimers and trimers [72]. The nonphenolic β-O-4 model **16** was treated with V catalyst **17** in CD_3CN at 80°C in air, leading to the formation of **18** and **19** in 82% and 57% yield respectively (Fig. 8). The reaction exhibits several interesting features. First, while the yield of cleavage products is enhanced in air, the reaction may be considered as non-oxidative, as **18** and **19** are formed in 60% and 64% yield, respectively, when the reaction is carried out under N_2. Under anaerobic conditions, the catalyst dimerizes to **20**, which can be reconverted to the active **17** upon exposure to oxygen. The reaction fails in the absence of a free OH group on the model's side chain, and thus, a mechanism proceeding through initial ligand exchange between the benzylic OH on the model and the OMe on the catalyst followed by formation and cleavage of an intermediate benzylic radical is suggested.

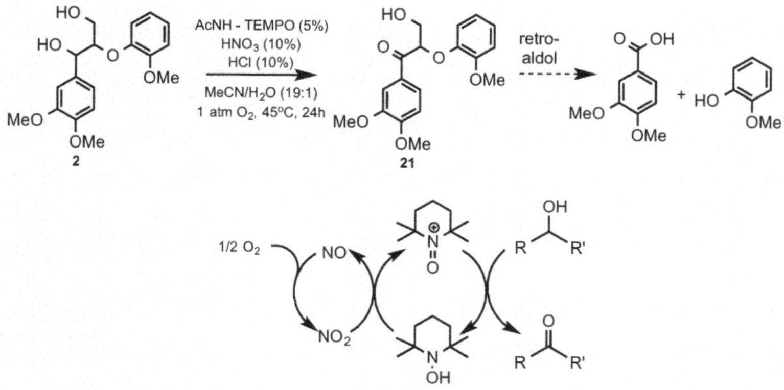

Fig. 9 Organocatalytic cleavage of lignin model β-aryl ethers with TEMPO

The optimal V catalyst has been examined for its reactivity with whole *Miscanthus* lignin. Although the very mild lignin extraction and isolation process employed is likely not representative of a commercial biorefinery operation, treatment of ethanol, acetone, and dioxane organosolv lignin with catalyst **17** led to significant cleavage of the β-O-4 linkages as shown by 2D-HSQC evaluation and a reduction in the lignin molecular weight. GC/MS analysis was also carried out to identify any monomeric materials formed during the oxidation, and revealed the presence of several aromatic monomers, including vanillin, syringic acid, and syringaldehyde, albeit in low yield [73].

Selective reaction along the side chain of lignin is the basis of a proposed two-step oxidative sequence for the conversion of lignin into low molecular weight aromatics. Stahl describes a process for the selective oxidation of side chain alcohol functionality to activate lignin for a subsequent cleavage via retro-aldol reaction (Fig. 9). Selective oxidation of the 2° alcohol group in β-O-4 model **2** can be carried out using a metal-free system and 4-acetamido TEMPO as an organocatalyst. High yields of the corresponding ketone are observed (96% on a 5-g scale), affording an intermediate susceptible to retro-aldol cleavage in the presence of a base. The organocatalyst employs a well-recognized alcohol oxidation process mediated by TEMPO derivatives. This process is made catalytic by the conversion of the N-hydroxy TEMPO to the catalytically active N-oxo cation by NO_2 generated from the reaction of HNO_3 and HCl.

The reaction was also applied to a sample of specialized cellulolytic enzyme lignin. HSQC analysis of the oxidized lignin revealed that most S units and all G units were converted to the corresponding ketone. Moreover, preliminary examination of the proposed retro-aldol cleavage of ketone **21** (2 M NaOH, HOOH in 1:1 THF/MeOH at 50°C) afforded an 88% yield of veratric acid and a 42% yield of guaiacol [74].

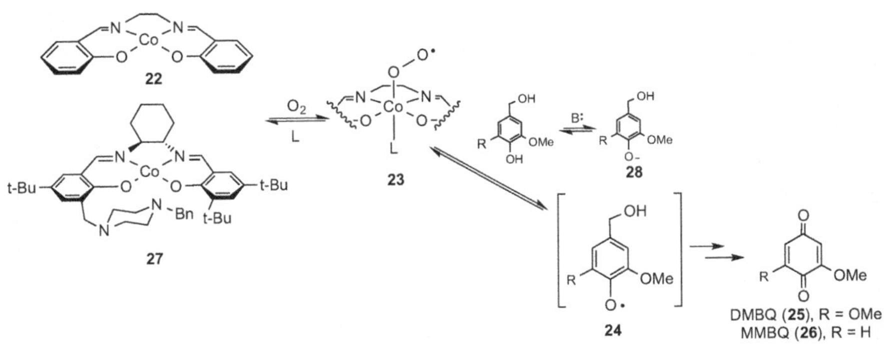

Fig. 10 Co-Schiff base catalyzed oxidation of *para*-substituted phenolics

3.3 Oxidation of Lignin's Phenolic Functionality

The recent increased interest in catalytic aryl ether cleavage has provided important insight regarding methodology that might provide more selective transformations of lignin. However, these studies have focused almost exclusively on the cleavage of dimeric, non-phenolic β-aryl ether models, representative of lignin's β-O-4 linkage. It is important to note that such processes are only effective if the starting lignin actually bears the interunit linkages being targeted. Although it is accurate that lignin as found in the plant may contain as much as 50–60% β-O-4 units, commercially viable methodology may significantly reduce that percentage [8–10]. Moreover, as these β-O-4 units are lost, lignin realizes a concomitant increase in free phenolic hydroxyl functionality [75]. Thus, catalytic processes targeting conversion of lignin-like phenols may provide a more realistic approach for the use of lignin as a chemical feedstock. Because of their electron-rich nature, such systems would be anticipated to undergo a wide range of oxidation processes. Development of new oxidation catalysts would ideally employ environmentally benign terminal oxidants such as O_2 or HOOH, and would demonstrate reactivity designed for the substructural units present in lignin as isolated by the biorefinery.

Co-Schiff base complexes catalyze the aerobic oxidation of *para*-substituted phenols under mild conditions (Fig. 10). In the presence of an external ligand such as pyridine, the Co catalysts [e.g., Co(salen), **22**] bind molecular oxygen to form a Co-superoxo complex **23** in solution [76]. Formation of **23** is significantly improved by addition of pyridine as an external axial ligand because **22** itself binds oxygen poorly [77, 78]. Complex **23** abstracts a phenolic hydrogen from the substrate to generate phenoxy radical **24**, and initiates a process that results in the production of *para*-benzoquinones from *para*-substituted lignin model phenols. Some of the first reported examples of this transformation converted S lignin models to dimethoxybenzoquinone (DMBQ, **25**) and G lignin models to monomethoxybenzoquinone (MMBQ, **26**) [79].

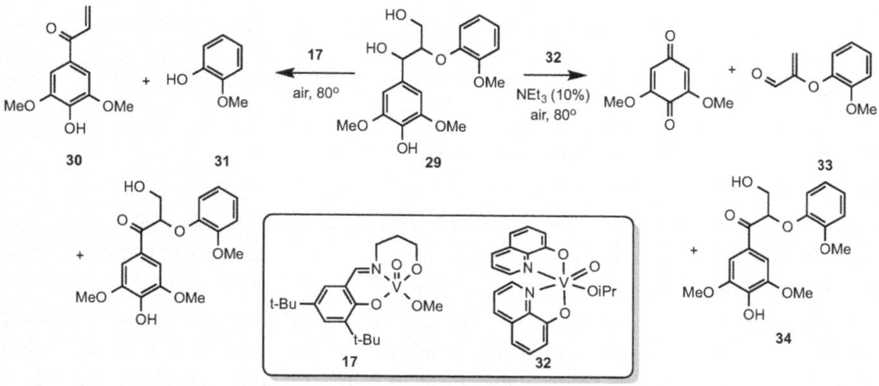

Fig. 11 Reactivity control as a function of catalyst in vanadium catalyzed β-aryl ether cleavage

Although the oxidation proceeded in good yield for the production of **25**, catalyst **22** gave low yields of **26** from oxidation of G model phenols. However, the yield of **26** from G models was improved when the Co/O_2/pyridine system was supplemented with a catalytic amount of a sterically hindered, non-coordinating aliphatic base [80]. By appending the hindered base to an aromatic ring of the Schiff-base ligand (e.g., the piperazine in complex **27**), improved yields, reaction times, and catalyst loadings for the oxidation of both S and G model monomers and dimers in the *absence* of an added external ligand were observed [81]. The results suggested that the presence of a bulky aliphatic base in close proximity to the Co-superoxo complex promotes the formation of the phenoxy radical **24** through formation of an easily oxidized phenoxide intermediate **28**. A computational evaluation of the reactivity of **22** with a family of substituted imidazoles as axial bases suggested that the steric environment around the Co could affect the geometry of **22**, and, thus, its reactivity [82]. When applied to biorefinery lignin samples, Co-Schiff base catalyzed oxidation induced quinone formation and the production of structurally related aromatics. Treatment of organosolv lignin isolated from mixtures of switchgrass and poplar with Co(salen) and O_2 affords roughly 10 wt % yield of low molecular weight products, most of which is DMBQ. Interestingly, oxidations of lignin are not improved by the addition of either an external basic ligand or a non-ligating hindered base, suggesting that the lignin itself may be acting in those capacities. Although the yields are still low, they are equivalent to similar lignin oxidation processes reported in the literature.

The impact of using phenolic lignin models has also been observed with V catalysts. The reactivity of two V catalysts was compared, and revealed a dependence on the ligands within the complex (Fig. 11). Reaction of phenolic β-O-4 model **29** with the Toste catalyst **17** afforded the expected products of benzylic hydrogen cleavage, **30** and **31**. In contrast, reaction with vanadium quinolate complex **32** under similar conditions induced the cleavage of the bond between the Cα and Cβ carbons of the model's side chain, affording dimethoxybenzoquinone (40%), acrolein derivative **33** (38%), and benzylic oxidation product **34**

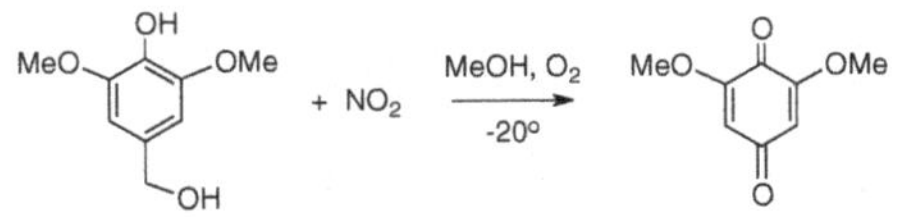

Fig. 12 Oxidation of syringyl alcohol with catalytic NO_2

(20%). The presence of a free phenolic OH group on the aromatic ring was suggested as the source of the C–C bond cleavage via formation of an intermediate phenoxy radical [83].

Recent work has examined the Co-catalyzed oxidation of several different lignins in ionic liquid medium using a simple catalyst. Alcell (organosolv) and soda lignin were treated with catalytic $CoCl_2{\cdot}6H_2O$ and NaOH in 1-ethyl-3-methylimidazolium diethyl phosphate under 0.5 MPa O_2 at 80°C. The lignin samples did not demonstrate conversion to low molecular weight aromatics, but infrared analysis showed an increase in the concentration of both –OH and aldehyde functionality. Model compound studies suggested that certain subunits in lignin were inert to these conditions, but that available –OH groups on the lignin side chain were converted to the corresponding aldehyde [84]. Subsequent spectroscopic and mechanistic investigation demonstrated that the NaOH played a key role in activating the Co precatalyst, and determining the distribution of products after reaction. The ionic liquid solvent was thought to stabilize several reactive intermediates, including a Co-superoxo complex analogous to **23** that is not observed from $CoCl_2{\cdot}6H_2O$ in conventional media [85].

Phenolic lignin models are also converted to the corresponding benzoquinone upon oxidation with stoichiometric NO_2 [86] or *catalytic* NO_2 in the presence of O_2 [87]. Reaction of syringyl alcohol with 10% $NaNO_2$ in MeOH and a small amount of concentrated HNO_3 or HCl (which generates NO_2 in situ), gives an 80–90% yield of DMBQ when carried out under 0.1 MPa of O_2. Reducing the $NaNO_2$ level to 5% gives DMBQ in 70–75% yield (Fig. 12). The mechanism is similar to Co-Schiff base catalyzed reactions, in that NO_2 initiates reaction through formation of an intermediate phenoxy radical, which is trapped by additional NO_2 [88, 89].

4 More Highly Selective Lignin Transformations

4.1 Vanillin

Despite the challenges to selectivity presented by lignin's heterogeneous structure, scattered examples of more highly selective transformations have been reported. The production of vanillin from lignin is perhaps the best example of an ongoing industrial effort for selective transformation of lignin. Borregard has been a commercial supplier of lignin-derived vanillin for decades, and carries out the conversion through the metal catalyzed oxidation of lignosulfonates (from sulfite pulping of wood) at high pH. The production of vanillin from lignin in the pulp and paper industry has a long history, with an initial observation of vanillin in lignin wastes appearing in 1875, and the first commercial scale production in the US starting in 1936 [90]. Yields are low (5–10%), but production has historically been used by the paper industry to generate additional revenue. Although 85% of vanillin today is now produced from guaiacol [91], production from lignin is still carried out on a limited scale by industry [92]. A recent mechanistic study of this process using sodium lignosulfonates and oxygen in the presence of a Cu^{+2} catalyst revealed the complexity of the process, and showed that vanillin resulted both from lignin hydrolysis in the absence of an oxidizing agent (~55% of the vanillin), while the remainder was formed oxidatively (~45%) [93].

Efforts have been made to improve this process through development of new oxidation catalysts. It has long been recognized that the yields of vanillin from sulfite liquors can be increased by the addition of metal catalysts such as Cu(II). Alkaline oxidation of sulfite waste liquor gave 22% vanillin in the presence of $CuSO_4{\cdot}5H_2O$, and 13.5% yield from sulfite waste liquor solids [94]. Organosolv lignins have also been reported as starting materials for vanillin synthesis. Organosolv lignin obtained from eucalyptus, sugarcane bagasse, and softwood via the Acetosolv or Organocell processes was converted to vanillin by oxidation in HOAc with O_2 catalyzed by $Co(OAc)_2$. The optimum yield reported for any of these systems was less than 6% [95].

Kraft lignin can also be converted to vanillin, although the yields remain uniformly low. Oxidation of black liquor from kraft pulping of *Pinus pilaster* with oxygen in alkaline medium at ~135°C gave a maximum vanillin yield of about 0.9 g of vanillin/100 g of black liquor solids. Catalyzed and uncatalyzed oxidation of isolated kraft lignin from eucalyptus with oxygen, or oxygen and added Cu(II) or Co(II) catalysts gave $<5\%$ yield of low molecular weight products as a mixture of materials [96]. Catalysis of kraft lignin oxidation with O_2 using a polyoxometalate ($H_3PMo_{12}O_{40}$) under acidic conditions in MeOH at 170° gave vanillin and methyl vanillate in a combined yield of 7–8%. The isolated material was accompanied by a large amount of oligomer, requiring additional purification of the reaction mixture to obtain pure monomers [92, 97].

4.2 *Other More Highly Selective Conversions*

If the lignin is retained in whole biomass, direct hydrogenolysis can give production of a relatively small number of lignin-derived phenols. The lignin in birch sawdust was converted into propylguaiacol (**10**) and propylsyringol (**35**) in 97% selectivity and a lignin conversion of 50% using a Ni/C catalyst in alcoholic solvents. The reaction process was demonstrated to occur via solvolytic fragmentation of lignin followed by hydrogenolysis to final products, with the alcohol solvents serving as the source of hydrogen in the reaction [98]. In a similar reaction, birch sawdust was treated with H_2 at 4 MPa and 200°C with a Rh/C catalyst in 1:1 dioxane/H_2O and

1% H_3PO_4 to give a 46 wt% yield of total monomers, 35% of which was **35**. Smaller amounts of **10** and alcohols **12** and **36** were also observed [99]. The resulting phenols were then transformed into a mixture of fuel-grade alkanes by hydrogenation at 4 MPa over a 5% Pd/C catalyst at 250°C.

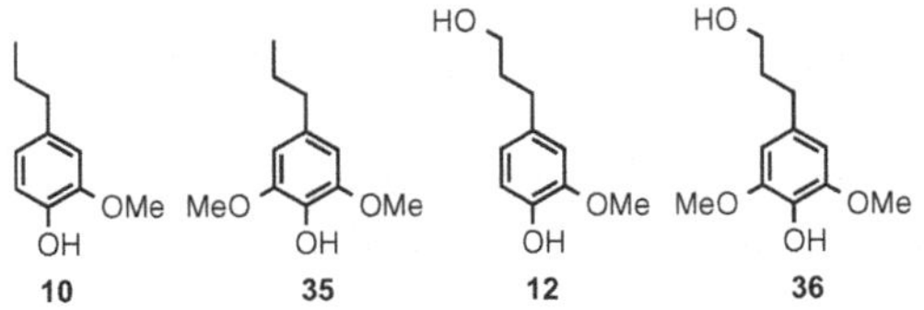

Catalytic oxidation of beech organosolv lignin in ionic liquid (1-ethyl-3-methylimidazolium trifluoromethanesulfonate) at 100°C and 8 MPa air in the presence of 20% $Mn(NO_3)_2$ formed DMBQ **25** in 11.5% isolated yield and 21% selectivity. The amount of **25** formed under these conditions depended on the catalyst level. At 2% catalyst, the product slate shifted to a mixture of syringaldehyde, syringyl alcohol, and vanillin, with only trace amounts of DMBQ observed. The authors propose that DMBQ results from syringaldehyde oxidation [100].

An interesting deconstruction of lignin occurs upon thermolysis in a large excess of a water/*p*-cresol mixture (1.8:2.5 g) at 400°C. Under these conditions, a single product **38** is formed (Fig. 13), accounting for as much as 80% of the carbon present in the starting lignin, and with no char formation being observed. The reaction is suggested to proceed through an initial depolymerization of the lignin, and then a rapid trapping of the reactive intermediates through reaction with the *p*-cresol [101]. Further mechanistic investigation suggests the intermediacy of a benzylic cation formed from the acid catalyzed lignin hydrolysis, and subsequent rapid electrophilic capping by the *p*-cresol leading to **37** [102]. The cresol adduct undergoes cleavage of the α,β-bond, affording **38** and the remainder of the lignin polymer. Deformylation of the aromatic ring is suggested to occur through the keto form of the phenol (**39**) via loss of formaldehyde.

Oxidation processes have been carried out on lignin models in the context of improving pulp bleaching processes, but also suggest methodology that, if optimized, might prove useful for transforming lignin into chemical products. Reaction of a family of lignin models (veratryl and vanillyl alcohol, β-O-4 dimers, and diphenylmethanes) with homogeneous and heterogenized methyltrioxorhenium and HOOH as the terminal oxidant afforded moderate yields of corresponding oxidation products. When applied to hydrolytic sugar cane lignin or red spruce kraft lignin, moderate yields of new functionality (as determined by ^{31}P NMR) in the lignin were observed, but no specific single compounds were isolated or reported. Interestingly, the heterogenized catalysts displayed significant reactivity toward the lignin, suggesting that there was no appreciable kinetic barrier to the interaction of two heterogeneous materials [16].

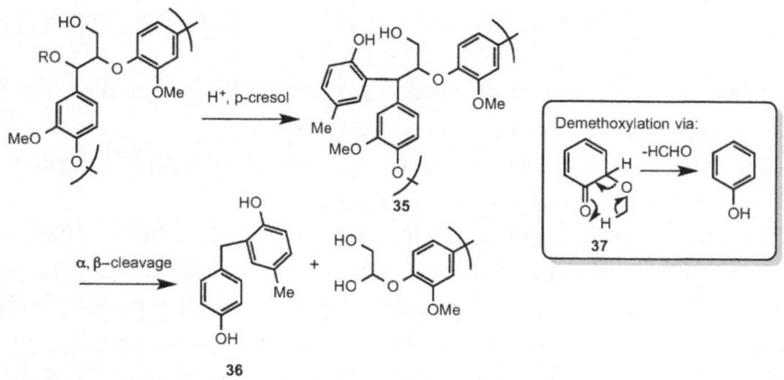

Fig. 13 Selective conversion of lignin to **38** by trapping in aqueous cresol

Oxidation of lignin and lignin models was investigated using a well-recognized industrial method based on aerobic oxidation catalyzed with metal bromide complexes in acetic acid. A group of lignin models was converted to the corresponding carboxylic acid upon treatment with Co/Mn/Br catalysts in HOAc solution under O_2 at 80–95°C. When an organosolv mixed hardwood lignin (Aldrich) was subjected to these conditions, 5–11% of similar, monomeric oxidation products were identified, and included vanillin, vanillic acid, syringaldehyde, and syringic acid. The reactivity of organosolv lignin was considerably higher than the other lignins tested in this study (acetylated organosolv lignin, sodium lignosulfonates, and hydrolytic lignin from a sugar cane biorefinery operation) [103].

5 Conclusions

Until recently, catalytic lignin conversion has been used primarily for the nonselective deconstruction of lignin into low molecular weight, easily removed fragments that can be separated from cellulose. Even more challenging to an effective coverage and understanding of the field is that much of this work remains within the more specialized journals of the pulp and paper industry. The *promise* of lignin conversion continues to be cited, but the *reality*, at least in the context of organic synthesis, is that lignin conversion still leads to the formation of complex mixtures enriched in certain functional groups representative of the starting lignin structure. *There is a vast difference between a valuable mixture of chemicals and a mixture of valuable chemicals.*

Nonetheless, the expanding interest in biorefining of lignocellulosics and full exploitation of all the biopolymeric components of biomass is moving lignin transformation more into the mainstream. Methodology for the selective transformation of lignin is beginning to emerge, but still faces several challenges inherent to lignin as a source of renewable carbon:

- Lignin's structure as biosynthesized in plants is not the structure normally isolated and available for conversion in the biorefinery.
- Lignin bears a large number of structurally different groups with similar reactivity, rendering selective transformations difficult.
- Upon isolation, lignin affords mixtures of materials with different molecular weight, further complicating its reactivity profile.
- Lignin's structure and composition can vary widely depending on source. ETEK lignin from dilute acid hydrolysis of biomass is actually about 46% cellulose and 29% lignin [104]. In contrast, organosolv lignins exhibit purities >95%.
- Accordingly, the reactivity of the lignin can also vary widely. In general, organosolv lignins (often prepared under acidic conditions) appear to offer greater reactivity and utility as raw materials than kraft lignins (prepared under basic conditions).

New catalytic transformations are being developed and reported for lignin models, and provide useful information regarding reactivity of specific functional groups under reductive and oxidative conditions. The opportunity in this field is the development of general, broad-based methodology that can accommodate *multiple* functional groups and still afford high yields of low molecular weight aromatics. By tailoring the tools of catalysis, long used for the selective transformation of petroleum based hydrocarbons, to the more highly oxygenated lignin polymers and substructures, and coupling this with new methodology for lignin fractionation and isolation, the potential for using lignin as a viable source of renewable carbon is increasing. Recognition that isolated lignin is available in a wide variety of forms, and, thus, a wide variety of functional group profiles, will enable the most effective catalyst design.

Acknowledgements This work was supported as part of the Center for Direct Catalytic Conversion of Biomass to Biofuels (C3Bio), an Energy Frontier Research Center funded by the U.S. Department of Energy, Office of Science, Office of Basic Energy Sciences, under Award Number DESC0000997.

References

1. Bozell JJ, Holladay JE, Johnson D, White JF (2007) Top value added chemicals from biomass. Volume II – results of screening for potential candidates from biorefinery lignin. Pacific Northwest National Laboratory, Richland
2. Bozell JJ, Black SK, Myers M, Cahill D, Miller WP, Park S (2011) Solvent fractionation of renewable woody feedstocks: organosolv generation of biorefinery process streams for the production of biobased chemicals. Biomass Bioenergy 35:4197
3. Rushton M (2012) Biochemicals and bioproducts: the new focus of biorefining, CanBio annual conference, Vancouver, BC, http://www.canbio.ca/upload/documents/van-12-presentations/rushton-mike.pdf
4. Bozell J, Dimmel DR, Power A (1994) Pulping catalysts from lignin. Ind Uses Agric Mat Sit Outlook Rep 27

5. Gluckstein J, Hu M, Kidder M, McFarlane J, Narula C, Sturgeon M (2010) Final report: Investigation of catalytic pathways for lignin breakdown into monomers and fuels. ORNL/TM-2010/281
6. Boerjan W, Ralph J, Baucher M (2003) Lignin biosynthesis. Ann Rev Plant Biol 54:519
7. Vanholme R, Demedts B, Morreel K, Ralph J, Boerjan W (2010) Lignin biosynthesis and structure. Plant Physiol 153:895
8. Bozell JJ, O'Lenick CJ, Warwick S (2011) Biomass fractionation for the biorefinery: heteronuclear multiple quantum coherence-nuclear magnetic resonance investigation of lignin isolated from solvent fractionation of switchgrass. J Agric Food Chem 59:9232
9. Samuel R, Pu YQ, Raman B, Ragauskas AJ (2010) Structural characterization and comparison of switchgrass ball-milled lignin before and after dilute acid pretreatment. Appl Biochem Biotechnol 162:62
10. Li JB, Gellerstedt G, Toven K (2009) Steam explosion lignins; their extraction, structure and potential as feedstock for biodiesel and chemicals. Bioresour Technol 100:2556
11. Morck R, Yoshida H, Kringstad KP, Hatakeyama H (1986) Fractionation of kraft lignin by successive extraction with organic solvents-1. Functional groups, C13 NMR spectra and molecular weight distributions. Holzforschung 40:51
12. Thring RW, Vanderlaan MN, Griffin SL (1996) Fractionation of Alcell(R) lignin by sequential solvent extraction. J Wood Chem Technol 16:139
13. Yuan TQ, He J, Xu F, Sun RC (2009) Fractionation and physico-chemical analysis of degraded lignins from the black liquor of *Eucalyptus pellita* KP-AQ pulping. Polym Degrad Stab 94:1142
14. Garcia A, Toledano A, Serrano L, Egues I, Gonzalez M, Marin F, Labidi J (2009) Characterization of lignins obtained by selective precipitation. Sep Purif Technol 68:193
15. Toledano A, Serrano L, Garcia A, Mondragon I, Labidi J (2010) Comparative study of lignin fractionation by ultrafiltration and selective precipitation. Chem Eng J 157:93
16. Lange H, Decina S, Crestini C (2013) Oxidative upgrade of lignin – recent routes reviewed. Eur Polym J 49:1151
17. Zakzeski J, Bruijnincx PCA, Jongerius AL, Weckhuysen BM (2010) The catalytic valorization of lignin for the production of renewable chemicals. Chem Rev 110:3552
18. Azadi P, Inderwildi OR, Farnood R, King DA (2013) Liquid fuels, hydrogen and chemicals from lignin: a critical review. Renew Sustain Energ Rev 21:506
19. Bu Q, Lei HW, Zacher AH, Wang L, Ren SJ, Liang J, Wei Y, Liu YP, Tang JM, Zhang Q, Ruan R (2012) A review of catalytic hydrodeoxygenation of lignin-derived phenols from biomass pyrolysis. Bioresour Technol 124:470
20. Calvo-Flores FG, Dobado JA (2010) Lignin as renewable raw material. ChemSusChem 3:1227
21. Doherty WOS, Mousavioun P, Fellows CM (2011) Value-adding to cellulosic ethanol: lignin polymers. Ind Crop Prod 33:259
22. Gasser CA, Hommes G, Schaffer A, Corvini PFX (2012) Multi-catalysis reactions: new prospects and challenges of biotechnology to valorize lignin. Appl Microbiol Biotechnol 95:1115
23. Laskar DD, Yang B, Wang HM, Lee J (2013) Pathways for biomass-derived lignin to hydrocarbon fuels. BioFPR 7:602
24. Crestini C, Crucianelli M, Orlandi M, Saladino R (2010) Oxidative strategies in lignin chemistry: a new environmental friendly approach for the functionalisation of lignin and lignocellulosic fibers. Catal Today 156:8
25. Dence CW, Reeve DW (1996) Pulp bleaching – principles and practice. TAPPI, Atlanta
26. Weinstock IA, Atalla RH, Reiner RS, Moen MA, Hammel KE, Houtman CJ, Hill CL (1996) A new environmentally benign technology and approach to bleaching kraft pulp. Polyoxometalates for selective delignification and waste mineralization. New J Chem 20:269
27. Perng YS, Oloman CW, Watson PA, James BR (1994) Catalytic oxygen bleaching of wood pulp with metal porphyrin and phthalocyanine complexes. Tappi J 77:119

28. Simmons BA, Logue D, Ralph J (2010) Advances in modifying lignin for enhanced biofuel production. Curr Opin Plant Biol 13:313
29. Vanholme R, Morreel K, Darrah C, Oyarce P, Grabber JH, Ralph J, Boerjan W (2012) Metabolic engineering of novel lignin in biomass crops. New Phytologist 196:978
30. Mansfield SD, Kang KY, Chapple C (2012) Designed for deconstruction – poplar trees altered in cell wall lignification improve the efficacy of bioethanol production. New Phytologist 194:91
31. Pandey MP, Kim CS (2011) Lignin depolymerization and conversion: a review of thermochemical methods. Chem Eng Technol 34:29
32. Yang HP, Yan R, Chen HP, Lee DH, Zheng CG (2007) Characteristics of hemicellulose, cellulose and lignin pyrolysis. Fuel 86:1781
33. Nowakowski DJ, Bridgwater AV, Elliott DC, Meier D, de Wild P (2010) Lignin fast pyrolysis: results from an international collaboration. J Anal Appl Pyrolysis 88:53
34. Britt PF, Buchanan AC, Cooney MJ, Martineau DR (2000) Flash vacuum pyrolysis of methoxy-substituted lignin model compounds. J Org Chem 65:1376
35. Chu S, Subrahmanyam AV, Huber GW (2013) The pyrolysis chemistry of a beta-O-4 type oligomeric lignin model compound. Green Chem 15:125
36. Thring RW, Katikaneni SPR, Bakhshi NN (2000) The production of gasoline range hydrocarbons from Alcell (R) lignin using HZSM-5 catalyst. Fuel Process Technol 62:17
37. Ma ZQ, Troussard E, van Bokhoven JA (2012) Controlling the selectivity to chemicals from lignin via catalytic fast pyrolysis. Appl Catal A 423:130
38. Zheng Y, Chen DY, Zhu XF (2013) Aromatic hydrocarbon production by the online catalytic cracking of lignin fast pyrolysis vapors using $Mo_2N/\gamma\text{-}Al_2O_3$. J Anal Appl Pyrolysis 104:514
39. Yu YQ, Li XY, Su L, Zhang Y, Wang YJ, Zhang HZ (2012) The role of shape selectivity in catalytic fast pyrolysis of lignin with zeolite catalysts. Appl Catal A 447:115
40. Fan MH, Jiang PW, Bi PY, Deng SM, Yan LF, Zhai Q, Wang TJ, Li QX (2013) Directional synthesis of ethylbenzene through catalytic transformation of lignin. Bioresour Technol 143:59
41. Huang WW, Gong FY, Fan MH, Zhai Q, Hong CG, Li QX (2012) Production of light olefins by catalytic conversion of lignocellulosic biomass with HZSM-5 zeolite impregnated with 6 wt.% lanthanum. Bioresour Technol 121:248
42. Mante OD, Rodriguez JA, Babu SP (2013) Selective defunctionalization by TiO_2 of monomeric phenolics from lignin pyrolysis into simple phenols. Bioresour Technol 148:508
43. Furimsky E (2000) Catalytic hydrodeoxygenation. Appl Catal A 199:147
44. Goheen DW (1966) Hydrogenation of lignin by the Noguchi process. Adv Chem Ser: 205.
45. Huibers DTA, Parkhurst HJ (1983) U. S. patent 4,420,644 to Hydrocarbon Research Institute
46. Oasmaa A, Johansson A (1993) Catalytic hydrotreating of lignin with water-soluble molybdenum catalyst. Energ Fuels 7:426
47. Oasmaa A, Alen R, Meier D (1993) Catalytic hydrotreatment of some technical lignins. Bioresour Technol 45:189
48. Ben HX, Mu W, Deng YL, Ragauskas AJ (2013) Production of renewable gasoline from aqueous phase hydrogenation of lignin pyrolysis oil. Fuel 103:1148
49. Jongerius AL, Bruijnincx PCA, Weckhuysen BM (2013) Liquid-phase reforming and hydrodeoxygenation as a two-step route to aromatics from lignin. Green Chem 15:3049
50. Jongerius AL, Copeland JR, Foo GS, Hofmann JP, Bruijnincx PCA, Sievers C, Weckhuysen BM (2013) Stability of $Pt/\gamma\text{-}Al_2O_3$ catalysts in lignin and lignin model compound solutions under liquid phase reforming reaction conditions. ACS Catal 3:464
51. de Wild P, Van der Laan R, Kloekhorst A, Heeres E (2009) Lignin valorisation for chemicals and (transportation) fuels via (catalytic) pyrolysis and hydrodeoxygenation. Environ Prog Sustain Energ 28:461
52. Ye YY, Zhang Y, Fan J, Chang J (2012) Selective production of 4-ethylphenolics from lignin via mild hydrogenolysis. Bioresour Technol 118:648

53. Kleinert M, Gasson JR, Barth T (2009) Optimizing solvolysis conditions for integrated depolymerisation and hydrodeoxygenation of lignin to produce liquid biofuel. J Anal Appl Pyrolysis 85:108
54. Holmelid B, Kleinert M, Barth T (2012) Reactivity and reaction pathways in thermochemical treatment of selected lignin-like model compounds under hydrogen rich conditions. J Anal Appl Pyrolysis 98:37
55. Gosselink RJA, Teunissen W, van Dam JEG, de Jong E, Gellerstedt G, Scott EL, Sanders JPM (2012) Lignin depolymerisation in supercritical carbon dioxide/acetone/water fluid for the production of aromatic chemicals. Bioresour Technol 106:173
56. Patil PT, Armbruster U, Richter M, Martin A (2011) Heterogeneously catalyzed hydroprocessing of organosolv lignin in sub- and supercritical solvents. Energ Fuels 25:4713
57. Kang SM, Li XL, Fan J, Chang J (2013) Hydrothermal conversion of lignin: a review. Renew Sustain Energ Rev 27:546
58. Constant S, Robitzer M, Quignard F, Di Renzo F (2012) Vanillin oligomerization as a model of side reactions in lignin fragmentation. Catal Today 189:123
59. Tymchyshyn M, Xu CB (2010) Liquefaction of biomass in hot compressed water for the production of phenolic compounds. Bioresour Technol 101:2483
60. Kang SM, Li XL, Fan J, Chang J (2011) Classified separation of lignin hydrothermal liquefied products. Ind Eng Chem Res 50:11288
61. Zhang B, Huang HJ, Ramaswamy S (2008) Reaction kinetics of the hydrothermal treatment of lignin. Appl Biochem Biotechnol 147:119
62. Yoshikawa T, Yagi T, Shinohara S, Fukunaga T, Nakasaka Y, Tago T, Masuda T (2013) Production of phenols from lignin via depolymerization and catalytic cracking. Fuel Process Technol 108:69
63. Hasegawa I, Inoue Y, Muranaka Y, Yasukawa T, Mae K (2011) Selective production of organic acids and depolymerization of lignin by hydrothermal oxidation with diluted hydrogen peroxide. Energ Fuels 25:791
64. Yuan ZS, Cheng SN, Leitch M, Xu CB (2010) Hydrolytic degradation of alkaline lignin in hot compressed water and ethanol. Bioresour Technol 101:9308
65. Zhang Y, Ye YY, Fan J, Chang J (2013) Selective production of phenol, guaiacol and 2,6-dimethoxyphenol by alkaline hydrothermal conversion of lignin. J Biobas Mat Bioener 7:696
66. Sergeev AG, Hartwig JF (2011) Selective, nickel-catalyzed hydrogenolysis of aryl ethers. Science 332:439
67. Sergeev AG, Webb JD, Hartwig JF (2012) A heterogeneous nickel catalyst for the hydrogenolysis of aryl ethers without arene hydrogenation. J Am Chem Soc 134:20226
68. Nichols JM, Bishop LM, Bergman RG, Ellman JA (2010) Catalytic C-O bond cleavage of 2-aryloxy-1-arylethanols and its application to the depolymerization of lignin-related polymers. J Am Chem Soc 132:12554
69. Wu A, Patrick BO, Chung E, James BR (2012) Hydrogenolysis of beta-O-4 lignin model dimers by a ruthenium-xantphos catalyst. Dalton Trans 41:11093
70. Chmely SC, Kim S, Ciesielski PN, Jimenez-Oses G, Paton RS, Beckham GT (2013) Mechanistic study of a Ru-xantphos catalyst for tandem alcohol dehydrogenation and reductive aryl ether cleavage. ACS Catal 3:963
71. Parsell TH, Owen BC, Klein I, Jarrell TM, Marcum CL, Haupert LJ, Amundson LM, Kenttamaa HI, Ribeiro F, Miller JT, Abu-Omar MM (2013) Cleavage and hydrodeoxygenation (HDO) of C–O bonds relevant to lignin conversion using Pd/Zn synergistic catalysis. Chem Sci 4:806
72. Son S, Toste FD (2010) Non-oxidative vanadium-catalyzed C-O bond cleavage: application to degradation of lignin model compounds. Angew Chem Int Ed 49:3791
73. Chan JMW, Bauer S, Sorek H, Sreekumar S, Wang K, Toste FD (2013) Studies on the vanadium-catalyzed nonoxidative depolymerization of *Miscanthus giganteus*-derived lignin. ACS Catal 3:1369

74. Rahimi A, Azarpira A, Kim H, Ralph J, Stahl SS (2013) Chemoselective metal-free aerobic alcohol oxidation in lignin. J Am Chem Soc 135:6415
75. Lai YZ, Guo XP (1991) Variation of the phenolic hydroxyl group content in wood lignins. Wood Sci Technol 25:467
76. Zombeck A, Drago RS, Corden BB, Gaul JH (1981) Activation of molecular-oxygen – mechanistic studies of the oxidation of hindered phenols with cobalt-dioxygen complexes. J Am Chem Soc 103:7580
77. Cozzi PG (2004) Metal-salen Schiff base complexes in catalysis: practical aspects. Chem Soc Rev 33:410
78. Chen D, Martell AE, Sun YZ (1989) New synthetic cobalt Schiff-base complexes as oxygen carriers. Inorg Chem 28:2647
79. Bozell JJ, Hames BR, Dimmel DR (1995) Cobalt-Schiff base complex-catalyzed oxidation of *para*-substituted phenolics – preparation of benzoquinones. J Org Chem 60:2398
80. Cedeno D, Bozell JJ (2012) Catalytic oxidation of para-substituted phenols with cobalt-Schiff base complexes/O_2-selective conversion of syringyl and guaiacyl lignin models to benzoquinones. Tetrahedron Lett 53:2380
81. Biannic B, Bozell JJ (2013) Efficient cobalt-catalyzed oxidative conversion of lignin models to benzoquinones. Org Lett 15:2730
82. Elder T, Bozell JJ, Cedeno D (2013) The effect of axial ligand on the oxidation of syringyl alcohol by Co(salen) adducts. Phys Chem Chem Phys 15:7328
83. Hanson SK, Wu RL, Silks LA (2012) C–C or C–O bond cleavage in a phenolic lignin model compound: selectivity depends on vanadium catalyst. Angew Chem Int Ed 51:3410
84. Zakzeski J, Jongerius AL, Weckhuysen BM (2010) Transition metal catalyzed oxidation of Alcell lignin, soda lignin, and lignin model compounds in ionic liquids. Green Chem 12:1225
85. Zakzeski J, Bruijnincx PCA, Weckhuysen BM (2011) In situ spectroscopic investigation of the cobalt-catalyzed oxidation of lignin model compounds in ionic liquids. Green Chem 13:671
86. Dimmel DR, Karim MR, Savidakis MC, Bozell JJ (1996) Pulping catalysts from lignin – 5. Nitrogen dioxide oxidation of lignin models to benzoquinones. J Wood Chem Technol 16:169
87. Bozell JJ, Hoberg JO, Dimmel DR (1998) Catalytic oxidation of *para*-substituted phenols with nitrogen dioxide and oxygen. Tetrahedron Lett 39:2261
88. Bosch E, Rathore R, Kochi JK (1994) Novel catalysis of hydroquinone autoxidation with nitrogen oxides. J Org Chem 59:2529
89. Rathore R, Bosch E, Kochi JK (1994) Selective nitration versus oxidative dealkylation of hydroquinone ethers with nitrogen-dioxide. Tetrahedron 50:6727
90. Hocking MB (1997) Vanillin: synthetic flavoring from spent sulfite liquor. J Chem Educ 74:1055
91. da Silva EAB, Zabkova M, Araujo JD, Cateto CA, Barreiro MF, Belgacem MN, Rodriques AE (2009) An integrated process to produce vanillin and lignin-based polyurethanes from kraft lignin. Chem Eng Res Des 87:1276
92. Voitl T, von Rohr PR (2010) Demonstration of a process for the conversion of kraft lignin into vanillin and methyl vanillate by acidic oxidation in aqueous methanol. Ind Eng Chem Res 49:520
93. Pacek AW, Ding P, Garrett M, Sheldrake G, Nienow AW (2013) Catalytic conversion of sodium lignosulfonate to vanillin: engineering aspects – part 1. Effects of processing conditions on vanillin yield and selectivity. Ind Eng Chem Res 52:8361
94. Pearl IA (1942) Vanillin from lignin materials. J Am Chem Soc 64:1429
95. Goncalves AR, Schuchardt U (1999) Oxidation of organosolv lignins in acetic acid – influence of oxygen pressure. Appl Biochem Biotechnol 77–9:127
96. Villar JC, Caperos A, Garcia-Ochoa F (2001) Oxidation of hardwood kraft lignin to phenolic derivatives with oxygen as oxidant. Wood Sci Technol 35:245

97. Voitl T, von Rohr PR (2008) Oxidation of lignin using aqueous polyoxometalates in the presence of alcohols. ChemSusChem 1:763
98. Song Q, Wang F, Cai JY, Wang YH, Zhang JJ, Yu WQ, Xu J (2013) Lignin depolymerization (LDP) in alcohol over nickel-based catalysts via a fragmentation-hydrogenolysis process. Energ Environ Sci 6:994
99. Yan N, Zhao C, Dyson PJ, Wang C, Liu LT, Kou Y (2008) Selective degradation of wood lignin over noble-metal catalysts in a two-step process. ChemSusChem 1:626
100. Stark K, Taccardi N, Bosmann A, Wasserscheid P (2010) Oxidative depolymerization of lignin in ionic liquids. ChemSusChem 3:719
101. Okuda K, Man X, Umetsu M, Takami S, Adschiri T (2004) Efficient conversion of lignin into single chemical species by solvothermal reaction in water-*p*-cresol solvent. J Phys Condens Matter 16:S1325
102. Takami S, Okuda K, Man X, Umetsu M, Ohara S, Adschiri T (2012) Kinetic study on the selective production of 2-(hydroxybenzyl)-4-methylphenol from organosolv lignin in a mixture of supercritical water and *p*-cresol. Ind Eng Chem Res 51:4804
103. Partenheimer W (2009) The aerobic oxidative cleavage of lignin to produce hydroxyaromatic benzaldehydes and carboxylic acids via metal/bromide catalysts in acetic acid/water mixtures. Adv Synth Catal 351:456
104. Nsimba RY, Mullen CA, West NM, Boateng AA (2013) Structure-property characteristics of pyrolytic lignins derived from fast pyrolysis of a lignin rich biomass extract. ACS Sus Chem Eng 1:260

Index

The manufacturer's authorised representative in the EU is Springer Nature Customer Service Centre GmbH, Europaplatz 3, 69115 Heidelberg, Germany. If you have any concerns regarding our products, please contact ProductSafety@springernature.com

Printed and bound by CPI Group (UK) Ltd, Croydon, CR0 4YY
30/09/2026
02225567-0011